Ethics and Sustainable Agriculture

Fabio Caporali

Ethics and Sustainable Agriculture

Bridging the Ecological Gaps

Fabio Caporali
DAFNE
Università degli Studi della Tuscia
Viterbo, Italy

ISBN 978-3-030-76685-6 ISBN 978-3-030-76683-2 (eBook)
https://doi.org/10.1007/978-3-030-76683-2

This Springer imprint is published by the registered company Springer Nature Switzerland AG
The registered company address is: Gewerbestrasse 11, 6330 Cham, Switzerland

To
all people
who work
with faith, hope and care

The necessary transformation of agriculture will come only because of a transformation of culture, when humanity recognizes the surpassing value of a world that transcends the species [...] Indeed we do need a land ethic, expanded into a world ethic.
Stan Rowe—, 2002.

Home Place: Essays on Ecology

Preface

A recall to ethics of responsibility is today a deontological urgency on the part of all components of the human activity system we call agriculture. Starting from farmers that govern fields and farms at local level, a hierarchical range of institutional bodies is involved which governs science, technology, economy and policy applications to agriculture at regional, national and international levels. The world of agriculture is part of a greater socio-ecological system that includes it, affects it and is affected by it. Today, the extent of affection is alarming and demands prompt remedies.

Recognition, *obligation* and *action* are the keywords inspiring a new ethics of responsibility for rendering human activity systems, such as agriculture, biophysically and socioeconomically sustainable. These three words summarise what is the content of this book that first assumes *recognition* as the starting point, or ecological understanding, which implies redefining the environment according to the new transdisciplinary science of ecology and its epistemological instrument, the ecosystem concept. Ecological understanding facilitates an integrated vision of reality where components are interdependent, and cooperation, instead of competition, is crucial for ecosystem maintenance and evolution. An unexpected result of ecological enquiry is the recognition that sustainability is an inherent quality of nature organisation.

Nature's work provides ecological services (support, provision, regulation and culture) for a living community, including human activities. *Obligation* implies to be consequentially responsible to the recognition achievements and be able to think of and design sustainable land use systems, such as those for multifunctional and sustainable agriculture, that comply with the principles of eco-development (intensive use of solar radiation, biodiversity and re-cycling). *Action* implies organising the whole agro-food system in practice, according to the previous steps of recognition and obligation, by coherently influencing policy, economy, education, science and technology for an *ecological intensification* of their means and goals. Ecological intensification concerns the strengthening of ecological services provided by agroecosystems, with positive implications for both human and environmental health, and social and environmental justice. The author regards this book as a tool for ecological intensification of both human understanding and behaviour for the benefit of the planetary community.

The originality of the book resides in its narrative grounded on the use of the ecosystem concept as the key for understanding, designing and managing agricultural systems at any spatial and temporal scales (plot, field, farm and landscape) for better balancing trade-offs between commodity production and environment protection. Ecosystem epistemology ensures a more truthful representation and organisation of agroecosystems in order to make them more responsive to the current ethical expectations for food security, human health, social equity, care for livestock and natural biodiversity.

No book until now bridges the ecological gaps between ethics and sustainable agriculture using an ecological narrative that highlights the best practices grounded on principles of ecological intensification. This is the main reason why I decided to write this book focusing on the historical steps signing the cross-fertilisation between ecology and agriculture that are now starting the conversion process towards an ethic of responsibility. The void of transdisciplinary knowledge and culture may have been only temporary, and there is a plausible reason to retain that the progressive expansion of agroecological research, and education can help pave the way towards a more multifunctional and sustainable agriculture.

There are at least two other elements of novelty in the book for enlarging the ethical horizon of the agro-food system through an ecological perspective: ecolinguistics and ecotheology. They both contribute to the enrichment of the ecological narrative as it relates to spirituality. It is through language that reality takes a mental form and acquires meaning. It is through language that we can transcend materiality and interpret reality as a sign for creation, opening a window for a dialogue between science and religion, touching the new field of enquiry of ecotheology, and extending it out to embrace sustainable agriculture.

Pisa, Italy
March 2021

Fabio Caporali

Acknowledgements

To fully acknowledge the contributions to this book, I should mention all authors of the cited references. Some of them appear at the onset of each chapter with their inspiring sentences. Having said that, what I really want to acknowledge is the whole process of thought movement that, as a spiritual driver, has emerged and developed in favour of sustainable human activity systems. Among the latter, agriculture should be the first one to undergo a process of ecological conversion. Considering that writing a book is a communicative act, I want to acknowledge the institutions, both public and private, and the people who participated in the process.

First of all, the European Union has been able to establish a far-sighted policy for a higher education open to student and teacher mobility, building up of common curricula development and participated research among stakeholders. This was the context where I had the opportunity to mature the experiential learning needed to write a book on ethics and agriculture. In this context, I want to acknowledge the contribution of all academicians involved in the ENOAT association (European Network of Organic Agriculture Teachers), with their dedicated competence and commitment to innovative contents and methods for jointly developing an agroecological paradigm to face the challenge of sustainability in agriculture and society.

To publish a book, the role of the publishing editor is pivotal, and I am grateful to Springer for having sustained me in both previous publications and the present one. I am particularly indebted to two anonymous reviewers who offered inspiring suggestions on the original book proposal, whereby I decided to write an afterword.

In the end, I am very grateful to all virtuous farmers that I met and their traditions, experience, innovative ideas and tireless work, which taught me the silent ethics of agriculture.

Contents

Chapter 1
Introduction: Agriculture as if Ethics Mattered

> *What must be indelibly marked by agricultural scientists is that science was not, is not, and can never be value-free, or even ethics-free…*
> *Thus the very logic of science is modulated by social ethical concerns*
>
> *(Rollin BE 1996, p. 536)*

To re-establish a meaningful bridge between ethics and agriculture, a new science (ecology) with both a new epistemological tool, that of the ecosystem concept, and a novel narrative, that of sustainable development, is growing necessary. This evidence stems from the recent history of *the human predicament* (Meadows et al. 1972) and the present era of *Anthropocene* (Crutzen 2002), where the challenge of sustainable development needs confrontation in every field of human activity systems. The main aim of this book is to focus on ethics as a liver for raising scientific, technical, social, economic and political solutions to adopt in agriculture as a model of symbiotic relationships between man and nature.

Current international initiatives aimed at establishing a qualitative and quantitative framework for global risks analysis deploy a picture of an unsettled world (WEF 2020), affected by climate threats and accelerated biodiversity loss, risks to economic stability and social cohesion, and health systems under new pressures. All these risks show interconnections (Beck 2008), as we unfortunately deduce from the latest devastating Covid-19 pandemic, which undermines at the same time human health and human economy at both regional and global levels. Interestingly, *The Global Risks Report 2020* (WEF 2020) also provides a survey ranking of risks likelihood where the top 10 risks fall into five categories: economic, environmental, geopolitical, societal, and technological. Interestingly, the order of the first five risk positions is the following: extreme weather, climate action failure, natural disaster, biodiversity loss, human-made environmental disasters. According to the voice of the WEF President (Børge Brende), "respondents to our Global Risks Perception Survey are sounding the alarm, ranking climate change and related environmental

F. Caporali, *Ethics and Sustainable Agriculture*,
https://doi.org/10.1007/978-3-030-76683-2_1

issues at the top five risks in terms of likelihood - the first time in the survey's history that one category [the environmental one] has occupied all five of the top spots". Furthermore, *The Global Risks Report 2020* provides a Global Risks Interconnections Map that shows the linkage and complexity of global challenges and associated risks, including food crises, water crises, unemployment, involuntary migration, infection diseases, and social instability, all social-environmental processes that include agriculture and land management as human activity systems. The web of intricate relations between the environment and its management through human agency is felt mounting in a globalised world both in material and immaterial terms, with energy and matter flows channelled by informational flows. All this complex and dynamic scenario requires capacity of governance inspired by systems thinking and systems practice integration at each level of spatial and temporal organisation.

A recent document of the Scientific and Technical Advisory Panel (STAP) of the Global Environment Facility (GEF)[1] points out to *systems thinking and integration* to solve complex environmental problems (Bierbaum et al. 2018). STAP encourages the GEF to continue pursuing integrative projects and recommends a sequence of systems rules based on the general assumption that applying systems thinking is addressing "interconnected environmental, social, economic, and governance challenges across sectors with an eye towards resilience and transformational change". If change is required to tackle the drivers of environmental degradation, the first rule is *developing a clear rationale and theory of change*. The second rule is providing *stakeholder engagement* in the entire project for identifying diverse needs and managing trade-offs. The third rule is devising "*adaptive implementation pathways*", i.e. "a logical sequence of interventions, which is responsive to changing circumstances and new learning".

Scientific support to systems theory and practice for solving social and environmental problems is growing as reported in influential review papers concerning the challenge of global sustainability (Liu et al. 2015). The potential of holistic approaches and systems integration in the process of better understanding of reality appears in a statement like the following:

> One major advance has been recognizing Earth as a large, coupled human and natural system consisting of many smaller coupled systems linked through flow of information, matter, and energy and evolving through time as a set of interconnected adaptive systems. (Liu et al. 2015)

The mental attitude to consider reality as a hierarchical construction is at the hearth of both the *process philosophy* and the *science of ecology*. There is no surprise that this knowledge innovation has penetrated human activity systems like agriculture in the recognised form of the transdisciplinary *science of agroecology*

[1]GEF was established on the eve of the 1992 RIO Earth Summit to help tackle our planet most pressing environmental problems and has become an international partnership for investing in integrative projects of countries, civil society organisations and the private sector.

(Caporali 2010), agriculture being in its own essence the trophic link between the social systems and the natural system.

The general evolving framework of *the human predicament* is worth to be recognised and made known as a new cultural achievement for creating awareness and responsible action (*ethics*) at both global and local levels in each field of human activity systems. This is one of the main aims of the present book, the author of which personally has lived and participated in the process of agroecology evolution.

Agriculture is a human activity system that has an ancient origin and more than others has to do with both the natural and the social environment in terms of use or abuse of their resources (Caporali 2008). Already in the Latin literature concerning agriculture (Caporali 2015), the first book of the Latin tradition[2] formulated both principles for evaluating the biophysical and social context for agricultural activities, and suggestions of best practices to carry out at the farm level. All recommendations were made with and ethical savour, where the terms *good* (for environment, climate, soil, water, buildings, labour, etc.) and *care* (for biodiversity of both cropland and woodland, livestock, management, etc.) were the most recurrent in the narrative.

All these recommendations were summarised by Cato under the general assumption that "it is from the farming class that the bravest men and the sturdiest soldiers come". Commenting on these historical findings, Caporali (2015, p. 9) envisages linkages with modern agroecology as follows:

> We can hardly contest that these points are still today relevant for a good modern farming carried out on a family-based organisation, even if labour as human slaves has been replaced by machine-driven labour. Indeed, agriculturalist tradition has kept these general recommendations valid for most countries in the world because their success has been assured by the experiential learning cumulated along continuing generations of farmers. Agroecology in general recognises the value of tradition in agriculture, establishes the scientific reasons of traditional agriculture's best practices and recognises their values as foundation for a sustainable agriculture.

However, this traditional attitude to regard agriculture as a service of common utility for both society and environment has progressively lost his weight with the increasing industrialisation of society, the involvement of agriculture in the *business-as-usual* economy and the consequent environmental and social impacts. Biologist Rachel Carson was the first one to raise the moral appeal against the use of pesticides in agriculture and the widespread destruction of wildlife in America, when she published her brilliant and controversial book "Silent Spring" (Carson 1962). The subtitle in the cover prophetically advices that "no single book on our environment has done more to awaken and alarm the world". Indeed, with a specific reference to the science of ecology that is about the relation of plants and animals to their environment and to another, the book introduction reports the following acknowledgment:

> Ecologists are more and more coming to recognize that for this purpose man is an animal and indeed the most important of all animals and that however artificial his dwelling, he

[2] Cato's *De Agri Cultura* (On Agriculture), which dates back to the second century BC.

> cannot with impunity allow the natural environment of living things from which he has so recently emerged to be destroyed. (Carson 1962, p. 11)

Since the publication of "Silent Spring", enormous developments have occurred in every field of human activity systems in order to better understand and manage the conditions of human existence as a part of a larger living world. This big adventure is revealing itself as the most relevant unprecedented challenge in the entire history of both humanity and the planetary life. Special responsibility rests with agricultural scientists that produce the knowledge flowing from institutional bodies of research and education, such as the Universities, to diffuse in society through extension services, communication facilities and participatory initiatives involving local stakeholders (farmers, traders, local administrators and consumers). A seminal paper of Jane Lubchenco (1998)[3] reminds scientists about their fundamental obligation to take an ethics of responsibility due to their mission in society which includes "more comprehensive information, understanding, and technologies for society to move toward a more sustainable biosphere—one which is ecologically sound, economically feasible, and socially just". To meet this challenge, there is a strong request for" new fundamental research, faster and more effective transmission of new and existing knowledge to policy- and decision-makers, and better communication of this knowledge to the public".

Trusting on ethics to alleviate *the human predicament* is a recipe that involves the entire systems of human values and not only science and technology. Starting from ethics and ecology in agriculture is a plausible way to try to get the final goal of sustainability, in that agriculture is the human activity system that historically has much to do with all components of nature, be they biotic or abiotic (Bardgett and Gibson 2017; Weiner 2017; Bullock et al. 2017). However, the challenge of agricultural sustainability is very serious, the 'limiting factor' being "not our scientific knowledge, but the political and economic structures within which agriculture is practised" (Weiner 2017), whereby someone calls for "a socio-ecological revolution in agriculture" (Norton 2016). In broad terms, the ecological crisis of agriculture reflects the global ecological crisis of human civilisation (Gare 2017). In modern industrialised societies, the philosophical roots of the Cartesian *Weltbild* (anthropocentrism, mechanistic conception of the natural world, and metaphysical dualism between humanity and the rest of natural world) "continue to serve as the foundation and the encompassing horizon of much contemporary philosophical thought, science and technology, neo-liberal economy and political and educational institution"(Kureethadam 2017).

[3] The text is an extension of her Presidential Address at the Annual Meeting of the American Association of the Advancement of Science, 15 February 1997.

References

Bardgett RD, Gibson DJ (2017) Plant ecological solutions to global food security. J Ecol 105:859–864

Beck U (2008) Weltrisikogesellschaft. Suhrkamp, Frankfurt am Mein

Bierbaum R et al (2018) Integration: to solve complex environmental problems. Scientific and Technical Advisory Panel (STAP) to the Global Environment Facility (GEF), Washington, DC

Bullock JM et al (2017) Resilience and food security: rethinking an ecological concept. J Ecol 105:880–884

Caporali F (2008) Ecological agriculture: human and social context. In: Clini C, Musu I, Gullino ML (eds) Sustainable development and environmental management. Springer, Dordrecht, pp 415–429

Caporali F (2010) Agroecology as a transdisciplinary science for a sustainable agriculture. In: Lichtfouse E (ed) Biodiversity, biofuels, agroforestry and conservation agriculture. Springer, Dordrecht, pp 1–71

Caporali F (2015) History and development of agroecology and theory of agroecosystems. In: Monteduro et al (eds) Law and agroecology. A transdisciplinary dialogue. Springer, Berlin, pp 3–29

Carson R (1962) Silent Spring. Penguin Books

Crutzen J (2002) Geology of mankind. Nature 415(3):23

Gare A (2017) The philosophical foundation of ecological civilization. Earthscan Routledge

Kureethadam JI (2017) The philosophical roots of the ecological crisis. Cambridge Scholars Publishing, Newcastle upon Tyne

Liu J et al (2015) Systems integration for global sustainability. Science 347(6225):963

Lubchenco J (1998) Entering the century of the environment: a new social contract for science. Science 279:491–497

Meadows DH et al (1972) Limits to growth. A Report of The Club of Rome's project on the predicament of mankind. Universe Books, New York

Norton LR (2016) Is it time for a socio-ecological revolution in agriculture? Agric Ecosyst Environ 235:13–16

Rollin BE (1996) Bad ethics, good ethics and the genetic engineering of animals in agriculture. J Anim Sci 74(3):535–541

WEF (World Economic Forum) (2020) Global risks report 2020. An unsettled world. Geneva, Switzerland. www.weforum.org

Weiner J (2017) Applying plant ecological knowledge to increase agricultural sustainability. J Ecol 105:865–870

Chapter 2
Ethics Structure and Goals

It is not clear to me where ecology ends
and the study of the ethics of nature begins,
nor is it clear to me where ecology ends
and human ecology begins

(Golley FB 1993, p. 205)

To define ethics is not a simple task in that definitions can change according to the evolution of human knowledge, institutional organisations and subjective factors. To overcome the problem, I would like to identify a few relevant components that build up a framework of reference for ethics as a system of values that guides behaviour. Figure 2.1 shows a representation of such an ethics structure.

Science, technology and economy are ethics components of the advanced industrial societies that nowadays pertain to the dominant field of *technocratic ethics* (Blok 2018), a paradigmatic area of interest and behaviour for progressive technological development, global market, and increase of both corporate profit and consumption of material goods. In this kind of technocratic ethics, the nonhuman world is valued in purely instrumental terms as the ground of human activity under a regime of nature domination. Sociologist Max Weber was the first one to rise this question in his book "Protestant ethics and the Spirit of Capitalism"[1] (Weber 2009). A strong criticism against this technocratic ethics started around the 1940s with the influential work by the members of the Frankfurt Institute of Social Research ("the Frankfurt School"). First, under the leadership of prominent sociologists and philosophers like Theodor Adorno and Max Horckheimer; successively, with Herbert Marcuse, and since the 1960s, with the substantial contribution of Jurgen Habermas under the heading of "Critical Theory":

[1]The original German text (1904–1905), was translated into English by American sociologist Talcott Parsons in 1930.

F. Caporali, *Ethics and Sustainable Agriculture*,
https://doi.org/10.1007/978-3-030-76683-2_2

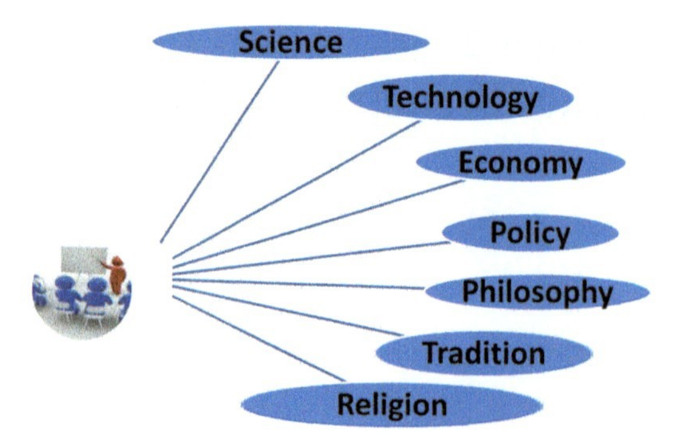

Fig. 2.1 Components of ethics as a system of values that guides behaviour

> This general critique of instrumental rationality has been carried forward and extensively revised by Jurgen Habermas, who has sought to show, among other things, how political decision-making has been increasingly reduced to pragmatic instrumentality – serving the system while "colonizing the life-world" [or cultural tradition]. According to Habermas, the "scientization of politics" has resulted in the lay public ceding ever greater area of system-steering decision-making to a technocratic elite". (Eckersley 1990, p. 741)

The main critical point raised by Habermas is the suggestion that, in a rational society, instrumental reason should "be made subservient to the norm established by practical reason"[2], adopting "a concomitant rationalization of social norms in the sphere of communication". In his "Theory of Communicative Action", Habermas (1984) advocates the establishment of a participatory democracy based upon social learning practices in a dialogical society for the achievement of a rational and normative consensus. In practice, *technical* knowledge and *practical* knowledge should converge in an *emancipatory* knowledge, whose way of inquiry is critical reflection or a *communication ethics*.

A distinguished contemporary economist as Amartya Sen[3] (Sen 1987), admits that "in the usual economic literature a person is seen as maximizing his utility function, which depends only on his own consumption, and which determines all his choices" (p. 80). This dominant paradigm of the contemporary "welfare economics" is the driver for an ethics of "self-interested behaviour" as condensed in Fig. 2.2.

Self-interested behaviour is an outcome of the reasons reported below (summarized from Sen 1987), which call for a need of bringing back economics in the more inclusive field of ethics (point v):

(i) the importance of the ethical approach has weakened as modern economics has evolved (p. 7);
(ii) the neglect of both deep normative analysis and ethical considerations in the actual human behaviour is continuously progressing (p. 7);

[2] Practical reason has its ground on language and is directed toward interpretive understanding in a dialogical community of speech (cfr. Eckersley 1990, cit.).

[3] Professor of Economics and Philosophy at Harvard University. He was awarded the Nobel Memorial Prize in Economic Sciences in 1998.

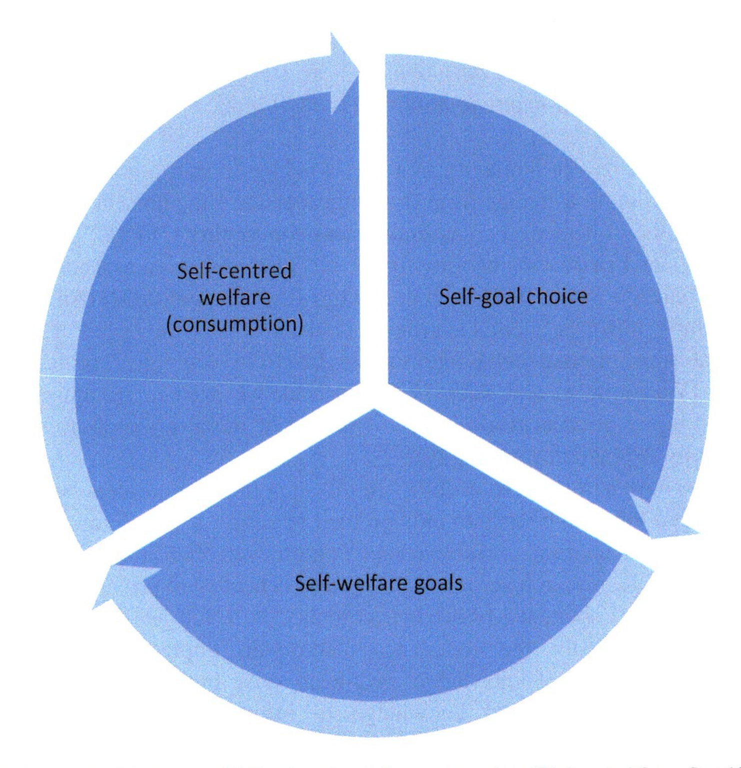

Fig. 2.2 Loop of self-interested behaviour in modern economics. (Elaborated from Sen 1987)

(iii) the distance between economics and ethics has grown impoverishing modern economics (p. 7);
(iv) economics can be made more productive by paying greater attention to the ethical considerations that shape human behaviour and judgement (p. 9);
(v) a closer contact between ethics and economics can be beneficial to both (p. 78).

Hodges (2003) claims that ethics is a relational and community issue of paramount importance in a world where billions of people are "shrinking to a village community". However, it is paradoxical that "the current focus in Western society upon the rights of the individual blurs our vision of community" in a way that a peril emerges:

> Excessive emphasis upon individual and national competition distorts progress into ruthless self-interest. A human population without ethics means the end of civil society with civilization declining into primitive lifestyles.

The advent of a transdisciplinary science like ecology has radically changed the way we understand nature and, consequently, the way we should manage nature in comparison with the apparent rationality of the instrumental reason. With the emergence and affirmation in science, philosophy and religion of an *ecological* rationality (Dryzek 1987), a revision of ethical stances has become necessary and urgent for both human interests and nature itself, i.e. for achieving a *harmony-with nature development* or a *sustainable development* (Orr 1991). According to Dryzek, the most appropriate human-nature relationship is a symbiotic one rather than an instrumental one.

The big challenge for humanity as a whole is how to start a process of *cultural conversion* leading to a change from the Anthropocene – the present Era of human dominion on the biosphere seen as limitedness commodity – to the 'Ecozoic Era' (Berry 2006), an Era of sustainability for the biosphere regarded as a community to be kept in balance within its limited context of life (Crutzen 2002). An epistemological failure known as 'mental apartheid' is likely to be at the origin of the dominant ill-informed human behaviour that is far from ecological principles and outcomes (Wackernagel and Rees 1997). Instead, we need to recognise that 'to understand better in order to do better' is a kind of moral principle to share and implement by the human community.

Scientists have a major responsibility in fulfilling this commitment because they are "privileged today to be able to indulge their passion for science and simultaneously to provide something useful to society"; moreover, "because the environment is so broad a topic, research across all disciplines is needed to provide the requisite knowledge base" (Lubchenco 1998). In accordance with Noorgard and Baer (2005), we have to recognise that the modern world shows an "unprecedented fragmentation and specialisation of knowledge", while to solve problems "scientists must bring together the dispersed knowledge to inform collective deliberation". Transdisciplinarity in human knowledge and action should facilitate the inescapable connections for achieving sustainable development patterns, in a way that unity of knowledge, unity of judgement and coherent action can proceed in tune. The science of ecology can help in this connecting process as clearly stated by Keller and Golley (2000) in the following passage:

> From ecology to ethics: the step is inevitable. When the issue of human behaviour arises, it is difficult –and may be impossible– not to ask: Is there any difference between how humans are acting, and how humans should act? Now, at the end of the second millennium, we live on a planet where the activities of one species have an impact on all processes of the biosphere. The hegemony of *Homo sapiens* constricts the freedom of all organisms [...] Ecologists cannot, and ought not, refrain from making moral judgments. Yes, ecology is political. (Keller and Golley, p. 321)

Accordingly, the quest for an *ecological ethics* is emerging as "not only desirable but urgently needed" (Curry 2011, p. 269). For legitimation, it needs a science, such as ecology, which has its ground on both value-free and value-laden research. The role of values in scientific research has become an important topic of discussion in both scholarly and popular debates as recently documented in the book "A Tapestry of Values" (Elliott 2017). Values play an important role in science by limiting unethical forms of research and by deciding what areas of research have the greatest relevance for society. Elliott's book shows cases where values have necessary roles to play in identifying research topics, choosing research questions, determining the aims of inquiry, responding to uncertainty, and deciding how to communicate information. According to a thoughtful reviewer (Nash 2017), *Tapestry of Values* will help people learn to recognise (a) the presence of values in science, (b) assess their impact and legitimacy, (c) how science can help achieve better society. Elliott suggests that scientific reasoning can be assimilated to a tapestry because it consists of two kinds of intertwined threads that scientists need to bring together to set up a research and draw a conclusion. Some of these threads are relatively rule governed and free of values, such as logical principles and mathematical techniques. The others are value-laden threads, such as background assumptions, terminological choices, and methodological decisions. Elliott supports scientists' choice to adopt value-laden research ethics if three conditions are satisfied, (a) *transparency* about the use of values, (b) *social priorities* about the common good, and (c) *engagement* for selecting and steering those values. Moreover, he identifies five key roles for values in science to play in the sequential chain of research steps reported in Fig. 2.3.

Each step entails an appropriate scrutiny of the values involved for legitimating choices and decisions. Today, ecological ethics appears as a further step in the process of rational evolution (Savulescu 2019), in that we have both to balance individual interests against those of society and to promote "some conception of justice at the cost of some aspects of individual wellbeing".

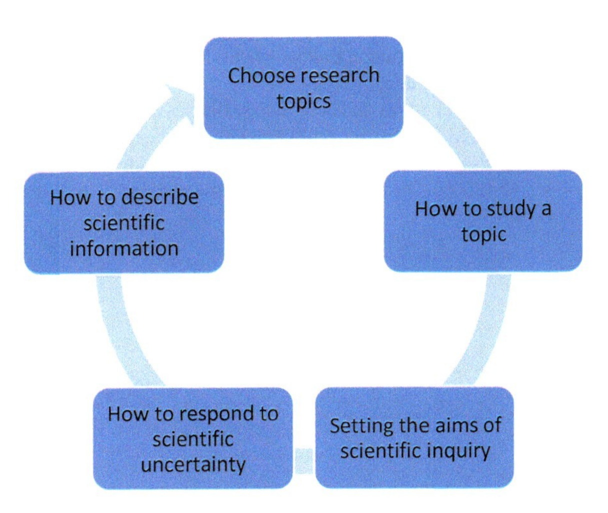

Fig. 2.3 Value-dependent research stepping stones. (Adapted from Elliott 2017; Nash 2017)

References

Berry T (2006) Evening thoughts. Reflecting on earth as sacred community (ed. Tucker ME). Sierra Club Books, San Francisco
Blok V (2018) Technocratic management versus ethical leadership. Redefining responsible professionalism in the Agri-Food Sector in the anthropocene. J Agric Environ Ethics 31:583–591
Crutzen J (2002) Geology of mankind. Nature 415(3):23
Curry P (2011) Ecological ethics. An introduction. Polity Press, Cambridge
Dryzek J (1987) Rational ecology. Basil Blackwell Ltd, Oxford
Eckersley R (1990) Habermas and green political thought: two roads diverging. Theory Soc 19(6):739–776
Elliott KC (2017) A tapestry of values: an introduction to values in science. Oxford University Press, New York
Golley FB (1993) A history of the ecosystem concept in ecology. Yale University Press, New Haven
Habermas J (1984) Theory of communicative action, vol I–II. Beacon Press, Boston
Hodges J (2003) Livestock, ethics, and quality of life. J Anim Sci 81:2887–2994
Keller DR, Golley FB (2000) The philosophy of ecology. From science to synthesis. The University of Georgia Press, Athens
Lubchenco J (1998) Entering the century of the environment. Science 279:491–497
Nash EJ (2017) Are values in science like a tapestry or a patchwork quilt? Sci Educ 26:1063–1069
Noorgard RB, Baer P (2005) Collectively seeing complex systems: the nature of the problem. Bioscience 55(11):953–960
Orr DR (1991) Ecological literacy. SUNY Press, Albany
Savulescu J (2019) Human Enhancement. In: Edmond D (ed) Ethics and the contemporary world. Routledge, London, pp 319–334
Sen A (1987) On ethics and economics. Blackwell Publishing, Oxford/New York
Wackernagel M, Rees WE (1997) Perceptual and structural barriers to investing in natural capital: economics from an ecological footprint perspective. Ecol Econ 20:3–24
Weber M (2009) Die protestantische Ethik und der Geist des Kapitalismus. Anaconda Verlag GmbH, Köln

Chapter 3
The Emergence of Ecological Awareness

It is the system as a whole that evolves,
and we can hope to establish a definition of the direction of
evolution only in terms of the system as a whole.

(Lotka AJ 1925, p. 22)

I chronologically assume to divide the development of ecological awareness in two separate periods, the first or *emergent* one and the second or *operational* one. The first one started with the birth of the new term *ecology.* The point of transition from the first to the second one was the formulation of the new term *ecosystem* by the ecologist Arthur Tansley, which dates back to the year 1935.

The word "ecology" originated in German language (Oecologie) from Ernst Haeckel (1866), a theoretical morphologist, with the aim to provide a neologism to identify 'relations' among organisms within their environment of life. Originally, *relations* as key word reveals the epistemic need to overcome the fallacy of considering a thing, be it an organic or inorganic one, as a separate entity. A thing exists and coevolves within its context of relations. The focus of interest for both ecological inquiry and management is not so much an individual entity, but the *web of relations* in which he/she/it is involved. This epistemological stance to 'see' *webs* instead of separate points opens up a new window to appreciate reality from an *ecocentric* perspective instead of an *anthropocentric* one, where man is alienated from nature and regards it as a separate ground for resources and wastes. This epistemological change is a step forward to recognising humanity's embeddedness in the natural world. Such as cultural transition is still in progress and involves new ethical implications, as shown in the following quote:

> Ecocentric theorists are concerned to emphasize our continuity with and relatedness to the nonhuman world, rather than our separation and differentiation from it, and to cultivate an orientation that recognises how the development and fulfilment of the part can only proceed from its complex interrelationship and unfolding within the larger whole. Such an empathic orientation should imbue all of human activity – not only art, play, and contemplation but also work, science, and technology. (Eckersley 1990, p. 764)

F. Caporali, *Ethics and Sustainable Agriculture*,
https://doi.org/10.1007/978-3-030-76683-2_3

Haeckel's inspiring term *ecology* has likely part of its roots in the emergence of chemistry, a science developed in the nineteenth century. Justus von Liebig, a celebrated German soil chemist, made an important contribution to the biogeochemistry approach (Box 3.1) as early as 1840, in his book *'Chemistry in its application to Agriculture and Physiology"*[1] as this passage shows:

> An inquiry into the conditions on which the life and growth of living beings depend, involves the study of those substances which serve them as nutriment, as well as the investigation of the sources whence these substances are derived, and the changes which they undergo in the process of assimilation. (Liebig 1847, p. 9)

The fundamental ecological insights concerning *trophic relations* and their local and global effects, as reported in Box 3.1, were due to Liebig's commitment to investigating agricultural activity with a transdisciplinary attitude, where "the power and knowledge of the physiologist, of the agriculturalist and chemist, must be united for the solution of problems" (Liebig 1847, p. 49).

Box 3.1: Biogeochemical Relations in the Web of Life and Planetary Metabolism (From Liebig 1847, Summarised)

1. A beautiful connection subsists between the organic and inorganic kingdoms of nature.
   Inorganic matter affords food to plants and they, on the other hand, yield the means of subsistence to animals (p. 9).
2. The leaves and other green parts of a plant absorb carbon acid, and emit an equal volume of oxygen (p. 16).
3. [Oxygen] is essential for the support of the important vital process of respiration (p. 16).
4. Animals on the other hand expire carbon, which plants inspire; and thus the composition of the medium in which both exist, namely, the atmosphere, is maintained constantly unchanged (p. 16).
5. Plants thus improve the air by the removal of carbonic acid, and by the renewal of oxygen, which is immediately applied to the use of man and animals (p. 17).

PLANETARY METABOLISM

6. The proper, constant, and inexhaustible sources of oxygen gas are the tropics and warm climates, where a sky, seldom clouded, permits the glowing rays of the sun to shine upon an immeasurably luxuriant vegetation (p. 17).
7. The temperate and cold zones, where artificial warmth must replace deficient heat of the sun, produce, on the contrary, carbonic acid in superabundance, which is expended in the nutrition of the tropical plants (p. 17).
8. The same stream of air, which moves by the revolution of the earth from the equator to the poles, brings to us in its passage from the equator, the oxygen generated there, and carries away the carbonic acid formed during our winter (p. 17).

Another supposed source of inspiration for the fortunate term *ecology* was Haeckel's major interest for Darwin's theory of *evolution*, as advanced by Stauffer (1957), who cites Haeckel himself referring on his "Generelle Morphologie" as follows:

> A somewhat comprehensive work, which constituted the first attempt to apply the general doctrine of development to the whole range or organic morphology (Anatomy and Biogenesis), and thus to make use of the vast march onwards which the genius of Charles Darwin has effected in all biological science by his reform of the Descent Theory and its establishment through the doctrine of selection.

Providing a series of citations from Darwin's *Origin of Species* (1859) in order to demonstrate the Darwinian derivation of the term *ecology*, Stuffer ends up with this last citation:

> We can let Darwin conclude in the terms which seem to have been most influential on Haeckel: "Let it be borne in mind how infinitely complex and close-fitting are the mutual relations of all organic beings to each other and to their physical conditions of life". (Darwin1859, p. 69)

It is very likely true that both the emergence of *chemistry* and the conceptual innovation in biology carried out by the term *evolution* brought about the inspiring conditions for Haeckel to coin the term *ecology*. To confirm this statement suffices considering the following and more detailed definition of ecology issued by Haeckel in the second volume of his "Generelle Morphologie"[2]:

> By ecology, we mean the whole science of the relations of the organism to the environment including, in the broad sense, all the "conditions of existence". These are partly organic, partly inorganic in nature; both, as we have shown, are of the greatest significance for the form of organisms, for they force them to become adapted. Among the inorganic conditions of existence to which every organism must adapt itself belong, first of all, the physical and the chemical properties of its habitat, the climate (light, warmth, atmospheric conditions of humidity and electricity), the inorganic nutrients, nature of the water and of the soil etc. As organic conditions of existence we consider the entire relations of the organism to all other organisms with which it comes into contact, and of which most contribute either to its advantage or its harm. Each organism has among the other organisms its friends and its enemies, those which favour its existence and those which harm it [...] In our discussion of the theory of selection we have shown what enormous importance all these adaptive relations have for the entire formation of organisms, and specially how the organic conditions of existence exert a much more profound transforming action on organisms than do the inorganic.[3]

In this extended Haeckel's definition of *ecology*, the term *relations* constitutes the very core of a new science able to investigate the interactions between organisms and their contest of life. The web of relations accounted for by the language of chemistry, either organic or inorganic, and the language of biology, either as morphology or physiology, are also functional for explaining evolution of organisms in their community of life.

[2] Haeckel, E 1866, Vol.II, p. 286.

[3] Translation from German in Stauffer 1957.

The concept of evolution was also important to identify and investigate the relations between man and nature at the level of geographic and geological scale. A celebrate geographer of the XIX century, Alexander von Humboldt, reports nature as "that which is ever growing and ever unfolding itself in new forms"(von Humboldt 1849, p. 21) and as a conclusion of his "Cosmos" expresses a clear concept of co-evolution between man and nature in the following terms:

> In this fragmentary sketch of the phenomena of organisation, I have ascended from the simplest cell – the first manifestation of life – progressively to higher structures [...] The general picture of nature I have endeavoured to delineate, would be incomplete, if I did not venture to trace a few of the most marked features of the human race, considered with reference to physical gradations – to the geographical distribution of contemporaneous types – to the influence exercised upon man by the forces of nature, and the reciprocal, although weaker, action which he in his turn exercises on these natural forces [...] By activity of mind and the advance of intellectual cultivation, no less than by his wonderful capacity of adapting himself to all climate – man everywhere becomes most essentially associated with terrestrial life. (von Humboldt 1849, pp. 360–361)

From a geological perspective, Austrian geologist Edward Suess (1875) introduced a more adequate terminology to express the interaction among physical and biological forces in determining the evolution of life. He coined the term *biosphere* in order to situate the right place where the phenomenon life happens on Earth. According to Suess, biosphere is "the region of interaction" between the atmosphere and the lithosphere, where plants live to breath into the air and feed into the soil. A few years earlier, Italian geologist Antonio Stoppani (1871) had provided a more detailed conception of what happens at a planetary level as effect of the interaction among geological, physical, chemical, biological and anthropic forces. He had defined the entire planet as "a telluric organism" (from Latin *tellus* – *Earth*), having Earth a proper metabolism, where production and consumption happen in a circular and complementary patterns on a spatiotemporal scale and allow life to prosper, regenerate and change in a continuous flow of biological forms. Concerning the anthropic influence, he had made a prophetic claiming by asserting that humanity is "a telluric force" that does compare with the other geological forces because of its technological power and universal effect. He had announced that "the anthropozoic era" (today defined "Anthropocene") had just began and that the geologist was unable to foresee its end. His conclusion had been extremely truthful: "Earth will be never getting out from the hands of humanity until completely moulded by human footprint" (Stoppani 1873, p. 740).

Philosophers at that time did not fail to challenge the conversion of society to rapid industrialisation under the push of a rampant capitalism. A fine collection of essays on key concepts in Critical Theory edited by Carolyn Merchant (Merchant 1994) presents the contribution of marxist philosophers to alleviate the emergent rush toward the domination of nature. Notably, Friedrich Engels and Carl Marx put forward interesting insights with ecological meaning, a few of which are in Box 3.2. This essay collection reveals Merchant's intention of both bringing together critiques and alternatives to mainstream social theory, science, technology, and addressing ethical challenges:

Domination is one of our century's most fruitful concept for understanding human-human and human-nature relationships. When the domination of nonhuman nature is integrated with the domination of human beings and the call for environmental justice, Critical Theory instils the environmental movement with ethical fervour (Merchant 1994, p. 1).

Box 3.2: Engels and Marx on Ecology and Society (From H.L. Parsons 1977, Reported in Merchant 1994, pp. 28–43, Modified)

Friedrich Engels, *Dialectics of Nature*

1. For in nature nothing takes place in isolation. Everything affects every other thing and vice-versa, and it is mostly because this all-sided motion and interaction is forgotten that our natural scientists are prevented for clearly seeing the simple things (p. 40)
2. Thus at every step we are reminded that we by no means rule over nature like a conqueror over a foreign people, like someone standing outside nature – but that we, with flesh, blood, and brain, belong to nature and exist in its midst, and that all our mastery of it consists in the fact that we have the advantage over all other creature of being able to know and correctly apply its laws (pp. 41–42).

Karl Marx, *Capital*

CAPITALISM RENDS THE UNITY OF AGRICULTURE AND INDUSTRY, DISTURBS MAN'S RELATION TO THE SOIL, WASTES WORKERS, AND ROBS LABORERS AND SOIL (pp. 39–40)

1. Capitalist production, by collecting the population in great centres, and causing an ever increasing preponderance of town population, on the one hand concentrates the historical motive power of society; on the other hand, it disturbs the circulation of matter between man and the soil, i.e., prevents the return to the soil of its element consumed by man in the form of food and clothing; it therefore violates the conditions necessary to lasting fertility of the soil.
2. In modern agriculture, as in in the urban industries, the increased productiveness and quantity of the labour set in motion are bought at the cost of laying waste and consuming by disease labour-power itself.
3. Moreover, all progress in capitalistic agriculture is a progress in the art, not only of robbing the labourer, but of robbing the soil.

As to the biosphere concept, it would have had a bright future as a pillar of ecology that persists until now. It took strength in the 1920s, especially with the contributions of Russian biogeochemist Vladimir Vernadskij (1945) and French priest and paleontologist Pierre Teilhard de Chardin (1959). Their personal acquaintance, developed during a common period of permanence at the Sorbonne in Paris in 1922–1923, was fruitful in yielding a common interest for scientific and philosophical speculation about the process of coevolution between man and nature, intended at planetary scale and identified with the new term *noosphere* (the sphere of "reason"). While Vernadskij was able to provide the concept of noosphere with scientific evidence, Teilhard de Chardin emphasized its spiritual content and the eschatological meaning, casting the base for a new field of theological and ecological speculation recognised today as *ecotheology* (ecological theology). The cultural potential of the biosphere concept would have reached its maximum later, with the introduction of the *ecosystem* concept, where biodiversity is one of the most meaningful components.

Vladimir Vernadskij left in the last years of his life an important manuscript in the hands of his son George, who translated it from the Russian at Yale University and let it published under the editorship of Professor G.E. Hutchinson in 1945. The paper shows all steps he made to perform the transition from the concept of biosphere to the concept of noosphere and its relative ethical implications, as summarised in Box 3.3.

Vernadskij mentions biophysicist Alfred Lotka as a scholar that would have tried to solve the last challenge in Box 3.3 (reconstruction of the biosphere). Indeed,

Box 3.3: Vernadskij's Transition from the Biosphere to the Noosphere Concept (After Vernadskij 1945, Summarised)

LIVING MATTER VS. LIFE

"Instead of the concept of 'life', I introduced that of 'living matter'[…] Living matter is the totality of living organisms […] The concept of 'life' always steps outside the boundaries of the concept of 'living matter'; it enters the realm of philosophy, folklore, religion and the arts.

The evolutionary process is a characteristic only of living matter. There are no manifestations of it in the inert matter of our planet. The history of living matter expresses itself as a slow modification of the forms of living organisms which genetically are uninterruptedly connected among themselves from generation to generation.

The whole of mankind put together represents an insignificant mass of the planet's matter. Its strength is derived not from its matter, but from its brain. If man understands this, and does not use his brain and his work for self-destruction, an immense future is open before him in the geological history of the biosphere".

MANKIND AS A GEOLOGICAL FORCE IN THE BIOSPHERE

"Mankind taken as a whole is becoming a mighty geological force […] Chemically, the face of our planet, the biosphere, is being sharply changed by man, consciously, and even more so, unconsciously, The aerial envelope of the land as all its natural waters are changed both physically and chemically by man […].

Man now must take more and more measures to preserve for future generations the wealth of the seas which so far have belonged to nobody".

MANKIND AS THE DRIVING FORCE IN THE NOOSPHERE

"There arises the problem of *the reconstruction of the biosphere in the interests of freely thinking humanity as a single totality*. This new state of the biosphere which we approach without our noticing it, is the noosphere. The noosphere is a new geological phenomenon on our planet. In it for the first time man become a large-scale geological force. He can and must rebuild the province of his life by his work and thought, rebuild it radically in comparison with the past.

The noosphere is the last of many stages in the evolution of the biosphere in geological history".

Lotka is credited for being the first scholars to write an ecological treatise (Lotka 1925) without mention the term ecology! (Caporali 2015). His insights are so deep in an ecological fashion to forerun those of the ecosystem approach. His endeavour starts from enquiry in energy transfer, trophic chains, biogeochemical cycles, biodiversity accounting and ends up with analysing human consciousness and its effects on planetary environment. As to the title and topic of the work, Lotka explains in the book's preface that he uses the expression *Physical Biology* to denote the broad application of physical principles and mathematical methods in the "contemplation of biological *systems*". That declaration unveils a transdisciplinary attitude of enquiry that is typical of a systems thinking approach, as confirmed in the first chapter (general principles) by the following statements:

> The several organisms that make up the earth's living population, together with their environment, constitute one system, which receives a daily supply of available energy from the sun [...] It is customary to discuss the "evolution of a species of organisms". As we proceed we shall see many reasons why we should constantly take in view the evolution, as a whole, of the system [organism plus environment] [...] It is not so much the organism or the species that evolves, but the entire system, species and environment. The two are inseparable. (Lotka 1925, p. 16)

The ascent of life evolution culminates with human consciousness, posing the critical question of thought development and its influence, as total humanity, on the evolution of the planetary system. Lotka advances the proposal to regard man as both a *Knower* and a *Willer*, considering *Knowing* and *Willing* as complementary thought manifestations that must yield beneficial outputs for both humanity and planetary life. As concerns the synthesis of knowing and willing, Lotka proposes an interesting framework for a classification of the Sciences, as reported in Table 3.1. Philosophy is the science with the task of connecting the self with the external world; Psychology is the science within the self that connects knowledge with will, and Ethics is the science of the Willer's choices and responsibility. In conclusion, for "man is part of Nature, the part that studies the whole", Lotka's final comments on the relation between man and nature are as follow:

> Thus, in the light of modern knowledge, man is beginning to discern more clearly what wise men of all ages have intuitively felt – his essential unity with the Universe [...] A race with

Table 3.1 Classification of the sciences

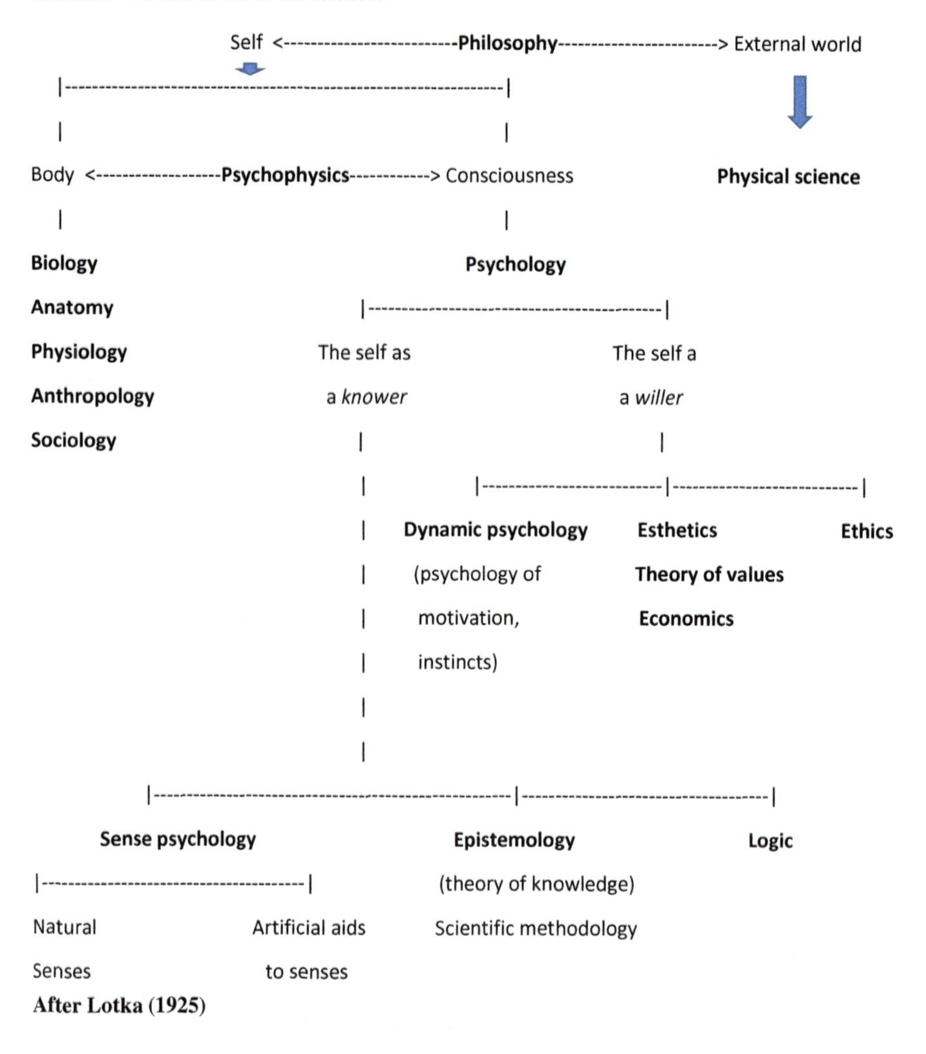

After Lotka (1925)

desires all opposed to nature could no longer endure; he that survives must, for that very fact, be in some measure a collaborator with Nature. With extending knowledge must come awakening consciousness of active partnership with the Cosmos. (Lotka 1925, p. 433)

A classification of the sciences as that reported in Table 3.1 reveals the hidden quest for a reunification of knowledge that is necessary, then as well as todays, to bridge the gap between to arbitrarily separated subjects, the human ego (self) and the human environment (external world).

Reconnecting mentally man with nature through a new formulated epistemological tool, such as the concept of ecosystem, was a necessary step for linking theory and practice on a point of more advanced ethical convergence, todays defined as sustainable development. The following extract from Lotka is extremely revelatory in that sense:

> We have noted elsewhere that, physically, no clear line of division can be drawn between the body and the environment. Psychologically too, we saw that the ego is not something that divides the world into separate fields, but is rather of the nature of a standard of reference in terms of which we find it most convenient to describe experience. But our system of coordinate reference frame, from a good servant, has threatened to become a bad master [...] We have estranged and objectified the world, and lost the sense that we are of it. (Lotka 1925, p. 433)

References

Caporali F (2015) History and development of agroecology and theory of agroecosystems. In: Monteduro et al (eds) Law and agroecology. A transdisciplinary dialogue. Springer, Berlin, pp 3–29
Eckersley R (1990) Habermas and green political thought: two roads diverging. Theory Soc 19(6):739–776
Haeckel E (1866) Generelle Morphologie der Organismen. Allgemeine Grundzüge der organischen Formen – Wissenschaft, meckanisch begründet durch die von Charles Darwin reformirte Decendenz-Theorie, 2 Vols. Reimer, Berlin
von Humboldt A (1849) Cosmos: a sketch of a physical description of the universe. Henry G, Bohn/London
Liebig J (1847) (English Edition) Chemistry in its application to agriculture and physiology. T.B. Peterson, Philadelphia
Lotka A (1925) Elements of physical biology. William & Wilkins Company, Baltimore
Merchant C (1994) Ecology. Key concepts in critical theory. Humanities Press, Atlantic Highlands
Stauffer RC (1957) Haeckel, Darwin and ecology. Q Rev Biol 32(2):138–144
Stoppani A (1871) Corso di Geologia, vol 1. Bernardini e Brigola, Milano
Stoppani A (1873) Corso di Geologia, vol 2. Bernardini e Brigola, Milano
Suess E (1875) Die Entstehung der Alpen. Inktanken
Teilhard de Chardin P (1959) The phenomenon of man (ed. and trans. Wall B). Collins, London
Vernadskij VI (1945) The biosphere and the noosphere. Am Nat 33(1):243–265

Chapter 4
Development of Ecological Awareness

Regardless of the level on which life is examined,
the ecosystem concept can appropriately be applied

(Evans FC 1956, Ecology, p. 1127)

The ecosystem concept is the theoretical underpinning of all modern science of ecology. It is worth noting that the "inventor" of the term "ecosystem", Arthur George Tansley, was a man of different interests, a biologist, a psychologist and a philosophy scholar, who "struggled throughout his career to open traditional botany to ecology and to have ecology equally accepted as part of the natural sciences" (Golley 1993, p 9). His biography and contribution to the science of ecology are well documented in the just mentioned Golley's book. However, in reconsidering the context that originated the ecosystem concept is very useful the information provided by Peder Anker (2002) on the intellectual underpinnings of Tansley's thinking on systems and ecosystem theory. An unpublished paper of Tansley (2002) prepared by Peder Anker for the scientific journal "Ecosystems"[1] shows details about his vision of the world and the way he managed to form it. Tansley first declares that his survey about the field of knowledge was inevitably biased by his training in natural sciences, with the end of using as a framework the history in time of the visible universe, the planet of earth, the appearance and development of life, the advent of man with his mind and specific activities. In practice, he reconstructs "the temporal genetic series", as he calls it, constituted by *inorganic matter – living organism – mind* as a succession that appears from scientific enquiry. The organisation level of living organism is a crucial one in this sequence and its characters are reported by Tansley as follows:

[1] Tansley, A.G. 2002. The temporal genetic series as a means of approach to philosophy. In "Ecosystems", 5, 614–624. This paper was presented before The Magdalen Philosophy Club of Oxford University on 5 May 1932.

F. Caporali, *Ethics and Sustainable Agriculture*,
https://doi.org/10.1007/978-3-030-76683-2_4

> Organisms are to be regarded as self-maintaining physico-chemical systems able to assimilate material from without, to reproduce themselves, to respond specifically to influences from without, and, within limits, of astonishing stability [...] Living organisms have given rise to the most extensive changes on the surface of this planet, and at the same time have created a distinct sphere of existence that has attained immense complexity and necessitates separate study and a distinct branch of science with its own method and "laws" [...] Organisms do seem to have "emerged" from an inorganic world, as something that we cannot refuse to call "new"(Tansley 2002, p. 617).

The emergence of mind constitutes the bridge connecting the gap between the physiological and the psychological, in a way that allows man to react to external or internal stimuli in an intentional way. "Mind emerged as organisms emerged; and gradually, not suddenly" and "the development of mind depends on the progressive extension of awareness,[2] both external and internal". Human intentionality, i.e. capability to make decisions, derives from a process of creation of values where "all the values we know are human values build up in human minds in relation to the most various aspects of the universe"(Tansley 2002). Moreover, Tansley admits a hierarchy of values based on "the ethical judgement of good", where good is regarded in general as harmony between means and ends, and in particular for man as harmony, both externally and internally, with our ideal of virtuous life.

In Fig. 4.1, I propose a tentative representation of Tansley's "view of the universe", i.e. the ascent from material to ethical values as suggested by his theory of "the temporal genetic series", where an apparent continuum exists among the

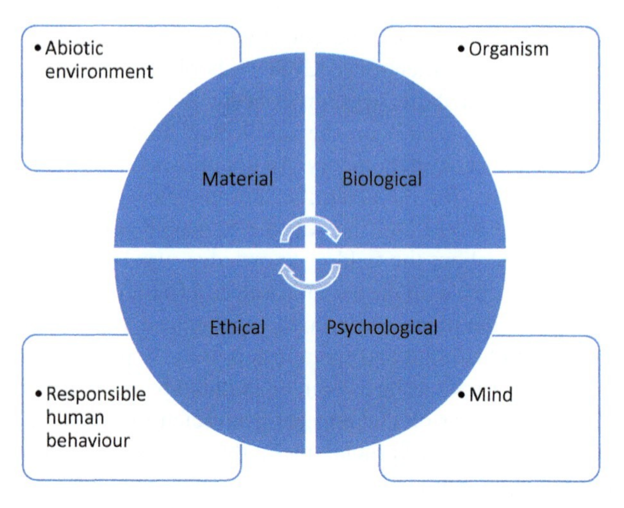

Fig. 4.1 Representation of Tansley's "view of the universe" as "a series of phenomena arising in temporal sequence"

[2] "The whole of the operations of mind, including conceptual thought, consists entirely in the interplay of external and internal awareness [...] Awareness must have arisen through combinations of physiological perceptions registered as wholes in the organism" Tansley 2002, p.618.

different steps, even if at each step a new gap needs to be overcome. The gaps between the material and the biological, and between the biological and the psychological are spontaneously bridged by nature itself in the process of evolution,[3] while the last gap between the psychological and the ethical is a cultural task to perform with progress in human understanding, will and sense of responsibility. Tansley made a decisive effort to bridge the last gap with the formulation of the ecosystem concept, which is to regard as both a powerful tool for epistemological and ethical advances.

As a conclusion of his "theory of temporal genetic series", Tansley proposes "a judgement of harmony and integration" as general criterion for ethical behaviour in that harmony and integration operate as "a common principle in the whole course of evolution". However, he also advises that there is a kind of physiological- psychological barrier to overcome in order to achieve a "judgement of harmony and integration":

> Because there is in the individual human being so intense a consciousness of his own unity, conditioned by his spatially limited and highly integrated organism and by his distinctness from the rest of the world, he abstracts and hypostatizes a sort of inner essence of himself that is represented as something distinct and separate fro the rest of the universe, sitting apart, as it were, and experiencing everything else (Tansley 2002, p 621).

To overcome this existential condition of "mental apartheid", as later defined by many scholars (Wackernagel and Rees 1997), a new powerful epistemological aid for human mind had to be envisaged in order to promote ecological understanding and behaviour, such a tool was the ecosystem concept.

4.1 The Ecosystem Theory

Tansley's merit concerns not only the definition of the ecosystem concept but also the construction of an ecosystem theory useful for better understanding the relation between man and nature. The new linguistic term *ecosystem* corresponds to the need of identifying with a unique term the organisation pattern of nature, as we see it. The term *ecosystem* provides a concept of organisation that is deducible from the observation of natural structures and processes. The occasion for its definition was a Tansley's paper devoted to clarify "the use and abuse of vegetational concepts and terms", where the main discussion point was the theory of vegetational development or *succession* as supported by scientific evidence and philosophical speculations (Tansley 1935). To define the entire development of the *successional series* and its laws until the final phase of *climax*, alternative terms coming from different scholars, such as Frederick E. Clements and John Philipps, were critically discussed along the sequence *organism -> quasi-organism -> complex organism -> biotic community*,

---

[3] "Organism" is the natural tool to bridge the gap between the material and the biological as "mind" is the natural tool to bridge the gap between the biological and the psychological.

before the final completion by Tansley with the term *ecosystem*. Scientific and philosophical grounds for the edification of the ecosystem concept can traced back to the bibliographical citations made by Tansley himself. The main scientific source is revealed just in his paper introduction where the "beautiful successional series" provided by Henry Chandler Cowles are mentioned as meaningful examples of "the reality and the universality of the process". It is worth examining the under reported extract from Cowles (1911) as an example of ecological holism derived by an inductive method:

> It is clear that vegetative cycles are not of equal value. Each climatic cycle has its vegetative cycle; each erosive cycle within the climatic cycle in turn has its vegetative cycle; and biotic factors institute other cycles, quite independently of climatic or topographic change. It is small wonder that within this complex of cycle within cycle, each moving independently of the others and at times in different directions, dynamic plant geography has accomplished so little in unravelling the mysteries of succession.

From a philosophical perspective, Tansley admits that the doctrine of "emergent evolution" in Smuts' holism (Smuts 1926) contains "some, though not all [...] acceptable and useful ideas", that he tries to adopt as shown in the following ecological assumption:

> Various "biomes", the whole webs of life adjusted to particular complexes of environmental factors, are real "wholes", often highly integrated wholes, which are the living nuclei of *systems* in the sense of the physicist. Only I do not think they are properly described as "organisms" (except in the "organicist" sense"). I prefer to regard them, together with the whole of the effective physical factors involved, simply as "systems".

The concepts of both hierarchy and integration, respectively expressed in the first and in the second of the two previous extracts, find their fusion point in the next synthetic concept of ecosystem, as shown in Box 4.1.

The definition reported in Box 4.1 shows that the word *ecosystem* is both a model of reality representation (a concept, an epistemological tool) and the reality itself

Box 4.1: *Definition of Ecosystem and Elements of an Ecosystem Theory (From Tansley 1935)*

DEFINITION

"The more fundamental conception is, as it seems to me, the whole system (in the sense of physics), including not only the organism-complex, but also the whole complex of physical factors forming what we call the environment of the biome – the habitat factors in the widest terms."

"It is the systems so formed which, from the point of view of the ecologist, are the basic units of nature on the face of the earth."

"These *ecosystems*, as we may call them, are of the most various kind and sizes. They form one category of the multitudinous physical systems of the universe, which range from the universe as a whole down to the atom."

METHODOLOGY AND PROPERTIES

"The whole method of science [...] is to isolate systems mentally for the purposes of study, so that the series of *isolates* we make become the actual objects of our study, whether the isolate be a solar system, a planet, a climate region, a plant or an animal community, an individual, an organic molecule or an atom."

"Some of the systems are more isolated in nature, more autonomous, than others. They all show organisation, which is the inevitable result of the interaction and consequent mutual adjustment of their components. If organisation of the possible elements of a system does not result, no system forms or an incipient system breaks up."

"The more relatively separate and autonomous the system, the more highly integrated it is, and greater the stability of its dynamic equilibrium."

"The *climax* represents the highest stage of integration and the nearest approach to perfect dynamic equilibrium that can be attained in a system developed under the given condition and with the available components."

"In fact the climatic complex has more effects on the organisms and on the soil of an ecosystem than these have on the climatic complex, but the reciprocal is not wholly absent."

MAN AS AN ECOSYSTEM COMPONENT

"Is man part of "nature or not? [...] regarded as an exceptionally powerful biotic factor which increasingly upsets the equilibrium of preexsisting ecosystems and eventually destroys them, at the same time forming new ones of very different nature, human activity finds its proper place in ecology."

"*Anthropogenic ecosystems* differ from those developed independently of man. But the essential formative processes of the vegetation are the same."

"Ecology must be applied to condition brought about by human activity. The "natural" entities and the anthropogenic derivates alike must be analysed in terms of the most appropriate concept we can."

that it refers to (the real, material world represented). It is a case where epistemology and ontology seem to coincide. From the perspective of the philosophical approach of the modern philosophy of science, the ecosystem concept may belong to the field of "scientific realism "(McGrath 2009, p. 214–215). The two basic pillars of *hierarchy* and *integration* that emerge as ecosystem properties methodologically determined have operated as pure "legs" on which the new ecosystem ecology has moved its first steps and is still standing and developing.

4.1.1 The Ecosystem Concept as a Meaningful Integrative Level in the Hierarchical Organisation of Reality

Looking at the astonishing scientific-philosophical debate that originated the term ecosystem, as historically presented by the same promoter (Tansley 1935), we can easily discover that the objective necessity which prompted the neologism was that of another word for the term "complex organism" suggested by Philipps and Clements as designing the development of the biotic community in its habitat. The term "complex organism" disliked Tansley because "the modern biologist means by an organism an individual animal or plant, and would usually refuse to apply the term to anything else". The matter of contention was properly a matter of defining appropriately with a term the state of organisation that characterises different hierarchical levels of reality. For the level of organisation of an individual plant or animal, the term organism was fine; for the level of organisation of the biotic community in its habitat was not, and "ecosystem" was the new term that Tansley proposed for defining that level of reality organisation.

Ecosystem was actually the missing term for conceptually identifying the *basic unity of organisation* between inorganic and organic components in nature. The link is through the *integration* of functional processes in a trophic dynamics. With the ecosystem concept, Tansley convincingly responded to the quest for an empirical knowledge of nature – i.e. with what we perceive through the senses – such as natural sciences are concerned with (Rowe 1961). A few year later, Lindemann (1942) made a clear illustration of the trophic-dynamic aspects of the processes of energy transfer and nutrient cycling that normally operate in nature, documenting logically and mathematically the functional integration between the inorganic and organic ecosystem components. After recalling the basic contribution of Thienemann (1918) to defining a new terminology with the language of "community economics",[4] Lindemann was able to demonstrate that life is an integrative process organised through a trophic chain. The first ring (*producers*) transforms solar energy and accumulates plant biomass (or chemical energy) available for further transformation (*consumers*, second ring as herbivors, carnivors, onnivors); the last ring (*decomposers*) dissipates energy while returning inorganic nutrients to the environment for fuelling new cycling. The bio-geo-chemistry perspective of reality clearly corroborates a theory of integrative life on the planet Earth. An iconographic representation of the ecosystem concept inspired by the insights expressed in the Lindemann's paper is shown in Fig. 4.2. According to his opinion, "the concept of the ecosystem is believed to be of fundamental importance in interpreting the data of dynamic ecology", and his definition of a real ecosystem is the following one:

> The *ecosystem* may be formally defined as the system composed of physical-chemical-biological processes active within a space-time unity of any magnitude.

[4] It refers to the introduction of the functional categories of *producers*, *consumers* and *reducers* (*decomposers*).

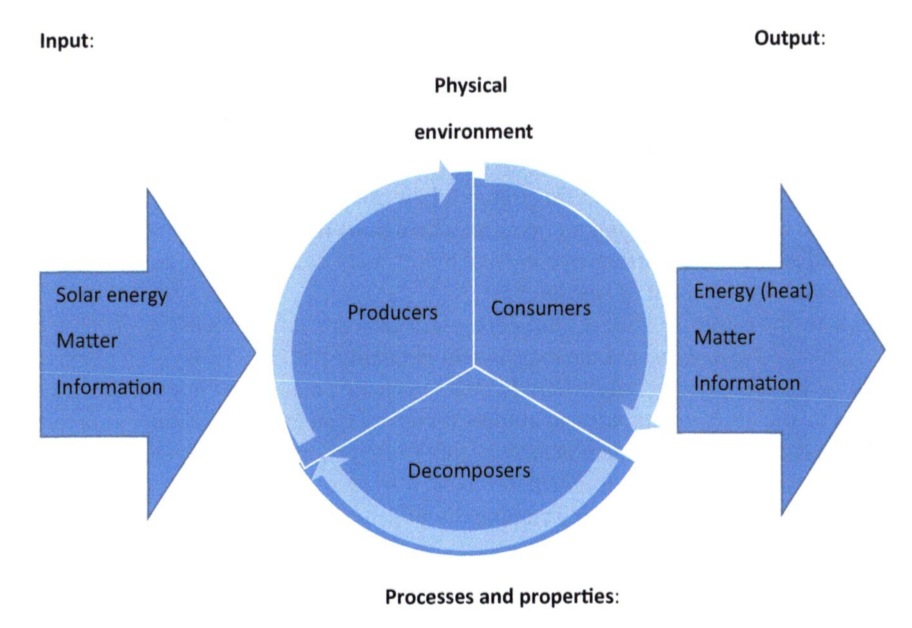

Fig. 4.2 Ecosystem organisation in a temporal trophic- dynamics perspective

The information content of Fig. 4.2 is relevant under multiple perspectives and unveils the enormous potential of the ecosystem theory. The concept of ecosystem is the rational instrument behind the conceptual map represented in Fig. 4.1 that integrates the relations among natural components in a functional web. In its turn, this relational web accounts for the existence, persistence and evolution of the whole ecosystem itself. The ecosystem concept is a broad concept that includes different meanings:

1. From a philosophical point of view, the ecosystem concept is an *epistemological tool*, i.e. a device of human mind useful to model a representation of reality in accordance with philosophical and scientific knowledge. As a rational instrument, it emerges from both natural and cultural roots and accounts for human capacity of representing the "voice" of nature, i.e. its self-consciousness.
2. Concerning the web of relations among ecosystem components in a definite place, the ecosystem concept is a *model of horizontal integration* for showing the functional complementarity of parts within a whole – a functional hierarchy in a trophic context.
3. As to the spatial extension, the ecosystem concept is dimensionless, i.e. it holds at all size scales from microsphere to biosphere (Reiners 1986). It implies a *model of vertical integration* of micro-ecosystems within macro-ecosystems- a nested integration in space and time of ecosystems in a hierarchical assemblage.

4. As to the temporal extension, the ecosystem concept is a *model of evolutionary integration*, spanning from developing processes at local scale (successional series) until change at planetary level (geosphere ---- > biosphere ---- > noosphere).
5. As to kinetics and dynamics, the ecosystem concept is a *structural-functional model* that identifies: a) discrete inorganic components and biological forms as entities of reality linked in functional relationships (autotrophic and etherotrophic components; grazing and detritus chains); b) functional processes such as transfer of energy, nutrient cycling and biodiversity creation as the basis for continuous development or *sustainability.*[5]
6. As to the systems science, the ecosystem concept is an *input/output model*, showing that *openness* is the most essential quality for a real ecosystem to exist, function and persist (Dale 1970). In general, energy, matter and information constitute the necessary inputs and outputs for an ecosystem to perform in its context of existence. An ecosystem is *inseparable* from its context of existence.
7. The ecosystem concept is a *heuristic model*, because it accounts for natural creative evolution and increasing complexity of nature toward more widespread and purposeful self-consciousness.
8. The ecosystem concept is an *aesthetic model*, because shows nature as a non-human marvellous construct of forms and relationships.
9. The ecosystem concept is an *ethical model*, because shows that: a) co-operation among ecosystem components is the driving force for sustainability in nature; b) nature is a common and collective good – a community instead of a commodity. The ecosystem concept provides basic principles for eco-development strategy such as integral use of solar radiation, material re-cycling and biodiversity enhancement.

Some more comments deserves point 3 (*vertical integration*) of the above reported list, because it has both logical and epistemological foundations, and practical meaning in the application of the ecosystem concept for management reasons. For those aspects, it is important to mention the following statement by Rowe (1961) on the basic proposition for "a logical, useful level-of-integration scheme":

> The object of study of whatever level must contain, volumetrically and structurally, the objects of the lower levels, and must therefore be itself a part of the level above. Each object will then constitute the immediate environment of the object at the level below while forming a structural-functional part of the object at the level above.

If the ecosystem concept is applicable "within a space-time unity of any magnitude"(Lindemann 1942), being it dimensionless (Reiner 1986), the critical points for correct ecological methodology, understanding and management, are:

---

[5]The term *sustainability* is an ecosystem property deducible from the following comment of Tansley (1939): "The equilibrium of an ecosystem is always false, but it is certain that its plant and animals, together with all inorganic factors on which they depend, work together to create a balanced structure which is self-maintaining".

1. to establish the position of the object of study on the hierarchical scale of the integrative levels of organisation;
2. to investigate the relationships of the object of study with both its lower and upper levels of organisation.

To clarify further about an inclusive definition of ecosystem in the hierarchy of levels of organisation, it is appropriate to mention the following one from Rowe (1961): "Any single perceptible ecosystem is a topographic unit, a volume of land and air plus organic contents extended areally over a particular part of the earth surface for a certain time". According to this definition, any piece of earth surface is susceptible of study and management within the framework of an ecosystem theory, with the result to obtain both scientific and philosophical knowledge. The under reported extracts from Rowe (1996) are illuminating on this respect:

> If there is to be a concise field of ecosystem science (ecosystemology) that includes, along with organisms, the air-water-land matrix of interest to earth scientists, then the concept of ecosystem – no matter how it is arrived at – must be that of a real structural-functional *volumetric system* occupying a relatively fixed earth space.
>
> An additional argument for the geo-ecosystem concept attributes the organizing principle we call "life" to ecosystems rather than to organisms *per se* (Rowe 1992). The fact that organisms are dead without their Earth matrixes means that ecosystems, not organisms are the vital bearers of life [...] Whatever "life" may be, it somehow blends both organic and inorganic.
>
> Planet Earth, the ecosphere, is the most important integrated unit, the largest and completest ecosystem. It is the creative entity, bearer of the organizing principle called "life".

The insights mentioned above justify the Rowe's conclusions reported in Box 4.2 in a way that blends ecological assumptions with ethical prescriptions.

Box 4.2: Rowe' s Assumptions as Outcomes from the Ecosystem Theory (Elaborated from Rowe 1996)

Basic assumption of Ecological Understanding

"Earth, the ecosphere, is a unified functional ecosystem"

Basic assumptions for Ecological Management

"The recognition of land/water ecosystems in a hierarchy of sizes provides a rational base for the many-scaled problems of protection and careful exploitation in the fields of agriculture, forestry, wildlife and recreation"

Ethical assumption based on Ecological Understanding

"Conceiving the world as comprising nested land/water ecosystems that are source of life, elevates the role of Earth-as-context, an antidote to destructive anthropocentrism"

The ecosystem level of organisation is the level where an organism like a human being makes his/her experience of life. It is the proper dimension for the experiential learning of every day. The ecosystem dimension is our natural dimension, i.e. the place where we actually feel and act. Our five natural senses and our mind are able to penetrate the proximate reality and construct what Popper (1959) called "knowledge of the common sense", distinguishing it from the scientific knowledge constructed through the continuous implementation of technical tools and appropriate technology. According to Popper, a major task for humanity is to transform the knowledge of the common sense into scientific knowledge. Scientific and technological advances are so evident today that it is easier for humanity to construct maps of reality for levels of ecosystem organisation both above and below the ecosystem level of the "knowledge of the common sense", letting the ecosystem theory recognised as inclusive, truthful, attractive and useful for the generations to come. On this point, Tansley himself in a later paper of 1939 prophetically foresees the necessary next steps for humanity to reach worldwide integration:

> In the phase through which humanity is now passing we see trend towards internationalism, with world federation as its ultimate goal – the establishment of a worldwide ecosystem – arising inevitably from the increased interdependence of the people, the multiplication of the bonds between them, and the immensely increased rapidity of every sort of intercommunication (Tansley 1939).

The challenge is to make a new ecosystem "which embraces the whole of humanity, retaining the more viable of the old partial systems, but subordinating them to the universal". This step will be affected "by the psychological factor, by the elements of self-consciousness, will, reason, the moral sense, and the power of deliberate action directed towards a conscious goal".

4.2 The Establishment of the Ecosystem Theory

The utility of the ecosystem concept resides in the smart connection of structural components and functional processes in a coherent unity of organisation for understanding nature according to scientific principles. Components and processes perform in relation to their specific role and their dynamics is a general pattern that yields energy transfer, nutrient cycling, biodiversity and ecosystem sustainability. Thompson (1996) and Alrøe et al. (2006) identify this ecological view of sustainability as 'functional integrity'.

The ecosystem concept subsumes disciplinary knowledge in a larger paradigmatic whole of transdisciplinary science. The ecosystem concept better that any other concept confirms the intelligibility of nature as an ordered construction comprehensible by human rationality. Energetics, cycling of nutrients, biodiversity development (including human beings), and consciousness development constitute elements of real ecosystems evaluation in order to establish principles of both self-organisation in nature and imposed-organisation (or control) by human interventions.

In 1988, the British Ecological Society (BES) – the world's oldest one – celebrated its 75th anniversary and organized symposia to review its ecology's intellectual achievements. BES sent all its 4376 member a questionnaire asking them what they thought were the most important concepts which ecology had contributed to our understanding of the natural world. The survey produced 236 recognisably different concepts; the most popular 50 were published with their mean score and choosing topic number (Cherret 1990). The first one in the list was the concept of *ecosystem* with 69.3% of citation frequency and a mean score almost double as compared with the second one, the concept of *succession*. It was recognised that the ecosystem concept "certainly provided the paradigm around which the International Biological Programme (IBP) (1968-1974), the largest and most expensive international research effort in ecology ever undertaken, was organised" (Cherret 1990). Commenting on those results, Burns (1990) provided the following outlook that the future would have revealed prophetic:

> The ecosystem concept and paradigm thus offer a hope of an enlightened public and a holistic Ecology for tomorrow. Most ecologists recognize the concept of ecosystem as being central to their discipline, and citizens all over the world are beginning to understand the ecosystemness of the biosphere of which they are integral part. Through this holistic conception of the unitary relationship of mankind to its environment, our species has the power to alter the course of history and achieve an improved quality of coexistence, a symbiosis with the rest of life in this planet.

4.2.1 *Research and Education*

At the outset of the 1970's, an enlightened group of people united under the name of "The Club of Rome" got started an influencing call for more justice and ethics among men and between man and the environment. That global question for urgent change took the name of "the predicament of mankind" (Box 4.3) and won a big resonance at global level. The Club of Rome published a report under the title

Box 4.3: Origin and Goals of "The Club of Rome" (Meadows et al. 1972, Excerpts from the Forward)

In April 1968, a group of 30 individuals from 10 countries – scientists, educators, economists, humanists, industrialists, and national and international civil servants – gathered in the "Accademia dei Lincei" in Rome. They met at the instigation of Dr. Aurelio Peccei, an Italian industrial manager, economist, and man of vision, to discuss a subject of staggering scope – the present and future predicament of man.

THE PROJECT ON THE PREDICAMENT OF MANKIND

The intent of the project is to examine the complex of problems troubling men of all nations: poverty in the midst of plenty; degradation of the environment; loss of faith in institutions; uncontrolled urban spread; insecurity of employment; alienation of youth; rejection of traditional values; and inflation and other monetary and economic disruptions. These seemingly divergent parts of the "world problematique," as The Club of Rome calls it, have three characteristics in common: they occur to some degree in all societies; they contain technical, social, economic, and political elements; and, most important of all, they interact.

The team examined the five basic factors that determine, and therefore, ultimately limit, growth on this planet – population, agricultural production, natural resources, industrial production, and pollution. The research has now been completed. This book [The Limits to Growth] is the first account of the findings published for general readership.

"Limits to Growth" (Meadows et al. 1972) which aroused more critics than consensus in an age of rampant industrialisation and use of natural resources.

That report, under the prestigious guide of MIT's[6] membership, was an earlier example of modelling in complex socio-ecological systems, as explicitly declared:

> Every person approaches his problems, wherever they occur on the space-time graph, with the help of models. A model is simply an ordered set of assumptions about a complex system. It is an attempt to understand some aspect of the infinitely varied world by selecting from perceptions and past experience a set of general observations applicable to the problem at hand (Meadows et al. 1972 p. 20).

The report structure examines growth as a metabolic activity of the complex system of coevolution between man and nature and treats the following contents: the nature of exponential growth, the limits to exponential growth in the world system, technology and the limits to growth, the state of global equilibrium. The report methodology consisted in analysing a series of indicators concerning:

(a) human perspectives, such as world population, world urban population, world industrial production, energy consumption, arable land, world fertilizers consumption, food production, protein and caloric intake, economic growth;
(b) environmental indicators of planetary metabolism, such as carbon dioxide concentration in atmosphere, DDT flows in the environment, lead in the Greenland ice cap;
(c) final modelling of a series of world scenarios with different intensity of human intervention and control in the range world model with unlimited or limited resources.

[6] MIT = Massachusetts Institute of Technology

As a conclusion, the report advances as final goal a proposal of transition from growth to global equilibrium, in order to establish an *equilibrium society* deemed to last permanently, generations after generations. To reach such an ideal *status*, a series of technological improvements would intervene, such as those reported in Box 4.4.

Box 4.4: Technological Steps to an Equilibrium Society (After Meadows et al. 1972, Modified)

THE EQUILIBRIUM STATE

Thus most basic definition of global equilibrium is that population and capital are essentially stable with the forces tending to increase or decrease them in a carefully controlled balance (p. 171).

TECHNOLOGICAL STEPS

- new methods of waste collection, to decrease pollution and make discarded material available for recycling;
- more efficient techniques of recycling, to reduce rates of resource depletion;
- better product design to increase product lifetime and promote easy repair, so that the capital depreciation rate would be minimized;
- harnessing of incident solar energy, the most pollution-free power source;
- methods of natural pest control, based on more complete understanding of ecological interrelationships;
- medical advances that would decrease the death rate;

contraceptive advances that would facilitate the equalization of the birth rate with the decreasing death rate (p. 177)

We can hardly sustain today that the technological steps recommended by "Limits to Growth" have had full implementation after 50 years from their advancement; however, at that time, international institution were ready in setting up research and education initiatives for bettering knowledge in the complex field of man-nature interactions at both global and local levels. An important initiative was UNESCO's "Programme on Man and Biosphere" (MAB 1971), an interdisciplinary vast survey to be implemented in close cooperation with the organisations of the United Nations concerned and the competent international non-governmental organisations focusing on:

(a) the general study of the structure and functioning of the biosphere and its ecological divisions;
(b) systematic observation and research on the changes brought about by man in the biosphere and its resources;
(c) the effects of these changes on the human species itself;
(d) the education and information to be provided on these matters.

The general objectives of MAB were:

(a) to develop the basis within the natural and social sciences for the rational use and conservation of the resources of the biosphere and for the improvement of the global relationship between man and the environment;
(b) to predict the consequences of today's actions on tomorrow's world;
(c) to increase man's ability to manage efficiently the natural resources of the biosphere;
(d) to promote environmental education in its broadest sense;
(e) to stress the interdisciplinary nature of environmental problems;
(f) to promote the idea of man's personal fulfilment in partnership with nature, and his responsibility for nature.

Methodologically, the informing principle of MAB was the ecosystem paradigm in that ecosystem research includes the description of abiotic variables, the inventory of biotic components and studies on the biology and physiology of selected species. It also involves the study of the various states of different types of ecosystems, the relationships between structure and functioning, the variability and magnitude of rates of energy flow and nutrient cycling, the ways in which social interactions affect ecosystem processes. Moreover, the use of modelling techniques was recommended on the fact that, if the state of an ecosystem or a complex of ecosystems at a given time can be expressed mathematically, there is greater opportunity to give quantitative expression to the effects of disturbs on the structure, functioning and management of ecosystems, and to evaluate their resilience capacity. Modelling of data had also credit to enhance their informative value and to widen the application of results.

An important ethical component partook in the scientific approach of MAB (1972) as revealed by the following extracts:

> Man, in modifying the biosphere of which he is an integral part, will produce new situations in which in turn may have a strong bearing on man himself. Man should be considered as being in partnership with nature. This means mutual taking and giving and thus use and conservation at the same time. This partnership includes all those qualitative values which man needs from nature for his physical and mental well being. Furthermore, this partnership is the expression of man's respect and responsibility for all other life on earth. In shaping his environment, man is in fact shaping his own future (p. 11).

> In order that man should learn to live in, and to maintain the quality of his environment – in order that the ecosystems of which he forms part should remain capable of supporting his continued presence and maintain their vital structure – it is essential for him to understand the driving forces of these ecosystems, their dynamics, and the mechanisms by which they are regulated (p. 40).

4.2.2 International Policy and Ecological Economics

The environmental question reached its apex at the international level with the United Nations Conference on the Human Environment held in Stockholm in 1972, the report of which (UN 1972) focuses on the "Action taken by the Conference "(Chapter I) and the "Action plan for the Human Environment" (Chapter II). The report provides an unprecedented information on the dramatic nexus connecting human beings and societies with their local and global environment, evidencing that both a turning point for humanity development was established and a new global policy was necessary to guide and implement it. This turning point, that started 50 years ago, is still in progress and the principles declared for the new development, defined 20 years later by UN as "sustainable development"(UN 1992a), are still valid. *Common outlook* and *common principles* constitute the pillars of the "declaration" that opens the report in order to "inspire and guide the *peoples of the world* in the preservation and enhancement of the human environment". These assumptions sound not rhetorical but ethical; they envisage humanity as both a teleological and theological category, the end of which is a duty – to maintain the whole context of life (or 'creation') healthy and regenerative.

The Stockholm Conference approved the "Declaration on the Human Environment" containing 26 broad principles that recognise the ecological quality of nature and the duties and rights of the whole humanity to preserve, use and enjoy

Table 4.1 Ethical values of rights and duties according to a few principles of Stockholm Declaration

Rights of man and nature	Duties of man
Principle 1	Principle 1
Man has the fundamental right to freedom, equality and adequate condition of life, in an environment of a quality that permits a life of dignity and well-being.	Man bears a solemn responsibility to protect and improve the environment for present and future generations.
Principle 3	Principle 2
The capacity of the earth to produce vital renewable resources must be maintained and, wherever practicable, restored or improved.	The natural resources of the earth, including the air, water, land, flora and fauna and especially representative samples of natural ecosystems, must be safeguarded […] through careful planning or management.
Principle 4	Principle 6.
[…] Nature conservation, including wildlife, must, therefore receive importance in planning for economic development.	The discharge of toxic substances or of other substances and the release of heat, in such quantities or concentrations as to exceed the capacity of the environment to render them harmless, must be halted in order to ensure that serious or irreversible damage is not inflicted upon ecosystems.
Principle 8	
Economic and social development is essential for ensuring a favourable living and working environment […]	

it. Table 4.1 proposes a re-organisation of a few principles of the Declaration according to their ethical value of rights and duties.

Another product of the Stockholm conference was an "Action plan for the human environment" containing 109 recommendations for international cooperation on the environment. The action plan assigned responsible leadership to the United Nations in order to assure assistance to the member states for implementing international cooperation on the environmental policy. This new alternative paradigm for a socio-ecological worldview and a sustainable management did not get practical effect immediately, although the conference recommendation led to the creation of the United Nations Environmental Programme (UNEP) for action and coordination activities within the UN system. According to later judgements (Porter and Brown 1991), policy and economy remained not affected:

> This new concern for the environment was not yet translated into alternative system of assumptions about both physical and social reality that could become a competing world-view. The essential assumptions of classical economics remained largely unchallenged. Confronted with evidence that existing exploitation of resources could cause irreversible damage, proponents of classical economics continued to maintain that such exploitation was still rational. Despite new rhetoric on the need to manage the global environment, the economic polices determining the rate of environmental degradation remained unchanged. (p. 28)

A severe critic toward mainstream economics arouse in the earlier 1970s by alternative economists, such as Nicholas Georgescu-Roegen (1971) and Ernst F. Schumacher (1973). The former provided a theoretical base for an ecological critic according to the laws of thermodynamics and the latter advanced technical and social solutions for the sake of the people of the world according to ethical principles.

The second law of thermodynamics, or the entropy law, states that any material transformation brings about a loss of energy and a generation of wastes in the environment. Therefore, increased production and consumption of goods generate increasing entropy, or environmental disorder, and eventually increasing pollution, which affects negatively both the quality of resources in the next economic cycle and the health of biological community, including human beings. As a result, economy-environment feedbacks are loaded with ethical consequences at both local and global levels. In mainstream economics, all this is miss-knowledge. A *linear* conception of technological processes holds, where inputs from environment serve to produce goods for the market and wastes for the environment. In these processes, there is a manifest asymmetry in the economic relations between man and environment, where inputs from the environment are resources with a market values and output to environment are wastes without a market value. In reality, wastes are *costs* of environmental degradation uploaded on the shoulder of both nature and human society as a whole. According to Georgescou-Roegen (1971), the entropy law "brings to the fore some fundamental yet ignored aspects of the two problem that now preoccupy the governed, the governments, and practically every scientist: pollution and the continuous increase of population"(p. 19).

He insists in charging mainstream economics of neglecting the entropic nature of the economic process:

> It is natural that the appearance of pollution should have taken by surprise an economic science which has delighted in playing around with all kinds of mechanistic models. Curiously, even after the event economics gives no signs of acknowledging the role of natural resources in the economic process. Economists still do not see to realize that, since the product of the economic process is waste, waste is an inevitable result of that process and *ceteris paribus* increases in greater proportion than the intensity of economic activity (p. 19).

Moreover, he claims that the entropy law is "the most economic of all physical laws" (p. 280) and that "the economic process is wholly *circular*" in that

> The economic process consists of a continuous transformation of low entropy [land resources] into high entropy, that is, into *irrevocable waste* or, with a topical term, into pollution (p. 281).

The economic process being dependent on the activity of humans, a sorting activity, there must be a reason behind that he explains in this way:

> True "output" of the economic process is not a physical outflow of waste, but the *enjoyment of life* [...]
>
> It is thus seen that we cannot arrive at a completely intelligible description of the economic process as long as we limit ourselves to purely physical concepts. Without the concepts of *purposive activity* and *enjoyment of life* we cannot be in the economic world. And neither of these concepts corresponds to an attribute of elementary matter or is expressible in terms of physical variables. [...] The economic process, to be sure, is entropic in each of its fibers, but the paths along which it is woven are traced by the category of utility to man (p. 282).

About the notion of *circular economy* at a larger scale level, an explicit example emerges from his description of the web of relations connecting different kind of human activity systems in an industrialised society:

> The manufacturing sector is completely tributary to the other two processes – agriculture and mining – in the sense that without the current input flows received from them it would have nothing to manufacture into industrial products. True, these other two sectors, in turn, are tributary to the industrial sectors for the tools they use and, implicitly, for a large measure of their technical progress. But this mutual dependence should not cause us to lose sight of the fact that it is the pace at which low entropy is pumped from the environment into the economic process that limits the pace of this process. (p. 292)

Concerning the two sources of low entropy that we find on our planet, the sun's radiation and the earth's own deposits, he sees an asymmetry that affects the economic development. The dispersed solar energy is associated primarily with husbandry while the concentrated mineral low entropy with industry. The asymmetry of the two sources generates peculiar forms of psychological reaction like these:

> With the stocks of low entropy in the earth's crust we may be impatient – as indeed we are – with their transformation into commodities that satisfy some of the most extravagant human wants. But not so with the stock of sun's energy. Agriculture teaches, nay, obliges man to be patient – a reason why peasants have a philosophical attitude in life pronouncedly different from that of industrial community (p. 297).

The question of the appropriate use of natural resources by man is also the topic discussed by Schumacher (1973) in his seminal book "Small is Beautiful. A Study of Economics as if People Mattered", where the subtitle denotes the importance of the human element as an end itself of the economic process. The book is a collection of essays and lectures given previously, between the 1960s and the early 1970s. The book content concerns social topics as peace, education, labour, technology, property and policy, all integrated within a general strategy of development, or social-ecological organisation, that preserves nature from degradation. According to Schumacher's first essay "The problem of production",[7] this kind of development – the definition of which is today "sustainable development"- sounds as "the possibility of evolving a new life-style, with new methods of production and new patterns of consumption: a life-style designed for permanence". In order to overcome a mainstream economics that disregards human interest for safeguarding nature, he advances a theory of nature recognition based on the following summarised assumptions:

(a) "modern man does not experience himself as a part of nature but as an outside force destined to dominate and conquer it";
(b) "we are estranged from reality and inclined to treat as valueless everything that we have not made ourselves";
(c) "far larger is the capital provided by nature and not by man";
(d) "let us take a closer look at this 'natural capital' ";
(e) "fossil fuels are merely a part of the 'natural capital' […] if we squander our fossil fuels, we threaten civilisation; but if we squander the capital represented by living nature around us, we threaten life itself".

The recognition of nature as "capital", using an economic term as a metaphor, means that nature has the power to provide "interests", usually identified today with the term "ecosystems services", i.e. benefits provided by nature for the sake of human beings. To reach an ideal compromise between exploitation and conservation of natural resources, there is the need to develop an appropriate technology, which Schumacher defines as "intermediate" in the essay "Technology with a human face"[8] . This kind of technology should be a means both to recover human creativity and to alleviate universal challenges, such as:

> Mass unemployment, mass migration into cities, rural decay, and intolerable social tensions. They [poor people] need, in fact, the very thing I am talking about, which we also need: a different kind of technology, a technology with a human face, which instead of making human hands and brains redundant, helps them to become far more productive than they have ever been before. As Gandhi said, the poor of the world cannot be helped by mass production, only by production by the masses.

The main synthetic characters of the two kinds of contrasting technology according to Schumacher appear in Table 4.2. He defines the technology of production by

---

[7] Based on a lecture given at the Gottlieb Duttweiler Institute, Rüschlikon, Nr. Zürich, Switzerland, 4th February 1972.

[8] Based on a lecture given at the Sixth Annual Conference of the Teilhard Centre for the Future of Man, London, 23rd October 1971.

Table 4.2 Technology for different socioecological performances (elaborated from Schumacher 1973)

Mass production technology	Production technology by the masses
The system of mass production has its base on highly capital- intensive, high energy-input dependent, and human labour-saving technology.	The system of production by the masses uses the priceless resources owned by all human beings themselves (clever brains and skilful hands supported by first-class tools).
EFFECTS	EFFECTS
"The technology of mass production is inherently violent, ecologically damaging, self-defeating in terms of non-renewable resources, and stultifying for the human person".	"The technology of production by the masses [...] is conducive to decentralisation, compatible with the laws of ecology, gentle in its use of scarce re-sources, and designed to serve the human person instead of making him the servant of machines".

the masses as "intermediate", i.e. "much simpler, cheaper, and freer than the super-technology of the rich. One can also call it self-help technology, or democratic or people's technology – a technology to which everybody can gain admittance and which is not reserved to those already rich and powerful".

Schumacher (1973) provides a more detailed description of the use of intermediate technology treating the problem of "The Third World".[9] He sees the need of an intermediate technology for solving the condition of the poor in the world, where "the poor are getting poorer while the rich are getting richer, and the established processes of foreign aid and development planning appear to be unable to overcome this tendency." To solve this problem, he proposes "that at least an important part of the development effort should by-pass big cities and be directly concerned with the creation of an agro-industrial structure in the rural and small-town areas". According to him, the primary necessity is to bring into existence millions of workplaces, in order to let everybody produce something, avoid migrations and urban unemployment, and overcome demographic unbalances at global, national and regional levels. "An unemployed man is a desperate man practically forced into migration. The provision of work opportunities [...] should be the primary objective of economic planning". The harmonisation of social policy, intermediate technology and economic development should follow four criteria as represented in Fig. 4.3.

According to Schumacher, to meet these requirements is possible first if there is a 'regional' approach to development and, second, if there is a conscious effort to develop and apply what might be called an "intermediate technology".

The complex relations between the socio-economic system and the natural system, denounced by both Georgescu-Roegen' s theoretical framework and Schumacher's socio-political perspective, can be summarised as represented in Fig. 4.4. Overlapping between the two systems is growing on time because of the increasing number of human population and the increasing power of its technological impact. On one hand, maintenance and growth of the economic system depends on material and energy extracted from environment low entropic sources; on the other hand, economic processes affect maintenance and

[9] Essay based on the Anniversary Address delivered to the general meeting of the Africa Bureau, London, 3rd March 1966.

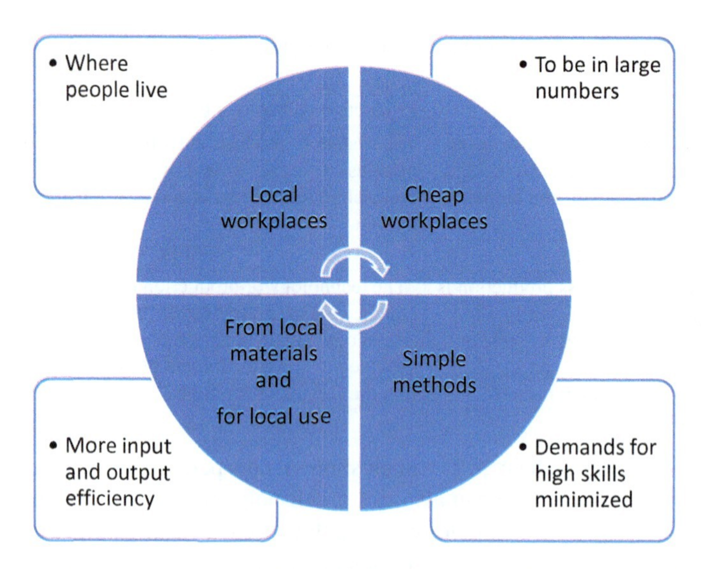

Fig. 4.3 Criteria for creating workplaces in poor countries (elaborated from Schumacher 1973, p.163)

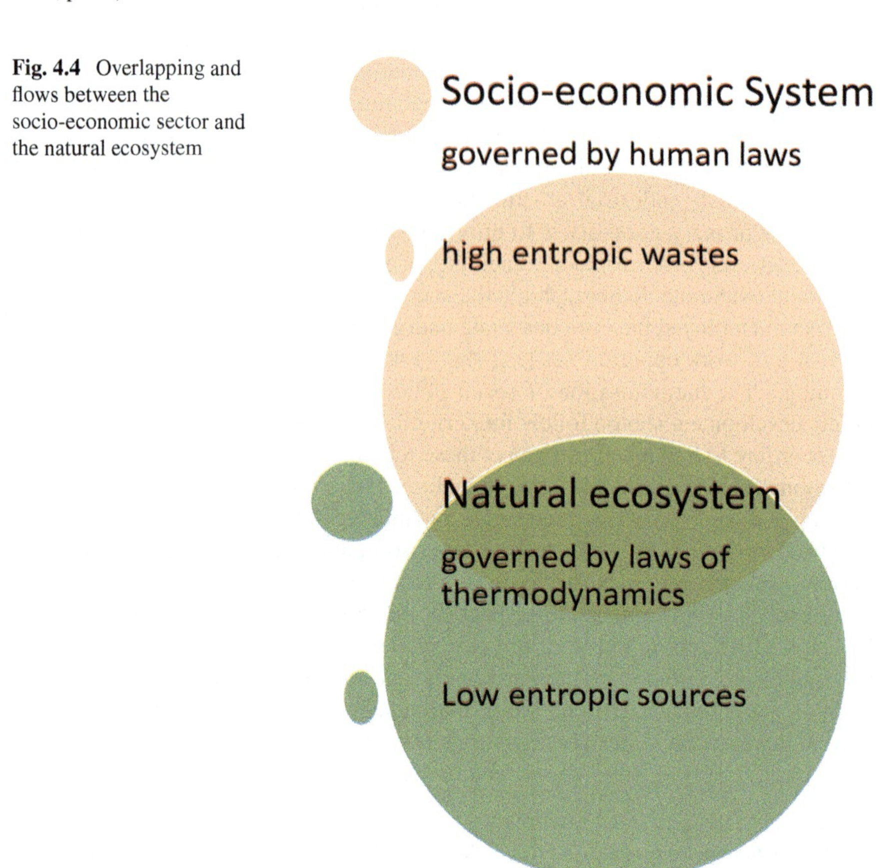

Fig. 4.4 Overlapping and flows between the socio-economic sector and the natural ecosystem

growth of natural ecosystems through emission of high entropic wastes. Altogether, the two systems generates a unique, dynamic, planetary socio-ecological system that denotes evolutionary quality and need governance by humanity as a whole.

Mainstream economics recognised environmental degradation as an economic issue, but under the unfortunate assumption that ecological and economic systems are separate entities (Bernstein 1981). The special branch devoted to study this problem took the name of *Environmental Economics*, major issues of which are the challenge of allocation of scarce resources and the market methods for pollution control (Solow 1974; Cropper and Oates 1992).

To appreciate the level of ecological awareness around the environmental degradation at the planetary level in the 1970s, it is interesting to consider the insights made by Charles Hall (1975), an ecosystem ecologist, between the effects of the two contrasting systems, *Biosphere* and *Industriosphere*, as reported in Table 4.3. The *Industriosphere,* according to Hall, is "the series of fossil-fuels powered patches of human industrial activity located within, and supported by the Biosphere".

As conclusive comments of his review paper on the ecology/economics interface and environmental degradation, Bernstein (1981) points out that:

1. the environmental degradation that accompanies expanding population and technological development results primarily from economic decisions about the most desirable uses of natural resources;
2. such outcomes make clear the conflict between the quest for short term gain and long-term concerns for sustained yield and ecosystem health;
3. such conflicts between ecological and economic priorities derive partly from a failure to realize that the links between the two systems are not just one-way flows of exploitable resources;
4. instead, human and natural systems are linked by a web of interconnected processes;
5. the challenge of economics is to redefine concepts of self- and public interest to include information about the long term effects of decisions for an age of ecological transition.

Following the lines traced by "Limits to Growth", "The Law of Entropy" and "Small is Beautiful", a World Bank economist, Herman Daly, attempted to connect the two separate fields of Economics and Ecology and developed a theory of a "Steady-State Economics" based on the ecosystem theory (Daly 1974 and 1991). He justifies his choice with these words (Daly 1991):

> Since "economics" as well as "ecology" come from the same Greek root (oikos), meaning management of the household," and since man's household has extended to include not only nations but also the planet as a whole, economics is probably the discipline that has least justification for taking a narrow view. Let us take a minute to consider the economy, environmental quality, food, energy, and adaptation as interrelated subtopics within the framework of economics viewed as management of the household of man (p. 7).

Table 4.3 Effects and interactions between *Biosphere* and *Industriosphere* (after Hall 1975, modified)

Effects of the biosphere	Effects of the industriosphere
☞ The biosphere contributes to human welfare by supplying man with food, fiber and woods as well as the fossil fuels, via ancients ecosystems, that power man's industrial systems.	☞ The development of the industriosphere has contributed to the management of the biosphere and the increase in the material wealth of Western man.
☞In addiction, and in conjunction with the atmospheric systems, the biosphere provides man with a stable physical and chemical milieu.	☞The industriosphere has not only elevated but also degraded human existence through pollution and industrial squalor, and it has leaked its toxic products into the biosphere.
☞As man has learned to manage the biosphere through agricultural and related activities, his ability to create cities, arts, armies and the other trappings of civilization has increased, as have his own numbers.	☞The industriosphere has grown so large as to over tax the capacity of biosphere to assimilate and disperse these by-products; the resulting and potential biotic degradations appear to be reaching critical dimensions on a global scale.
☞The biosphere performs a wide variety of *public service functions* that humans, often unwittingly, depend on.	☞As public services and primary functions of natural ecosystems are degraded, they must be replaced by more energy-intensive, less-efficient industrial substitutes".
☞The future well-being of man requires intelligent and sometimes intensive management of the biosphere using limited inputs from the industriosphere instead of the present, ever more intensive exploitation".	

> [Steady-state economics] at least seeks to integrate the two key disciplines of our time rather than further to subdivide each one into ever more arcane and irrelevant sub-sub-disciplines (p. xiii).

Starting from the assumption that economics has to do with ends and means, Daly represents the entire ends-means continuum as a hierarchical scale of subsequent achievements (Fig. 4.5).

Moreover, he states that the standard economists' attention has entirely focused on intermediate ends. The failure of economics to consider either ultimate ends (the persistence of the socio-ecological system and moral values) or means (sources of low entropy matter-energy) has originated the conflicts between short-term economic goals and long- term environmental concern:

> The lack of attention by economists to the ultimate extremes has been insulated by a relative lack of attention to ethics and technics (Daly 1991, p. 19).

What is necessary to do is changing "the defective structure of the semantic map used by economics to define human relationships with their environment" (Bernstein 1981). Better understanding of how perception of the environment influences self-interest and institutional decision- making is crucial for finding new ways to sustainable development.

The ecosystem theory appears to be an integral part of Daly's steady-state economics, as shown in Box 4.5 that presents a frame of its methods and contents from a systemic perspective.

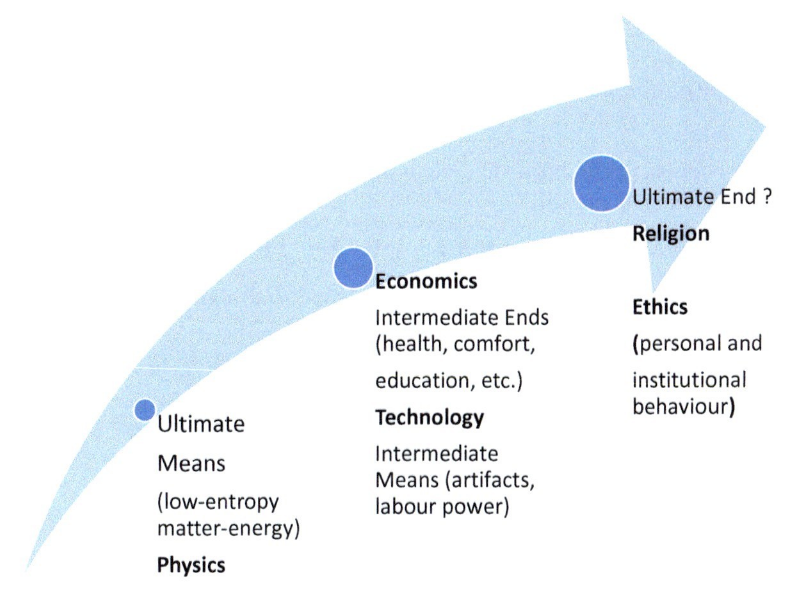

Fig. 4.5 The entire spectrum of the ends-means- continuum of humans relationship with their environment (Daly 1991, modified)

Methodologically, growth economics adopts the epistemic assumption that any analytical effort is of necessity preceded by a pre-analytic cognitive act (vision) that supplies the raw materials for the analytical effort. In steady-state economics compared with growth economics, such as vision is completely upside down in that it is the ecosystem that contains economy as a human subsystem and not vice versa. Therefore, economy expansion finds its proper limits within the frame of ecosystem resources, for both supplying law entropy input and absorbing high entropy output as waste. Economy cannot growth indefinitely within a finite space like the planetary ecosystem but needs to find an appropriate balanced organisation or tendency toward a steady- state that assures sustainability. Appropriate technology and policy changes that lead to personal involvement, control and creativity are also required to favour an ecological transition that invests ethical and spiritual values as well. However, a cultural barrier persists to this transition as a matter of fact:

> The temper of the modern age resists any discussion of the Ultimate End. Theology and purpose, the dominant concepts of an earlier age, were banished from the mechanistic, reductionistic, positivistic mode of thought that came to be identified with a certain phase of the of science. Economics followed suit by reducing ethics to the personal tastes: individuals set their own priorities, and economics is simply the mechanics of utility and self-interest" (Daly 1991, p. 20).

In the second edition preface of "Steady-state Economics" (1991), Daly confessed that his decision to publish again the book had two motivations:

Box 4.5: Basic Components of a Steady-State Economics According to Daly (1991, Modified)

VISION

The economy is an open subsystem of a finite and non-growing ecosystem (the environment) (p. xiii).

RATIONAL PRINCIPLES

Problems are not independent and sequential but highly interrelated and simultaneous: someone has to look at the whole (p. 7).

Any subsystem of a finite non growing system must itself at some point also become non-growing (p. xiii).

At some optimal, or at least sustainable, scale the economic subsystem should be maintained in a steady state as far as possible [p. xiii].

AUTOPOIESIS

The economy lives by importing low-entropy matter-energy (raw materials) and exporting high-entropy matter-energy (waste) (p. xiii).

FACTS

The biophysical facts have asserted themselves in the form of *increasing ecological scarcity*: depletion, pollution, and ecological disruption (p. 3).

The moral facts are asserting themselves in the form of *increasing existential scarcity*: anomie, injustice, stress, alienation, apathy and crime (p. 3).

TECHNOLOGY

Steady-state economics channels technical progress in the socially benign directions of small scale, decentralization, increased durability of products, and increased long-run efficiency in the use of scarce resources (p. 7).

POLICY

Human institutions should not be allowed to grow beyond the human scale in size and complexity (p. 4).

Lack of control by the individual over institutions and technologies is hardly democratic (p. 4).

FAITH

Only by returning to its moral and biophysical foundations and shoring them up, will economic thinking be able to avoid a permanent commitment to misplaced concreteness and crackpot rigor (p. 3).

(a) the foundation(1989) of a new interdisciplinary society, the *International Society for Ecological Economics* with its quarterly journal *Ecological Economics*;
(b) his conviction that cultural resistance would become weaker as the effects of the economy of growth on the environment be more severe.

Motivations for integrating ecology and economics are still increasingly important as humanity's impact on the natural world increases. An early collection of essays on this topic appeared on *Ecological modelling* (Costanza and Daly 1987) in order to make a hopeful first step toward a true synthesis of ecology and economics that could lead to better management of renewable and non-renewable natural resources and a sustainable future. In this perspective, sustainability is a concept that stresses responsibility for "maintaining our life support system":

> The most obvious danger of ignoring nature in economics is that nature is the economy's life support system, and by ignoring it we inadvertently damage it beyond repair. Several authors stress the fact that current economic systems do not inherently incorporate any concern about the sustainability of our natural life support system and the economies which depend on it (Costanza and Daly 1987).

In this vision, sustainability is a trans-disciplinary conception useful for integrating ecological theory with economic practice. The failure of standard economic organisation, like a free market, to guarantee sustainability is a topic discussed by Pearce (1987). In his vision, the issue of sustainability has connections with justice both within species, between species, and between present and future generations.

A complete outlook of goals, agenda and policy recommendations for Ecological Economics is in a subsequent seminal book (Costanza et al. 1991) with three parts. The first part is devoted to define the basic world view of ecological economics as compared with conventional economics; the second part focuses on accounting, modelling and analysis of ecological economic systems, under the perspective on incorporating natural capital and services into national income accounting; the third part deals with institutional changes necessary to achieve sustainability. Without discussing in details the book's topics, it is worth to report how the role of man is dealt with in the two contrasting paradigms of ecological and conventional economics:

> The basic world view of conventional economics is one in which individual human consumers are the central figures. Their tastes and preferences are taken as given and are the dominant determining forces. The resource base is viewed as essentially limitless due to technical progress and infinite substitutability.
>
> Ecological economics takes a more holistic view with humans as one component (albeit a very important one) in the overall system. Human preferences, understanding, technology and cultural organisation all co-evolve to reflect broad ecological opportunities and constraints. Humans have a special place in the system because they are responsible for understanding their own role in the larger system and managing it for sustainability (Costanza et al. 1991, p. 4).

To achieve sustainability, there is the need to recognise, incorporate and evaluate ecosystem goods and services into an economic accounting and modelling. Instead, Gross National Product (GNP), as well as other related conventional indicators of national, regional or local economic performances, ignores the contribution of nature to production and sustenance of the life support system. Without differentiating between costs and benefits for both human society and nature, there is the risk to attribute a harm to nature as a benefit to man. This is the case for pollution, in that it creates jobs for cleaning up and consumes resources, all of which adds to GNP and gives a false information on the state of the socioecological system's health.

To achieve safe, minimum sustainability standards, project implementation should meet the following criteria:

1. for renewable resources, the rate of harvest should not exceed the rate of regeneration (sustainable yield) and rates of waste generation not exceed the assimilative capacity of the environment (sustainable waste disposal);
2. for non-renewable resources, the rates of waste generation from projects shall not exceed the assimilative capacity of the environment and the depletion of the non-renewable resources should require comparable development of renewable substitutes for that resource (Costanza et al. 1991,p. 16).

To achieve sustainability, there is the need to develop both research and education. As to research, a special recommendation concerns the valuation of important non-market goods and services provided by ecosystems. Considering that "the role of nature in economics is that nature is the economy's life support system", there is the risk "to damage it beyond its ability to repair itself", just when "we have now entered a new era in which the limiting factor in development is no longer manmade capital but remaining natural capital (Costanza et al. 1991, p. 8). It is evident that in dealing with the ecological economic question of sustainability, scientific and technical knowledge is not enough, but it needs support by explicit ethical criteria for appropriate judgement and political decisions.

As to education, its institutional state is largely based on the criterion of disciplinarity, i.e. a fragmented assemblage of specialized fields of knowledge that methodologically prefer analysis to synthesis, making difficult for both teachers and students to reorganize a unitary body of knowledge. Daly and Cobb Jr (1994, p. 33–34) call this knowledge specialisation "disciplinolatry", in that has negative educational and ideological effects, inhibiting or preventing students from understanding connections not only among the disciplines, but also among the theories and objects of their interest. Instead, the need for an ecological economics core curriculum is a marked transdisciplinary character:

> We should promote at all levels education that weaves together fundamental understanding of the environment with human economic activities and social institutions, and promotes research that facilitates this interweaving process [...] We should promote education of broadly-trained environmental scientists, whose jobs will be to provide on-going environmental assessment as an addition to the decision-making process of various institutions (Costanza et al. 1991, p. 18).

Fragmentation of knowledge reflects also in the structure and functioning of public and private institutions. Their specialisation usually constitutes a barrier for change in that they lack of flexibility for dealing with ecologically sustainable development due to the following reasons:

(a) poor awareness and education about sustainability, the environment, and causes of environmental degradation;
(b) traditional assumption of continuous exponential growth;
(c) fragmented mandates and policies;
(d) many institutions do not possess market and non-market forces to resolve environmental problems;
(e) many institutions do not freely share or disseminate information;
(f) many institutions do not provide access to decision making.

Another problem to face in order to promote institutional change is de-bureaucratizing institutions so that they can effectively respond to the coming challenges of achieving sustainability (Costanza et al. 1991, p. 19).

Summing up all the previous considerations, Bormann and Kellert (1991) provide a general frame of man/nature relationship at the start of the 1990s, the picture of which appears in Fig. 4.6.

Ecology, the science of balance regulating man/nature relationship, should inform economics, the science of resources allocation, in order to fulfil ethical expectations. Ethics, the science of good behaviour, should inspire more appropriate ecological research and education. All tree scientific disciplines influence environmental policy. According to Bormann and Kellert (1991), the virtuous circle is unfortunately broken. The main reason is that the liver of on-going policy is short-term decision making, which ignores the ecological, economic, and moral burden that will be borne by future generations, if we continue to deplete and degrade the world's natural goods and services.

The challenge of restoring the balance among ecology, economy and ethics through adequate policy has become since then an urgent question but, paradoxically, often debated but never solved. It involves the notion of "common good", which ultimately refers to the health of the global ecosystem, and then concerns at

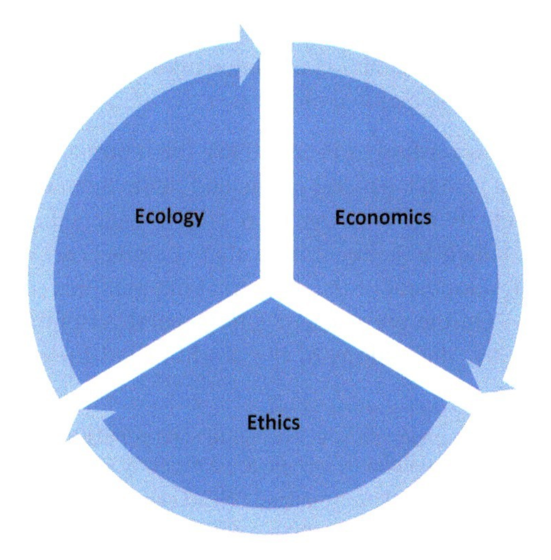

Fig. 4.6 The virtuous circle of human knowledge and behaviour (after Bormann and Kellert 1991)

the same time every man and the whole humanity. This is why Daly and Cobb Jr (1994, first edition 1989) invoke an "Ecological humanism" as end point of human predicament, in that "economics for the common good is what ecological humanism call for, and even more what stewardship for creation calls for" (Daly and Cobb Jr 1994, p. 21). They add that their chief task was "to provide an image or vision of an economic order[10] [...] that would be just, participatory and sustainable".

To the notion of "common good" was soon associated the notion of "common future" as a "brand" of international political credibility. This happened in October 1987, when the UN General Assembly welcomed the final report of the World Commission on Environment and Development (WCED 1987), the title of which was "Our Common Future", also known as the "Brundtland Report".[11]

A brief account of its content and suggestions appears in Box 4.6.

According to Starke," *Our Common Future* provided governments, private groups, and individuals with a powerful tool that helped them to reassess where they were headed, and rethink their approach to sharing this only one earth" (p. 1). The report's suggestions had successful acceptance worldwide, letting nation leaders to demand a major UN Conference on Environment and Development.

Together with *Our Common Future*, another report produced by international agencies, *Caring for the Earth* (IUCN/UNEP/WWF 1991), had a positive role in calling for such a conference. Pace (1996, p. 17) lists nine principles for sustainable development around which this report is structured:

1. respect and care for the community of life;
2. improve the quality of human life;
3. conserve the Earth's vitality and diversity;
4. minimise the depletion of non-renewable resources;
5. keep within the Earth's carrying capacity;
6. change personal attitudes and practices;
7. enable communities to care for their own environments;
8. provide a national framework for integrating development and conservation;
9. create a global alliance.

According to Pace (1996), the report addresses not just people in "high places", but also the general public and invite them to take concrete actions towards sustainable living. This approach, which renders the population active participants in environmental action rather than passive recipients of directives, characterised the development of the next great international initiative.

On the premises of UN General Assembly resolution 44/228 of 22 December 1989, the nations of the world called for a *UN Conference on Environment and*

[10] "An economy in which economic and population growth is halted, technology is controlled, and gross inequalities of income are done away with" (Daly and Cobb Jr 1994, p.21).

[11] Ms. Gro Harlem Bruntland, WCED chairperson, became also Norway's Prime Minister shortly after working on the report.

Box 4.6: Stepping Stones Toward the Goal of Sustainable Development According to "Our Common Future" (After Starke 1990, pp. 2-8-9, Modified)

OVERALL MESSAGE

Our future is undeniably threatened; the world can change; people can build a future that is more prosperous, more just, and more secure.

DEFINITION OF SUSTAINABLE DEVELOPMENT

Sustainable development is meeting the needs of the present without compromising the ability of future generations to meet their own need.

PARTICIPATORY METHODOLOGY

No single blueprint of sustainability will be found, as economic and social systems and ecological conditions differ widely among countries. Each nation will have to work out its own concrete policy implications.

Basic prerequisite for the pursuit of sustainable development is a political system that secures effective citizen participation in decision-making.

STRATEGY

For development to be sustainable, it must take account of social and ecological factors, as well economic ones; of the living and non-living resource base; and of the long term as well as the short -term advantages and disadvantages of alternative actions.

UNDERLYING PRINCIPLE

Recognition of the interdisciplinary nature of the world's environment and development problem as well as theirs solutions. Sectorial organizations tend to pursue sectorial objectives [...] Many of the environment and development problems that confront us have their roots in this fragmentation of responsibility.

Development, which took place in Rio de Janeiro, Brazil, 3 to 14 June 1992. One of the most important product of the conference was *Agenda* 21, which is a global plan for sustainable development, the most comprehensive and far-reaching programme of action ever approved by the world community (Pace 1996). It constitutes a turning point taken by the whole humanity for redirecting its thinking and behaviour with the aim of balancing the protection of the environment with social justice and economic development. The whole programme is a *global ethical plan* based on ecological principles. The exposition of the human predicament is clear in the first point of its preamble as follows:

> Humanity stands at a defining moment in history. We are confronted with a perpetuation of disparities between and within nations, a worsening of poverty, hunger, ill health and illiteracy, and the continuing deterioration of the ecosystems on which we depend for our well-

being. However, integration of environment and development concerns and greater attention to them will lead to the fulfilment of basic needs, improved living standards for all, better protected and managed ecosystems and a safer, more prosperous future. No nation can achieve this on its own; but together we can – in a global partnership for sustainable development.

Keywords in this preamble, from the bottom to the top, are *sustainable development, global partnership, ecosystem protection, improved living standards, hunger, disparities between and within nations, humanity*. Altogether, they form the effective opportunity for humanity, the only responsible category in the living system, to make a dramatic transition from the conventional culture of competition to the innovative culture of cooperation at every level of relationship, be it social or environmental.

Steps of implementation to reach this complex goal constitute the architecture of Agenda 21 with the following headings: Social and economic dimensions; Conserving/managing resources; Strengthening major groups; Means of implementation as articulated in Table 4.4

A few months after the Agenda 21 approval, the UN General assembly released the "*Rio* Declaration of Environment and Development", a kind of *press briefing* on how "establishing a new and equitable global partnership through the creation of new levels of cooperation among States, key sectors of societies and people". Its introduction makes clear the noble objective of "working towards international agreements which respect the interests of all and protect the integrity of the global environmental and developmental system recognizing the integral and interdependent nature of the Earth, our home". Its articulation in 27 Principles gives synthetic instruction on how to reach the shared end goal of sustainable development, as shown in Box 4.7.

Table 4.4 Main steps and items for the implementation of the Agenda 21 plan (UN 1992a)

Steps	Items
Social and economic dimensions	International cooperation; combating poverty; changing consumption patterns; demographic dynamics and sustainability; protecting and promoting human health; sustainable human settlements; integrating environment and development in decision-making.
Conserving/ managing resources	Protecting the atmosphere; protecting land, oceans, seas and fresh waters; combating deforestation, desertification and drought; sustainable mountain development; sustainable agriculture and rural development; conservation of biological diversity; environmentally sound management of biotechnology; environmentally sound management of hazardous waste and sewage.
Strengthening major groups	Women, children and young people; indigenous people; partnership with non-government organisations; local authorities; workers, trade unions, business and industry; scientists and technologists; farmers.
Means of implementation	Financial resources and mechanisms; cooperation and capacity building; science for sustainable development; education, public awareness and training; international institutional arrangements; international legal instruments and mechanisms; information for decision-making.

Box 4.7: Selected Order of the Rio Declaration's Principles for a Sustainable Development (UN 1992b)

PRECONDITIONS

Principle 25

Peace, development and environmental protection are interdependent and indivisible.

Principle 4

In order to achieve sustainable development, environmental protection shall constitute an integral part of the development process and cannot be considered in isolation from it.

MAIN GOAL

Principle 7

States shall cooperate in a spirit of global partnership to conserve, protect and restore the health and integrity of the Earth's ecosystem.

SECONDARY GOAL

Principle 8

To achieve sustainable development and a higher quality of life for all people, States should reduce and eliminate unsustainable patterns of production and consumption and promote appropriate demographic policies.

METHODOLOGY

Principle 10

Environmental issues are best handled with the participation of all concerned citizens, at the relevant level.

Principle 11

States shall enact effective environmental legislation. Environmental standards, management objectives and priorities should reflect the environmental and developmental context to which they apply.

Principle 17

Environmental impact assessment, as a national instrument, shall be undertaken for proposed activities that are likely to have a significant adverse impact on the environment and are subject to a decision of a competent national authority.

Principle 12

States should cooperate to promote a supportive and open international economic system that would lead to economic growth and sustainable development in all countries, to better address the problems of environmental degradation.

From this framework, it is easy to understand how complex and progressive may be the way towards sustainable development in that the whole human society, at any hierarchical level, must be involved in a process that requires time, education and financing. Sustainable development is a transformative, systemic paradigm with many components that evolve with different speed, crossroads, feedbacks, always running, and likely, never definitively achieved. It also needs a faith component that transcends facts and touch ideals, the spiritual reason that cast light on the obscure side of life. On this front, a valid contribution to theological insights came from Pope John Paull II in his message for the World Day of Peace in 1990(see Box 4.8). He defines the ecological crisis as "a moral problem" and "a common responsibility", whereby "the urgent need for a new solidarity" emerges.

Box 4.8: Pope John Paul II (1990) on the Ethical Value of the Ecological Crisis

INTRODUCTION

Many ethical values, fundamental to the development of a *peaceful society*, are particularly relevant to the ecological question. The fact that many challenges facing the world today are interdependent confirms the need for carefully coordinate solutions based on a morally coherent world view.

THEOLOGY OF CREATION

Biblical considerations help us to understand *better the relationship between human activity and the whole of creation.* When man turns his back on the Creator's plan, he provokes a disorder which has inevitable repercussions on the rest of created order. If man is not at peace with God, the earth itself cannot be at peace.

We must go to the source of the problem and face in its entirety that profound moral crisis *of which the destruction of the environment is only one troubling aspect.*

The earth is ultimately *a common heritage, the fruits of which are for the benefits of all.*

MORAL CONCERN

The most profound and serious indication of the moral implications underlying the ecological problem is the lack of *respect for life* evident in many of the patterns of environmental pollution.

Today, the dramatic threat of ecological breakdown is teaching us the extent to which greed and selfishness – both individual and collective – are contrary to the order of creation, an order which is characterised by mutual interdependence.

IN SEARCH OF A SOLUTION

The concepts of an ordered universe and a common heritage both point to the necessity of *a more internationally coordinated approach to the management of the earth's goods.*

The ecological crisis reveals the *urgent moral need for a new solidarity.*

An education in ecological responsibility is urgent: responsibility for oneself, for others, and for the earth.

Modern society will find no solution to the ecological problem unless it *takes a serious look at is life style.*

A few weeks after the Pope's message for the World Day of Peace, Mrs. Brundtland delivered a speech in Moscow at the "Global Forum on Environment and Development for Human Survival" referring to the spiritual challenge of sustainable development as follows (Starke 1990):

> A new environmental ethics must enter our consciousness, whether we are active in political or economic planning, whether we are bankers or industrialists, journalists or clergy, scientists or sales managers. The world's great spiritual and religious movements bear a profound responsibility because of the special influence, even authority, they hold in affecting people's most personal aspirations and motivations [...] They have a great role to play in redirecting human motivation in relation to future (p. 157).

Since the beginning of the 1990s, the concept of sustainable development has become a product of cultural development of humanity never abandoned until now.

4.3 The Ethics of Sustainable Development

Being ethics a system of values, an evolutionary process of reassessment of both facts and judgements in view of sustainable development goals is in progress. According to WCDE (1987, p. 308) "human survival and well-being could depend on success in elevating sustainable development to a global ethics". Jennings (2010) provides a conceptual system of four interdependent components as a philosophical representation of ethics, like that reported in Fig. 4.7.

As an evolutionary process, ethics is a decision-making process that needs a subject with a will to act in order to modify the context of life for his/her goals. The context modified feedbacks on the agent as a source of information for the next decision- making process. What counts for an ethical agent is the evaluation of the consequences that his/her action has upon the context in which it happens. This recognition of consequences means responsibility. According to Jennings, the evaluation of the consequences has the most direct connection with the common sense meaning of the concept of sustainability, which has to do with

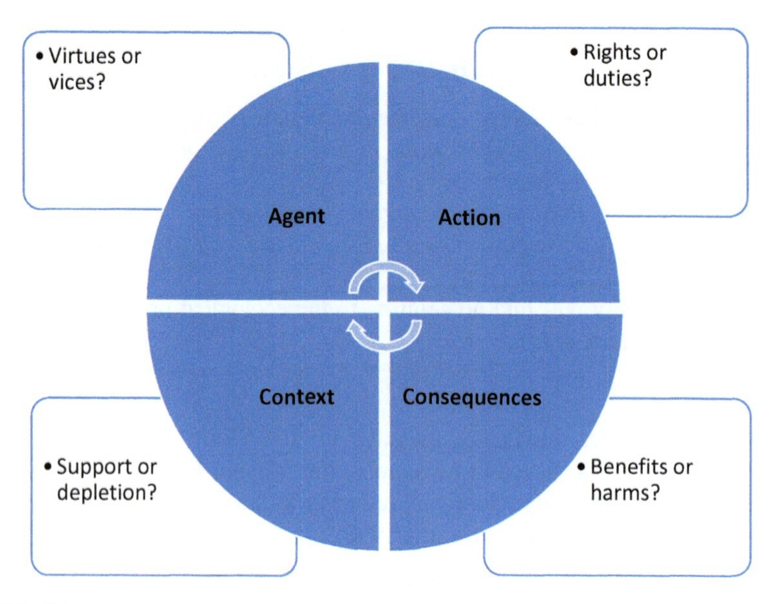

Fig. 4.7 Ethics as a conceptual system of four interdependent components (after Jennings 2010, modified)

> not undermining the prerequisite of what you are doing, living on the land without ruining it, using without using up, limiting how much you draw down reserves so that you do not deplete fast than you replenish (Jennings 2010).

However, a deeper judgement must consider all the elements included in the process as determinants of the concept of sustainability, in that virtue, rightness, motivation and moral commitment of the agent are sociological and psychological components of a sustainable society.

An agent can be an individual or an institution at every hierarchical level, family, association, State or corporation, up to the whole humanity. Important is to follow this prescription:

> a sustainable society lives within the carrying capacity of its natural and social system. It has a system of rules and incentives that promote replenishing and limit depletion and pollution. A sustainable society builds upon the commitment of its members to conform to these rules voluntarily (Jennings 2010).

From a philosophical perspective, a rational approach to an ethical analysis is the assumption of an ontological starting point of view, i.e. recognising facts as evidence from which recover values and disvalues. Facts emerge in the interface between ecosystems and human needs, whereby inevitably evaluation and judgements emerge about human actions concerning nature and about what we value in nature (Jax et al. 2013). Therefore, a comparison between the ecosystem performances and the nature of human needs seems appropriate to validate the concept of sustainable development.

The ecosystem concept opens an innovative approach to grasp the complex organisation of nature in a unitary framework, explaining both the functioning of the whole ecosystem and its specialised parts, i.e. organic functioning of a "body" and its "parts or organs". In this vision, a representation of man/nature interaction is shown in Fig. 4.8, where both biological and chemical-physical languages contribute to an ontology of the human predicament, from which the possibility of recovering judgements and values about sustainability and development emerges.

The ecosystem model is truly a heuristic one because unveils the real role that humanity plays in the ecosphere, man never being a true producer in thermodynamic terms, but always either a consumer or a decomposer. Real production is the accumulation of concentrated energy, matter and order within ecosphere, which is a consequence of photosynthesis. Through it, a steady stream of solar energy sustains essentially all biological activity and makes the earth a place where life thrives thanks to this transformation of sun's energy. The fossil-fuel era raised by man has dramatically changed the evolution pattern of natural development on Earth, because a further step of techno-metabolism has nested upon the original human bio-metabolism sustained by solar energy. Culture has given rise to an extra set of input and output of energy and materials, which are involved in all kind of technological processes and human artefacts. Most of human energy-matter consumption is an undirected consumption (techno-respiration), due to the construction and functioning of extra-somatic organs, such as any kind of machines used in human activities.

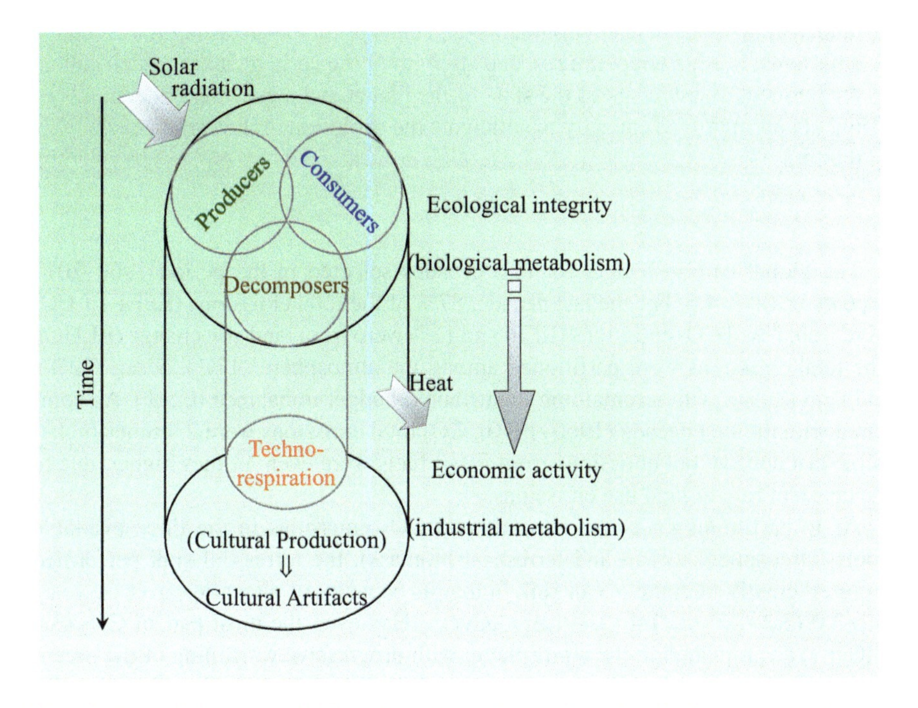

Fig. 4.8 Co-evolutionary metabolism of man/nature interaction at local and planetary level

Already in the 1990s, humans were using 120,000 times as much energy, mainly in the form of fossil fuels, as they were 400 generations ago when farming was first introduced (Boyden and Dovers 1992). Despite decades of warnings, agreements, and activism, human energy consumption, emissions, and atmospheric CO_2 concentrations all hit new records in 2017 (Quéré et al. 2018; Hagens 2020). The astonishing increase in human population in the last century, from 1 to 7.5 billions, combined with the astonishing increase in technology sustained by the use of fossil fuels, has yielded an increase of 35% in the carbon dioxide content of the atmosphere. Carbon dioxide is the main indicator of earth metabolism, i.e. the balance between its anabolic (photosynthesis) and catabolic (respiration) phases (Box 4.9).

Box 4.9: Relevance of the Carbon Budget Surveying (After Quéré et al. 2018)

"The delivery of an annual carbon budget serves two purposes:

First, there is a large demand for up-to-date information on the state of the anthropogenic perturbation of the climate system and its underpinning causes. A broad stakeholder community relies on the data sets associated with the annual carbon budget including scientists, policymakers, businesses, journalists, and nongovernmental organisations engaged in adapting to and mitigating human-driven climate change;

Second, over the last decade we have seen unprecedented changes in the human and biophysical environments (e.g. changes in the growth of fossil fuel emissions, Earth's temperatures, and strength of the carbon sinks), which call for frequent assessments of the state of the planet and a growing understanding of and improved capacity to anticipate the evolution of the carbon cycle in the future".

The global carbon budget for two decades selected in the period 1960–2017 appears in Table 4.5. For the last decade, 87% of the total emissions (EFF+ ELUC) were from fossil CO_2 emissions (EFF) and 13% were from land-use change (ELUC). The total emissions were partitioned among the atmosphere (44%), ocean (22%), and land (29%), with a remaining unattributed budget imbalance (0,5%). As compared with the first decade (1960–1970), CO_2 total emissions were 2.4 times higher in the last decade, but emissions from fossil fuels were even 3 times higher, due to the constant rate of land use emissions.

As to partitioning, i.e. the allocation of CO_2 emissions in the three available pools (atmosphere, ocean and terrestrial biomass), the terrestrial sink performed more efficiently than the ocean sink in that its capacity of allocation over 60 years was 2.0 (3.2–1.2) vs. 1.4 (2.4–1.0) $GtCyr^{-1}$. However, the most part of CO_2 (3.0 $GtCyr^{-1}$) accumulated in the atmosphere, with progressive worsening of the greenhouse effect. In the last 60 years, the atmospheric CO_2 concentration has increased from 310 ppm to 410 ppm, with an annual rate of 1.7 ppm. In the present stage of

Table 4.5 Decadal mean ($GtCyr^{-1}$) of the anthropogenic CO_2 budget in the first and last decades of the time period 1960–2017 (after Quéré et al. 2018, modified)

	1960–1969	2008–2017
Fossil CO_2 emission (EFF)	3.1 + – 0.2	9.4 + – 0.5
Land-use change emissions (ELUC)	1.5 + – 0.7	1.5 + – 0.7
Total emissions (EFF + ELUC)	4.7 + – 0.7	10.8 + – 0.8
Partitioning		
Growth rate in atmospheric CO_2 concentration	1.7 + – 0.07	4,7 + – 0.02
Ocean sink	1.0 + – 0.5	2.4 + – 0.5
Terrestrial sink	1.2 + – 0.5	3.2 + – 0.7
Budget imbalance (Total emissions – Partitioning)	0.6	0.5

ecosphere evolution, the total respiration rate overshoots the total production rate, inverting the trend of the past geological eras, when Carbon accumulated underground as fossil fuels. The present pattern of increasing resource-energy use and of waste production is not sustainable ecologically. Ominous changes in the biosphere are by now evident at regional as well global levels, and a transition to an ecological phase of human development is demanded and ever more urgent (Palumbi 2001; Foley et al. 2005; Hansen et al. 2005; Hagens 2020). International agreements, past and new, from Agenda 21 to the current Agenda 2030,[12] have produced some political effects and induced an internationally shared vision that should provide fruits at regional and local levels if a common commitment and responsibility by individuals and institutions will be developed and mutually reinforced. The future of humanity is in the humanity's hands and minds.

All these are facts that need confrontation with human needs and expectations in order to establish the concept of sustainable development as an ethical benchmark. *Development*, as a scientific concept derived from direct experience of changing forms in organisms, species, and ecosystems, has an analogous term in philosophy, *becoming*. The dialectic between Being and Becoming is a substantial part of human reflection in the psychology studies of the eminent scholar Abraham H. Maslow (1968), who advanced a theory of human needs that has received a large consensus in many disciplinary fields (Koltko-Rivera 2006). According to him, needs or values relate to each other in a hierarchical and developmental way, in an order of strength and priority. First come *physiological needs*, that concern the environmental relations for survival (provision of air, water, food, etc.), but "all these basic needs may be considered to be simply steps along the path to general self-actualization, under which all basic needs can be subsumed". Next, come *psychological needs*, concerning social relations for ensuring sense of belonging to community and requisites for personal and social security. At the top of the hierarchy stand *spiritual needs*,

[12] Resolution adopted by the UN General Assembly on 25 September 2015. Agenda 2030 is a plan of action for people, planet and prosperity, with 17 principles of sustainable development.

without which the human development process is not completed. On the meaning of both the whole hierarchy and each step, Maslow comments in this way:

1. "For one thing, it looks as if there were a single ultimate value for mankind, a far goal toward which all men strive. This is called variously by different authors self-actualization, self-realization, integration, psychological health, individuation, autonomy, creativity, productivity;
2. [the whole hierarchy must] be treated both as ends and as steps toward a single end-goal. It is true that there is a single, ultimate value or end of life and also it is just as true that we have a hierarchical and developmental system of values, complexly interrelated;
3. achieving basic-need gratifications gives us many peak-experiences, each of which are absolute delights, perfect in themselves, and needing no more than themselves to validate life;
4. the process of moment-to-moment growth is itself intrinsically rewarding and delightful in an absolute sense [...] Being and Becoming are not contradictory or mutually exclusive. Approaching and arriving are both in themselves rewarding" (p. 153–155).

According to Maslow (1968) "the tendency to grow toward full-humanness and health" can be agreed through a series of "objectively describable and measurable characteristics of the healthy human specimen" such as:

1. Clearer, more efficient perception of reality.
2. More openness to experience.
3. Increased integration, wholeness, and unity of the person.
4. Increased spontaneity, expressiveness; full functioning; aliveness.
5. A real self; a firm identity; autonomy, uniqueness.
6. Increased objectivity, detachment, transcendence of self.
7. Recovery of creativeness.
8. Ability to fuse concreteness and abstractness.
9. Democratic character structure.
10. Ability to love, etc. (p.156).

As to the relation man/environment, Maslow declares:

1. Man demonstrates *in his own nature* a pressure toward fuller and fuller Being, more and more perfect actualization of his Humanness;
2. creativeness, spontaneity, selfhood, authenticity, caring for others, being able to love, yearning for truth are embryonic potentialities belonging to his species-membership just as much as are his arms and legs and brain and eyes;
3. the role of the environment is ultimately to permit him or help him to actualize *his own* potentialities, not *its* potentialities (p.160–161).

Recent comments (Koltko-Rivera 2006) put emphasis on the utility of Maslow's psychological theory:

Recognizing self-transcendence as part of Maslow's hierarchy has important consequences for theory and research:

(a) a more comprehensive understanding of worldviews regarding the meaning of life;
(b) broader understanding of the motivational roots of altruism, social progress, and wisdom;
(c) a deeper understanding of religious violence;
(d) integration of the psychology of religion and spirituality into the mainstream of psychology;
(e) a more multi-culturally integrated approach to psychological theory.

The condition of health holds for both man's development and environment's development. Health is a value that derives from the observation of facts. Man's development, both as individual and species, and environment's development are mutually connected and it is insane to separate them. Both, man and environment are processes in development that must be harmonised, and sustainability is today the conceptual key for harmonisation. This assumption concerns the relation between the concept of health and the concept of integrity as well. We can start from the general definition of health provided by Rapport (2003) in relation to "regaining healthy ecosystems: the supreme challenge of our age":

> Clearly, the concept of health – which at root refers to the capacity of a system (whether biological, social, or mechanical) to perform normal functions – is not restricted to the hierarchical levels in which the particular system operate. There are healthy and unhealthy cells, tissues, organs, organisms, populations, biotic communities, ecosystems, and landscapes. When it comes to the biosphere, that all-encompassing dimension of life, its state of health also depends on the degree to which its functions are unimpaired (p.5).

There is no doubt that a vast group of indicators exist that provide an overview of signs commonly observed in ecosystems under stress, including the whole biosphere, such as altered climatic conditions, loss of biodiversity, altered primary productivity, altered nutrient cycling, pollution of air, water, soil, and food chains. In general, there is a diffuse state of *pathologies* denouncing a deviation from the state of normality recognised as *ecosystem health*. For looking at deeper, we can trace back the concept of deficient ecosystem health recovered by the ongoing pathological facts to the concept of ecosystems integrity. Such an operation has the meaning to discover what is missing, or compromising, to cause some impaired function of both ecosystem parts and ecosystem as a whole. Of great help in this effort is considering as a start the definition of the generic concept of integrity, as reported by Westra (2003):

> A valuable whole, 'the state of being whole, entire, or undiminished', or, 'a sound, unimpaired or perfect condition'. We begin with the recognition that integrity, in common usage, is an umbrella concept that encompasses a variety of other concepts.

On the base of Westra's account of a natural ecosystem unimpaired by human presence, it is possible to make a representation of the concept of ecosystem integrity like that reported in Fig. 4.9. The first point to work upon is that of "historicity of nature" (Pannenberg 2008), the meaning of which is well expressed by Gregersen (2008) in his introduction to Pannenberg's book as follows:

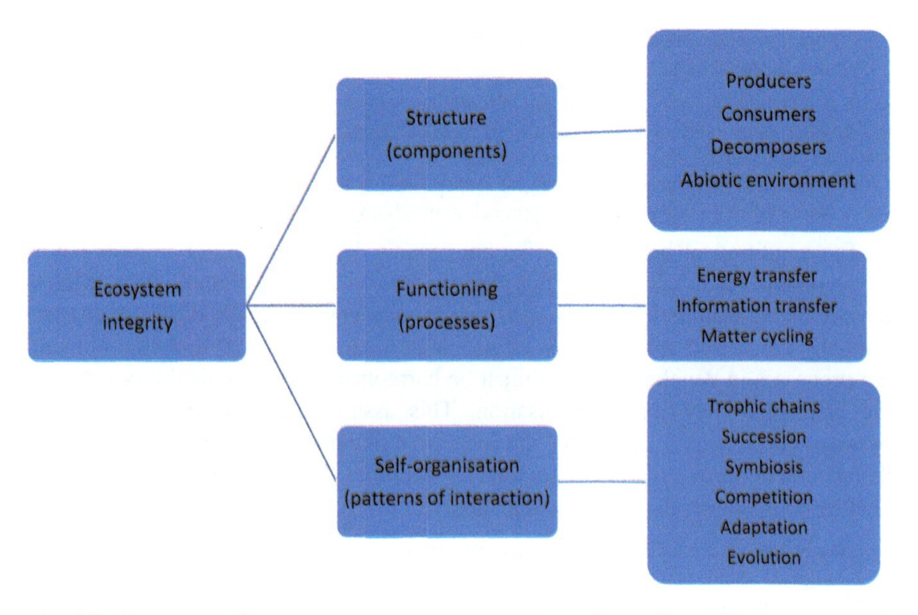

Fig. 4.9 The concept of ecosystem integrity as a means to look at its elemental composition and operational organisation on a spatiotemporal scale

> Not only is human existence shaped by historical decisions and cultural turning points, but nature also has an irreversible history in which virtually every event is unique and new complex structures are continually build up [p. vii]

The historical character of nature is a rational assumption resulting from both scientific and theological reasons, whereby Pannenberg (2008, pp.26–39) talks about "consonance between creation theology and natural science". From an ecological point of view, a pristine ecosystem is "a part of nature's legacy, the product of natural history" and its biota is the creature of evolutionary processes at work locally (Westra 2003, p. 32). A pristine ecosystem is an example of sustainable development in that it accounts for:

(a) the capacity of nature to persist by maintaining its fundamental auto-poietic processes of energy transfer, information transfer, and matter cycling;
(b) the capacity of nature to create different biological forms that are complementarily interwoven and prone to change due to environmental variations both induced or suffered;
(c) the capacity of nature to show locally the same principles that govern the self-organisation of the whole biosphere that is for man the total nature's legacy.

From Westra's reflexive approach, we can deduce that nature is a transmitter of important messages that are included in ecosystem structure, functioning and organisation, such as:

> The kinds of processes at work [in ecosystems] gave rise to the totality of life on Earth, including ourselves, and together maintain the conditions for the continuation of life as we

know it. Thus, natural ecosystems are valuable to and in themselves for their continuing support of life on Earth, for their aesthetic features, and for the goods and services they provide to humankind. Ecological integrity is thus essential to the maintenance of ecological sustainability as a foundation for a sustainable society (Westra 2003, p.32).

To confirm the validity of the concept of ecosystem integrity as an operational concept to make value judgements, Ulanowicz (2000) mentions the case of legislation of Canada and the United State where the notion of ecosystem integrity has repeatedly appeared. He identifies the concept of ecological integrity with the following four attributes; (1) system "health", i.e. the continued successful functioning of the community; (2) the capacity to withstand stress (resilience); (3) an undiminished "optimum capacity" for the greatest possible ongoing development options; and (4) the continued ability for ongoing change and development, unconstrained by human interruption.

Altogether, these attributes suggest the case for envisaging a map of sustainability (Fig. 4.10) that analogically represents them.

An ecosystem is sustainable if its parts co-operate appropriately, can resist to environmental stress, are able to balance input/output exchanges in order to maintain a relative stability, and manifest attitude to change in order to bear evolutionary development. The concept of sustainability, and its value, find a definite confirmation in the evolutionary process that has invested our planet and produced man, an entity able to value and change the future evolution process (Box 4.10).

Rolston III asserts that, axiologically, ecosystems have "systemic value" not only *instrumental* or *intrinsic*, as commonly attributed by standard economy or ethics, respectively. Sistemic values pertain to each ecosystem and the entire biosphere as well, in that ecosystems are "complex adaptive systems" sharing the properties of *aggregation*, *nonlinearity* and *diversity* (Levin 1998). Aggregation and hierarchical

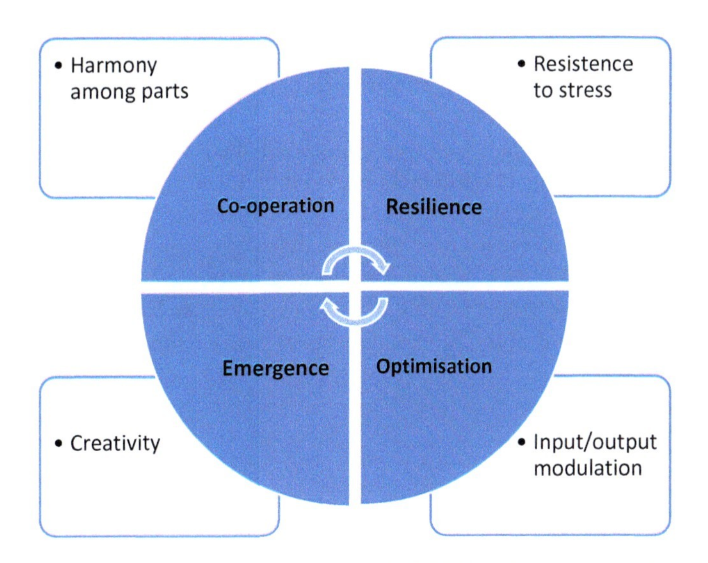

Fig. 4.10 Operational properties of ecosystem sustainability based on ecosystem integrity

assembly are not imposed, but emerge from local interactions of components and processes in nested hierarchical and evolutionary patterns (Pickett and Cadenasso 2001). Non linearity concerns change fluctuations due to chance events like environmental variations or genetic mutations that modify previous designs. Diversity within functional groups provides buffering and homestasis for critical ecosystem processes as diversity within a species provides resiliency and a hedge against extinction. Flows among components ensure exchange of energy, materials and information. The generation and maintenance of diversity is fundamental to adaptive evolution (Levin 1998). As result of aggregation, nonlinearity and diversity, complex adaptive systems possess the ultimate property of sustainability, i.e. the capacity of during through regeneration of their components, self-balance, and creative change.

On the front of human appreciation, ecosystem sustainability is self-revealing as a value emerging ontologically from real facts concerning nature's self-organisation. As such, ecosystem sustainability has potential for generating other values metaphorically connoted by it, such as *natural capital*, *ecosystem services*, and *ecological footprint* (Pickett and Cadenasso 2001). This metaphorical sequence of values allows their current social use in public discourse and contributes to the establishment of the paradigm of sustainable development.

Box 4.10: Steps of Earth's History According to a Philosophical Evaluation (After Rolston III 2006, Modified)

☺ Evolutionary ecosystems over geological time have increased the number of species on Earth from zero to five million or more.
☺ One- celled organisms evolved into many-celled highly integrated organisms.
☺ Photosynthesis evolved and came to support locomotion-swimming, walking, running, flight.
☺ Warm-blooded animals followed cold-blooded ones.
☺ Neural complexity, conditioned behaviour and learning emerged.
☺ Brains evolved, coupled with hands.
☺ Consciousness and self-consciousness arose.
☺ Persons appeared with intense concentrated unity.
☺ The products are valuable, able to be valued by these humans.
☺ Why not to say that the process is what is really value-able, able to produce these values?

4.3.1 *Natural Capital, Ecosystem Services and Ecological Footprint*

"In the application of natural capital to human use, the law of nature, or rather God's will promulgated by it, demands that right order be observed" (Pope Pius XI 1931). It is curious to note, from the above reported sentence of the encyclical letter "Quadragesimo Anno"of Pope Pius XI (1931), how the notion of "natural capital" appeared in a document of the Catholic Church before a scientific paper even mentioned it. Equally curious is the fact that the expression "natural capital" appears in the Italian version, but does not in the English translation by the Vatican, where is replaced by "natural resources". Instead, the first "scientific" mention of *natural capital* is due to the economist Schumacher (1973), as already cited. However, the ecologist Francis C. Evans (1956) made early important comments on the concept of ecosystem that recall at roots the meaning of natural capital:

> In the fundamental aspects, an ecosystem involves the circulation, transformation, and accumulation of energy and matter through the medium of living things and their activities.

These comments already show how the ecosystem organisation works. They show that nature has its own "economy" (circulation, transformation, accumulation), or a "great economy", where the human "small economy" is nested. The "great economy" of nature is the "natural capital", which the human "small economy" depends on. Economists have been able to explain some aspects of the natural capital with appropriate economic language and terminology, such as stock, income, renewable and non-renewable resources, discounting, and so on, but they have failed in extracting and valuing the role of biodiversity, which unfortunately escape the market's mesh network. Ecological economists, like Costanza and Daly (1992), explained plainly what natural capital is and why accounting for natural capital is so important. They were also right in criticising conventional economics for having considered human- made capital as a near perfect substitute for natural resources. They suggested the best methods for an economic evaluation of natural capital (cost/benefit analysis, methods based on human utility or preferences as "willingness to pay", and so on). They even were able to perform a total accounting for the world's ecosystem services and natural capital (Costanza et al. 1997). Nevertheless, something is still missing in a valuation process if it does not account for values that touch the soul and open to an interspecific ethics, "an Earth ethics, one that discovers a global sense of obligation to this whole inhabited biosphere", as Rolston III (2006) explains:

> The production of value over the millennia of natural history is not something subjective that goes on in the human mind. In that sense, a valuable Earth is the foundational value. The creativity within the natural system we inherit, and the values this generates, are the ground of our being, not the ground under our feet. Earth could be the ultimate object of duty, short of God, of God exist.

The simplest way to understand how nature accumulates its 'wealth', i.e. natural capital, is considering the process of *ecological succession*. The ecologist Eugene Odum (1969) smartly did it in a seminal paper devoted to understanding how the strategy of ecosystem development provides a basis for resolving man's conflict with nature. He suggested that the development of ecosystems has many parallels with the development of both organisms and human society. Starting from the definition of an ecosystem as a "unit of biological organisation made up of all the organisms in a given area", he accounts for ecological succession as a process (Box 4.11) with the following characters:

(a) It is an orderly process of community development that is reasonably directional and, therefore, predictable;
(b) it results from modification of the physical environment by the community (succession is biologically-controlled);
(c) the physical environment determines the pattern, the rate of change, and often sets limits to how far development can go;
(d) it culminates in a stabilized ecosystem in which maximum biomass (or high information content) and symbiotic functions between organisms are maintained per unit of available energy flow.

Odum adds that the strategy of ecosystem development at local scale is similar to that of biosphere development at global scale, with increased control on the physical environment in the sense of achieving maximum protection against its perturbations. Moreover, looking at the extreme stages on the axis of ecosystem development, there are contrasting characteristics between young and mature-type ecosystems, such as production/growth/quantity for the former and protection/stability/quality for the latter. He concludes that the tabular model (Box 4.11) presented for the ecosystem development has many parallels with the development of human society:

> In the pioneer society, as in the pioneer ecosystem, high birth rates, rapid growth, high economic profit, and exploitation of accessible and unused resources are advantageous, but, as the saturation level is approached, these drives must be shifted to consideration of symbiosis (that is, "civil rights", "law and order", "education", and "culture"), birth control, and the recycling of resources".

Box 4.11: Tabular Model of Ecological Succession According to the Concept of Ecological Integrity: Main Trends in the Development of Ecosystems (After Odum 1969, Modified)

Ecological succession is an integration process involving biotic and abiotic components in setting up more complex communities and resilient ecosystems.

Community features	Ecosystem attributes	Developmental stages	Mature stages
Community structure	Stratification and heterogeneity	Poorly organised	Well-organised
	Species diversity	Low	High
	Biochemical diversity	Low	High
	Total organic matter	Small	Long
Community energetics	Gross production/ community respiration (P/R ratio)	Greater than 1	Approaches 1
	Gross production/standing biomass (P/B ratio)	High	Low
	Biomass supported/unit energy flow (B/E ratio)	Low	High
	Net community production (yield)	High	Low
Community nutrients	Mineral cycles	Open	Closed
	Nutrient exchange rate, Organisms/environment	Rapid	Slow
	Role of detritus in nutrient regeneration	Unimportant	Important
	Inorganic nutrients	Extra-biotic	Intra-biotic
	Nutrient conservation	Poor	Good
Community self-organisation	Food chains	Linear, grazing	Web-like, detritus
	Internal symbiosis	Undeveloped	Developed
	Stability	Poor	Good
	Niche specialisation	Broad	Narrow
	Entropy	High	Low
	Information	Low	High

His final recommendation concerns the future of socioecological systems, in which a balance between youth (productive) and maturity (protective) developmental stages is "the really basic goal that must be achieved" in the transition to "the ultimate equilibrium-density stage".

The work of nature meant as "natural capital" has obviously contributed to valuing *nature's services*[13]in economic terms and to measuring them in monetary currency. The first written witness of this trend was an influential paper of the ecologist Walter E. Westman (1977) in which he expressed the following motivation about his work:

In the inexorable quest to rationalize the activities of the civilizations, policy makers in Western societies have increasingly asked the monetary value of items and qualities formerly regarded as priceless: clean air and water, untamed wildlife, wilderness itself. Behind this search has been the hope that, by weighing the benefits to society of nature in the undeveloped state against the benefits of resource development, an objective basis for decision-making will be achieved. Commonly, policy analysts further seek to estimate the equivalence in currency of the values lost by damaging ecosystems.

The most ambitious attempt of this trend of studies was the evaluation of the global world's ecosystem services and natural capital in comparison with the gross national product total (Costanza et al. 1997). In this kind of studies, a marked "anthropocentric bias" prevails, in that nature's services appear as a flow of benefits enjoyed exclusively by human society. Within this primary bias, a second one nests, that we can define as an "economic bias", which powerfully reinforces the first one and confirms the underlying motivation, i.e. the intentional, instrumental use of nature by man. A mitigation of this unbalanced ecosystem services value was present in some ecologists, like Ehrlich and Mooney (1983), who vindicated the *life-support-value* of ecosystem services instead of the human well-being value:

> Clearly, the diminution of solar-powered natural systems and the expansion of fossil powered human systems are currently looked in a positive feedback cycle. Increased consumption of fossil energy means increased stress on natural systems, which in turn means still more consumption of fossil energy to replace lost natural functions if the quality of life is to be maintained.

An important step for the development of the ecosystem services concept was an international, concerted initiative of the policy and science communities, completed in 2005, in order to make a massive synthesis of scientific knowledge about global ecosystems and their relation with human society, the Millennium Ecosystem Assessment or MEA (2005) (www.MEAweb.org). MEA grounded the basis for a sustainability science (Kates et al. 2001) in connection with policy relevant objectives, such as ecosystem services (Daily 1997; Daily and Matson 2008), land use dynamics (Turner II et al. 2007), governance of common property resources

[13] *Nature'services* is the first documented version of the concept that later has taken other spellings, such as *life-supporting-services*, *ecological services* and, lastly, *ecosystem services*.

(National Research Council 2002), connections of human and earth system history (Costanza et al. 2007) and earth system modelling (Steffen et al. 2004). As Carpenter et al. (2009) explain:

> The Millennium Ecosystem Assessment used a new conceptual framework for documenting, analysing, and understanding the effects of environmental change on ecosystems and human well-being. It viewed ecosystems through the lens of the services that they provide to society, how these services in turn benefit humanity, and how human actions alter ecosystems and the services they provide. The focus on ecosystem services has been adopted widely among the scientific and policy communities and has resulted in new approaches for research, conservation, and development.

As to the ecosystem services concept, MEA was able to enlarge its meaning in a way to include values that transcend the instrumental use of nature, signalling new dimensions as reported in Fig. 4.11.

Indeed, the MEA report on biodiversity synthesis assumes that "human well-being is the central focus for the MEA, but biodiversity and ecosystems also have intrinsic value. People make decisions concerning ecosystems based on considerations of well –being as well as intrinsic value"(MEA 2005, p. IV).

Biodiversity holds a systemic value connecting every level of organisation just in its definition: "biodiversity refers to diversity at multiple scales of biological organisation (genes, populations, species and ecosystems) and can be considered at any geographic scale (local, regional, or global)" [MEA 2005, p. 2]. This high ranking of biodiversity in the ecosystem organisation recalls the role that Evans (1956) assigns to it in regulating ecosystem services:

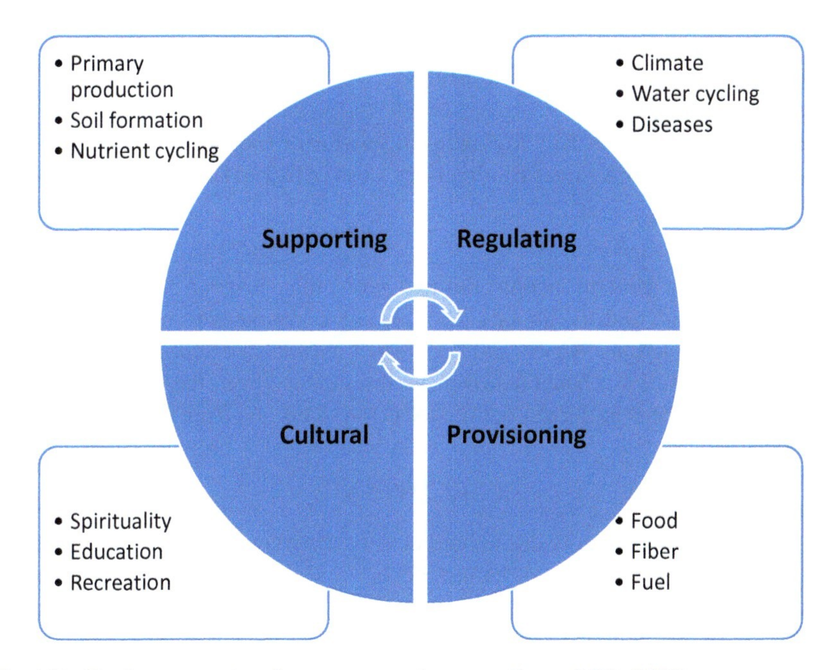

Fig. 4.11 The four categories of ecosystem services according to MEA (2005)

> Ecosystems are further characterized by a multiplicity of regulatory mechanisms, which, in limiting the numbers of organisms present and in influencing their physiology and behaviour, control the quantities and rates of movements of both matter and energy. Processes of growth and reproduction, agencies of mortality (physical as well as biological), patterns of immigration and emigration, and habits of adaptive significance are among the more important group of regulatory mechanisms. In the absence of such mechanisms, no ecosystem could continue to persist and maintain its identity.

Moreover, since "ecosystemness" operates as a universal pattern, ecosystem services can be investigated, monitored and evaluated at any scale of reality:

> Each population can be regarded as an entity in its own right, interacting with its environment [...] In turn, each individual animal or plant, together with its particular microenvironment, constitutes a system of still lower rank. Or we may wish to take a world view and look upon the biosphere with its total environment as a gigantic ecosystem. Regardless of the level on which life is examined, the ecosystem concept can appropriately be applied (Evans 1956).

For all above- mentioned reasons, ecosystem services "are a powerful lens through which to understand human relationships with the environment and to design environmental policy" (Brauman et al. 2007). MEA reports a list of key messages to inform society about the value of biodiversity in improving ecosystem sustainability (Box 4.12).

As to the current evaluation of ecosystem services according to the proposed frame (Fig. 4.11), MEA reveals the state of their incomplete assessment as follows:

> Only provisioning ecosystem services are routinely valued. Most supporting, and regulating services are not valued because the willingness of people to pay for these services – which are not privately owned or traded – cannot be directly observed or measured. In addition, it is recognised by many people that biodiversity has intrinsic value, which cannot be valued in conventional economic term (p.7).

Box 4.12: Selection of Key Messages in Favour of Biodiversity Improvement and Against Biodiversity Loss (After MEA 2005, p. VI, Modified)

BENEFITS

Biodiversity benefits people through more than just its contribution to material welfare and livelihoods. Biodiversity contributes to security, resiliency, social relations, health, and freedom of choices and actions.

Many people have benefited over the last century from the conversion of natural ecosystems to human-dominated ecosystems and from exploitation of biodiversity.

CHANGES

Changes in biodiversity due to human activities were more rapid in the past 50 years than at any time in human history and the drivers of change that cause biodiversity loss and lead to changes in ecosystem services are either steady, show no evidence of declining over time, or are increasing in intensity.

DRIVERS

The most important direct drivers of biodiversity loss and ecosystem services changes are **habitat change** (such as land use change, physical modification of river and water withdrawal from rivers, loss of coral reefs, and damage to the see floors due to trawling), **climate change**, **invasive alien species**, **overexploitation**, and **pollution**).

GOALS

Short-term goals and target are not sufficient for the conservation and sustainable use of biodiversity and ecosystems.

Improved capability to predict the consequences of changes in drivers for biodiversity, ecosystem functioning, and ecosystem services, together with improved measures of biodiversity would aid decision-making at all levels.

Science can help ensure that decisions are made with the best available information, but ultimately the future of biodiversity will be determined by society.

Moreover, MEA recognises the paradox of the present inconsistency of main indicators of economic performance when dealing with environmental health and wealth:

> A country's ecosystems and its ecosystem services represent a capital asset, but the benefits that could be attained through better management of this asset are poorly reflected in conventional economic indicators. A country could cut its forests and deplete its fisheries and this would show only a positive gain to GDP despite the loss of the capital asset (p.6).

The most innovative message transmitted by the new classification of ecosystem services provided by MEA is not in view of benefiting only human-well- being, but in showing that, with the flow of supporting and regulating ecosystems services, the entire living community is being benefited. Ecosystem health and wealth depend on the "right order" and harmonisation of its components, including human beings. This is why the inclusion of cultural services in the frame of ecosystem services is of crucial meaning, in that confirms the acceptance of man's belongingness in a living community. In practise, this means that man is potentially not only a user but also a provider of ecosystem services. Indeed, the truthfulness of this assumption accords with the following consideration:

> Ultimately, most biodiversity will be conserved if ethical, equitable distribution and spiritual concerns are taken into account than if only the operation of imperfect and incomplete markets is relied on (MEA 2005, p.8).

Indeed, many authors (Peterson et al. 2010; Kosoy and Corbera 2010; Peterson 2012) contest that the economic recognition of ecosystem services is the ultimate level of recognition. It easily opens the way to a process of "commodification" of nature and an "ethics of commodification" as that described by Plumwood (2002):

> It assumes a moral dualism between the group taken to be morally considerable ("persons") and the rest – which are "things" (and, potentially at least, property), and are assumed not to matter or count ethically at all, hence to be open to rational instrumental use (p. 144).

Such an ethic disregards the fact that "human dependence is multidimensional as our layered being":

> We are dependent on physical, biological and ecological, psychological, cultural and political institutions, and these multiple agencies themselves condition our own conceptions of dependence. [...] Replacing the economistic conception of quantifiable ecosystem services with a conception of diffuse and multiple agencies upon which human and nonhuman lives are diffusely dependent will require changes not merely in ontological categories, but also in political and economic institutions themselves (Peterson 2012).

Plumwood (2002) advances a tentative framework of "communicative virtues"[14] as foundations of a new "interspecific ethics", some of which appear in Table 4.6. This kind of re-framings has potential for: a)" recognising continuity with the non-human to counter dualistic construction of human/nature difference as radical discontinuity", b) "reconstructing human identity in ways that acknowledge our animality, decentre rationality and abandon exclusionary concept of rationality",

c)"opening the way for a culture of nature that allows [...] mutual adjustment between species, starting with our own, in what could become a liberatory blending or mingling of nature and culture"(p. 194–195). Such an "ecological virtue language" has potential for becoming and sharing a moral guide that expresses reason and emotions, commitment and hope, self-development and self-sacrifice, in "holistic ways consistent with ecologically informed worldviews" (van Wensveen 2000, p.161). Such a language can change radically our appreciation of humanity in the world in that "not only are we in the Earth envelope, we are part of it, participants in it, born from it, sustained and reproduced by it" (Rowe 2002, p.3). Such a language in favour of a community of life and physical environment is not only desirable but urgently needed for the foundation of "an ecological ethics" (Curry 2011).

Table 4.6 Ethical stances and communicative virtues (after Plumwood 2002, p.194, modified)

Ethical stances	Communicative virtues
Ontological stance	Decentring the human/nature contrast to allow a more inclusive, interspecies ethics
Intentional recognition stance	Openness to the non-human other as potentially an intentional and communicative being
Attentiveness stance	Listening to the other, active invitation to communicative interaction
Non-ranking stance	Minimising interspecies ranking and ranking context
Generosity stance	Redistribution, consideration without closure directed toward an excluded class
Mutual adjustment stance	Negotiation between parts
Self-critical stance	'Studying up' in problem context

[14] Those traits that sustain and deepen relationship.

Human activity systems of any kind have an environmental impact generating either ecosystem services or disservices, whereby this impact takes the name of *ecological footprint.* According to Wackernagel and Rees (1997), the ecological footprint in a concrete economy of a community is "the aggregate area of land and water necessary to produce all the resources consumed and to absorb all the wastes emitted by the participants".

Estimates of the ecological footprint indicate whether current or projected consumption levels can be sustained by available ecological productivity. In other words, "the difference between the size of a region (adjusted by its ecological productivity) and the footprint of this region's population must be covered by imports of ecological surpluses or the depletion of natural capital". The ecological footprint can be reduced by lowering population and consumption, or bettering technological efficiency and ecological productivity. According to Wackernagel and Rees (1995), the ecological footprint of all industrialized nations, representing less than 20% of the world population, is larger than the available ecological productivity land on Earth.

In the current organisation of the world economy driven forth by neoliberal principles, the instrumental use of nature is predominant. The global human economy functions as a great machinery that extracts a great deal of resources from nature giving back a great deal of wastes that impair the functioning of the natural system that support and sustain the global economy. On one hand, the instrumental use of nature depletes the "natural capital"; on the other hand, it undermines the release of the "ecosystem services" flow from the natural capital. The paradoxical trends between "economic rationality"[15] and "ecological reality" appear in Fig. 4.12.

The "ecological footprint" of humanity is progressively enlarging in the current state of economic organisation, causing a retrogression on the path of sustainable development and a worsening of the human predicament. This kind of assessment is certainly in line with the trends foreseen by Meadow et al. (1972). Recent surveys at planetary levels show that a safe operating space for humanity is already impaired for a series of biophysical thresholds such as climate change, rate of biodiversity loss, and nitrogen cycle (Rockstrom 2009; Rockstrom et al. 2009). According to Ehrlich et al. (2012), conceptual and analytical approaches of the human predicament and ecological footprint reveal both states of facts and prospects. As to the former, the great disparity between rich and poor nations in the per capita appropriation of Earth's capacity to support human activities, and in vulnerability to natural catastrophes. As to the latter, increasing well-being in poor countries will require significant reduction in the deleterious environmental impacts by rich and poor countries alike, and greatly narrowing the rich-poor gap will be key to achieving sustainability advances in many areas.

[15] "By economic rationality we understand subjecting everything to the laws of the market, whereby human beings act in accordance to the profit motive. Economic rationality assumes that all members of society are organised in the interest of profit and that their action can be understood by the production and distribution of goods while perceiving nature as a commodity to be traded in terms of market values" (Giraldo 2019, p.18).

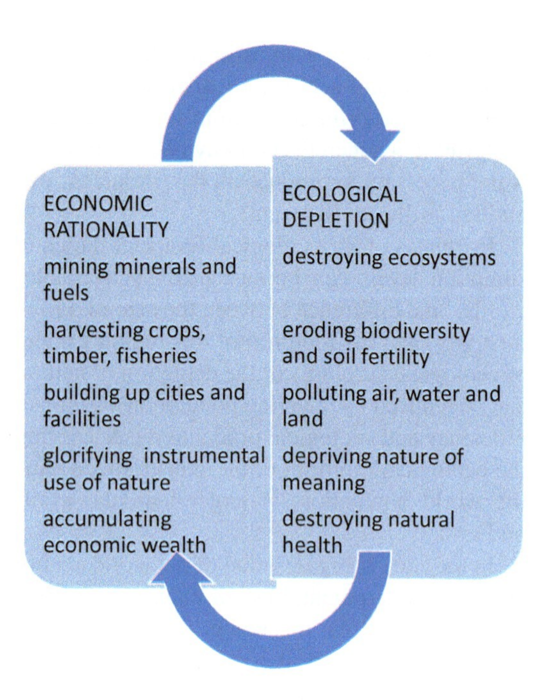

Fig. 4.12 Paradoxical trends causing the enlargement of the *ecological footprint* and the worsening of *the human predicament*

Arguing that the connectivity of the world is increasing and many of the most important environmental problem are global, Erlich et al. (2012) warn that "it is in everybody's interest to reduce ethically both the size of the population and our per capita impacts". This ethical challenge concerns avoiding "the amount of irreversible damage that overshoot inflicts on Earth's life-support systems", through a responsible action in the following direction:

> Just as climate change is speeding the flow of glaciers, [a responsible action] should speed the transition of the human enterprise towards a sustainable scale – at which care for all human beings and the natural capital upon which they depend is a the top of the political agenda.

To implement such a programme requires a cultural regeneration that implies an effort in many directions, especially in the field of education for reframing the current conception of the environment.

4.3.2 The Role of Education

According to Schumacher (1973), "the key factor of all economic development comes out of the mind of man", therefore, it is of fundamental importance to know how we think. George Lakoff's brilliant competence in cognitive science and linguistics helps a lot in explaining how we think in terms of "frames", i.e. typically unconscious structures physically realized in neural circuits in the brain. Frames include semantic roles, relations between roles, and relations to other frames (Lakoff 2010), as reported in Box 4.13.

Box 4.13: Potentialities and Limitations of "Framing" (After Lakoff 2010, Modified)

GENERALITIES
1. All thinking and talking involves "framing".
2. Frames come in systems, a single word activates its defining frame and the system its defining frame is in.
CONNECTIONS
3. Emotions are an inescapable part of normal thought.
4. You cannot be rational without emotions. Without emotion, you would not know what to want, since would be meaningless to you.
CHANGING
5. There are limited possibilities for changing frames.
6. Introducing new language is not always possible.
7. The new language must make sense in terms of existing system of frames.
8. It must work emotionally.
9. It must be introduced in a communication system that allows for sufficient spread over the population, sufficient repetition, and sufficient trust in the messengers.
MEANING
10. Facts must make sense in terms of people's system of frames.
11. A person must have a system of frames in place that can make sense of the facts.
12. Such frame systems have to build up over a period of time.

As well explained a century ago by the sociologist Max Weber (2009, first edition 1920), *der Geist des Kapitalismus* (the spirit of capitalisms) inspired by *die protestantische Ethik* (the protestant ethics) is still present today and provides, as then, the "frames" in which we are borne and brought up. In these frames, there is no limit for the instrumental use of nature, and money is the only driver of human economy. How to change the current "frames" is a challenge for the future ethics of sustainability. Lakoff suggests a methodology of communicative approach aimed at understanding the "real crises" (that we recognise as *the human predicament*), after taking account of the barriers reported in Box 4.14. The goal is building up conceptual structures to understand environmental issues. The next step after understanding "the real crisis" is to understand what to do about it. That means choosing the right policies and understanding how they can work. Again, it is possible to analyse the current political system with the communicative approach methodology in order to grasp its basic rationale and the possibility of changing it.

All this work has the first goal in bridging the cultural gap that Lakoff calls "environmental hypo-cognition: the tragedy of the absence of frames". The second and final goal is to establish a "progressive moral system" which has at its heart "the values", such as "empathy, responsibility (personal and social), and the ethic of excellence (make the world better, starting with yourself)".

As to the analysis of the current political system, Lakoff identifies it as a "conservative moral system" with a series of characters that contrast the emergence of a sustainable development paradigm, as summarised below:

Box 4.14: Communicative Approach Methodology: Barriers to Environmental "Framing" (After Lakoff 2010, Summarised)

GOALS
1. Building up conceptual structures to understand environmental issues.
2. Frames communicated via language and visual imagery.
3. The right language is necessary for communicating "the real crisis".
BARRIERS
1. Many people have in their brain circuitry the wrong frames for understanding "the real crisis".
2. They have frames that would either contradict the right frames or lead them to ignore the relevant facts.
3. What is needed is a constant effort to build up the background frames needed to understand the crisis, while building up neural circuitry to inhibit the wrong frames.
4. That is anything but a simple, short-term job to be done.

(i) man is above nature in a moral hierarchy and nature is there for human use and exploitation;
(ii) the natural world is a resource for short-term private enrichment;
(iii) the market is both natural and moral (Let-the-Market-Decide ideology);
(iv) conservatives tend to think more in term of direct rather than systemic causation;
(v) present-day market fundamentalism assumes that greed is good;
(vi) market principles should govern our conflicts between environmentalism and economics;
(vii) one such principle is cost-benefit analysis (CBA);
(viii) the basic math of CBA uses subtraction: the benefits minus the costs summed up over time indefinitely;
(ix) benefits and costs are seen in monetary terms, as if all values involving the future of the earth were monetary;
(x) the Equivalent Value Metaphor that compute ecosystems services as costs that a private enterprise should bear to provide the equivalent services.

According to Lakoff, the result of a policy based on the conservative moral system is the progressive destruction of the environment health and its sacrifice to short–term profit. What needs to do is "to activate the progressive frames on the environment (and other issues) and inhibit the conservative frames. This can be done via language (frame the truth effectively) and experience". Via language, he suggests personal recommendations, such as talking at the level of value, framing issues in terms of moral value, telling stories that exemplify values and rouse emotions, and dealing with general themes or narratives. Via experience, a topic issue is food, because it is "central to our existence as individuals and the politics of food is central to our existence on the planet". Food is a global need that easily can be associated with local environment, whereby the idea of *Globalizing Localism* emerges as a topic for framing development:

> Localism is the idea that food, energy, housing, and many other necessities of life can be made available locally in most of the world, that the third-world development depends on it, and that the control of carbon dioxide in our atmosphere may also depend on it (Lakoff 2010).

The topic of "frames and framing" appears as a chapter of a recent book on "ecolinguistics" (Stibbe 2015). Starting from the assumption that language influences how we think about the world, "ecolinguisics is about critiquing forms of language that contribute to ecological destruction, and aiding in the search for new forms of language that inspire people to protect the natural world". It contributes to the creation of a deeper connection between two areas of life that appear separate in two distinct fields of knowledge, ecology in the field of science and language in the field of humanities. Instead, there is no ontological discontinuity between the two areas of life, ecology being nested on language as a simple terminological refinement of experiential knowledge explained in terms of a transdisciplinary science. As mental models that influence behaviour, frames constitute powerful tools of cybernetic control of ecosystem management, even if their setting up demands long-term assemblage by promoters and long-term assimilation by users. They are prone to a selective pressure in a participatory community and their affirmation depends on attaining public consensus. From an educational point of view, ecolinguistics represents the most effective public means for creating the basis for a culture and an ethic of sustainability (Kates et al. 2001; Jenning 2010). As such, it should permeate all channels of communicative activity.

What we are trying to do in this heading 4.3. is just framing an ethic for sustainable development by setting up a circular sequence of meaningful metaphors via language (Fig. 4.13).

Frames are cognitive explanations based on conceptual systems done within cognitive science and cognitive linguistic tools, like metaphors. Lakoff (1995) argues that "the most important part of any real moral system is the system of metaphors for morality". He recalls that in the past most of explanations have tended to be given on economics, or class, or models of power. Currently is ecolinguistics that can support an innovative education based on real facts concerning the organisation of the planet economy, made up by the work of both nature and culture:

> It is through language that economic systems are build, and when those systems are seen to be lead to immense suffering and ecological destruction, it is through language that they are resisted and new forms of economy brought into being (Stibbe 2015, p.2).

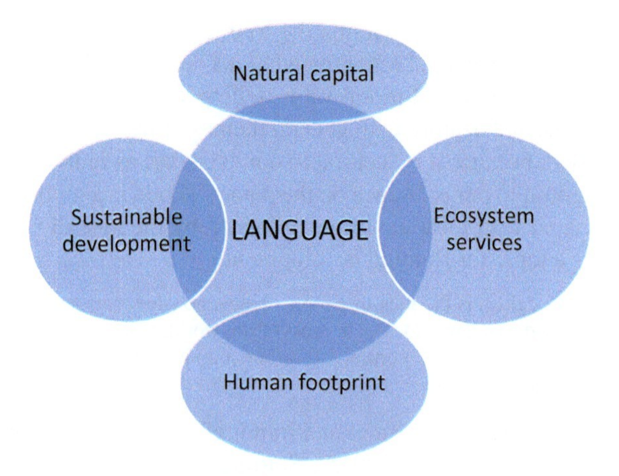

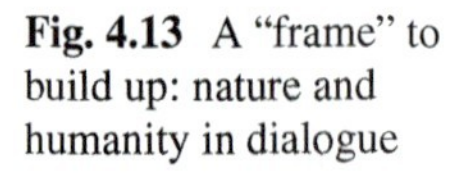
Fig. 4.13 A "frame" to build up: nature and humanity in dialogue

Figure 4.13 is a circular frame to be read clockwise starting from natural capital (or 'wealth' of nature) that provides ecosystem services (benefits for humanity and for the entire living system); humanity's using up of natural capital (human footprint) compromises the flow of ecosystem services and imposes a revision and resetting of the whole socio-ecological system toward a sustainable development. The frame entails: a) resistance toward all frames that facilitate humanity's using up of natural capital for instrumental use of nature and short-term profit; b) promotion of frames that value harmonisation between human needs, both material and immaterial, and nature provision of ecosystem services, including the cultural ones.

Nested on the dialogical frame of man/nature dialogue, another frame can be set up considering the dominant role assigned to human beings as market "consumers" (Fig. 4.14).

In this frame, a double opposition emerges, that between 'natural workers and market consumers', and that between 'natural resources and human care'. The expression "natural workers" mimics the original metaphor *ecosystem workers* due to Peterson et al. (2010) in this context:

> When ecosystem services to humanity become commodities, the biotic components of ecosystems become the workforce whose labour and energy is purchased. We use the phrase *ecosystem workers* as shorthand for the organisms that produce services in an ecosystem service marketplace.

The opposition 'natural workers and market consumers' is clear in defining the ecological role of the living community and, as a separate one, the economic role of man as user of commodified resources provided by gratuitous natural forces, often treated as mechanical slaveries. The right side of Fig. 4.14 is a sub-frame to reject (Wick 2010), recovering the dignity of human beings by opposing (left side) the economic metaphor of 'resources' with the ethic metaphor of 'care', a behavioural attitude guided by respect, gratitude and well-managed practises.

In literature, there are well-known examples of environmental ethics, like Aldo Leopold's land ethics, that contains important insights, such as "the land ethics

Fig. 4.14 Dialogic frame between 'consumerism and care'

simply enlarges the boundaries of the community to include soil, waters, plants, and animals, or collectively, the land" (Leopold 1949, p. 204). Yet, it is locally dimensioned; therefore, it is "a poor fit with the most urgent and dire environmental concern of our time", as recently reported by J. Baird Callicott (2013).[16] However, Leopold wrote in 1923 an essay unpublished over his life, which dealt with an earth ethic. In this essay, he traced a sketch where Callicott recognises important elements to inspire a frame for an ethic of sustainability, such as "personal, professional and social virtue (self-respects)", and two distinct kinds of responsibility to future generations – to immediate posterity and to the "Unknown Future".

An example of ecolinguistics *sui generis*, the case of "ecotheology", is rather relevant today for the development of a culture and an ethic of sustainability. The definition of ecotheology reported in Encyclopedia of Science and Religion (van Huyssteen 2003, p.247) explains the science of ecology as an informant of theology as follows:

> The term *ecotheology* came into prominence in the late twentieth century, mainly in Christian circles, in association with the emergent scientific field of ecology. Ecotheology describes theological discourse that highlights the whole 'household' of God's creation, especially the world of nature, as an interrelated system (*eco* is from the Greek word for household, *oikos*).

The text goes on with "ecotheology arose in response to the widespread acknowledgment that an environmental crisis of immense proportions was threatening the future of human life on the earth". It is likely true that the first input that originated the interest in connecting ecology with theology was the paper of the historian Lynn White (1967) who made the following conclusion:

[16] He taught the world's first course in environmental ethics in 1971 at the University of Wisconsin-Stevens Point.

> Since the roots of our trouble [a worsening ecological crisis] are so largely religious [rightful mastery of nature], the remedy must also be essentially religious, whether we call it that or not. We must rethink and re-feel our nature and destiny. The profoundly religious, but heretical, sense of the primitive Franciscans for the spiritual autonomy of all parts of nature may point a direction: I propose Francis as a patron saint for ecologists.

White's article gave birth to a fervent discussion of the role of Christianity in perpetrating the environmental crisis (Sponsel and White Jr 2016). Such a discussion has continued through to the present day. Meanwhile, the thesis regarding the environmental guilt of Christianity passed into the subject literature as the *Lynn White thesis* (Whitney 2013; Sadowski 2017). In practise, all major religions have passed a scrutiny about their potential in favouring or contrasting the socio-environmental crisis and this matter is in continuous development (Sadowski 2017). Catholic Church has given its own contribution since the seminal document of Pope John Paul II (1990) on "Peace with God the creator, Peace with all of Creation"(see Box 4.8). The global message of this document is an educational warning with and ecotheological content:

> Today, the dramatic threat of ecological breakdown is teaching us the extent to which greed and selfishness – both individual and collective – are contrary to the order of creation, an order which is characterised by mutual interdependence (p.8).

This kind of approach to the human predicament has been maintained to this day as a thread in developing ecotheology such as a complex reframing of the theology of creation through the information derived by the science of ecology. In doing that, a new kind of ethic has emerged, an ecotheological ethic. For instance, most of Franciscan literature developed since the *Lynn White thesis* has given origin to a vast body of knowledge to constitute a real, new transdisciplinary field of survey, as well-documented by the Australian ecotheologist Denis Edwards (2006):

> Learning about science and theology, I found myself spending months reading Bonaventure and discovering in his Trinitarian theology abundant resources for an ecological theology that takes Christology and the Trinity as central (p.2).

The task of recovering the teaching of patristic and medieval fathers for a theological foundation of environmental ethics has been carried out by ecotheology scholars like Jame Schaffer (2009), and has touched its apex with the recent Pope Francis' encyclical letter "Laudato si' – On Care For Our Common Home"(2015) [from now on LS]. This monumental encyclical letter has received a vast resonance, recognition and appreciation by religious authority and political institutions all over the world. Nowadays, it continues to rise dialogue on the interrelated issues of theology, ecology, economy and equity. Soon after its publication, a collection of essays representing multiple fields and faiths appeared with an explicit dedication to Pope Francis, recognising his moral leadership in guiding to "a complete transformation in how we teach, how we govern, how we do business, how we think, and who we include" (Cobb Jr and Castuera 2015). From this collection, Rosemary Radford Ruether's essay summarises at best the ecotheological content of LS, as briefly reported in Box 4.15.

Box 4.15: Ecotheology Contents of Pope Francis' Laudato si' (Elaborated from Ruether 2015, pp. 19–20)

Theological contents
1. The earth is God's, not ours.
2. We are called to repentance and conversion from our exploitation of one another and for the earth, into a restored attitude of care.
3. We are called to facilitate the fulfilment of creation in that goodness which is our and creation's calling.
4. The ecological crisis is not "natural" [...] but reflects human sin.
5. Religions need to dialogue with each other, as well as with sciences.
Ecological contents
6. The exploitation of nature by humans, and the misuse of human power, especially by the richest ones.
7. Technology needs to be guided by ethics.
8. Local cultures should be respected, especially indigenous cultures that have values of harmony with and care for creation.
9. Francis concludes by calling for an ecological conversion of humanity.

The final part of LS deals with "Education and ecological spirituality" (Chapter 6), that is, more than a logical conclusion of the previous parts, a window open toward a new phase of human civilisation. Chapter 6 presents 9 headings, 5 concerning suggestions for civic ecological duties and 4 concerning the relative connections of theological meaning. A key word for the whole chapter is *ecological conversion*, where conversion means both "profound interior conversion" (LS, 217) and "community conversion" (LS, 219). Box 4.16 reports the LS five headings and a few points of civic interest for an ecological education.

To deepen the relationship between "Religion and Ecology", it is highly recommendable to consult the 2001 special issue on this topic of *Dædalus*, a journal published by the MIT Press on behalf of American Academy of Arts & Sciences. In the "Introduction", Tucker and Grim (2001) report this important information:

> This issue of *Dædalus* brings together for the first time diverse perspectives from the world's religious traditions regarding attitudes toward nature with reflections from the fields of science, public policy, and ethics. The scholars of religion in this volume identify symbolic, scriptural, and ethical dimensions within particular religions in their relations with the natural world. They examine these dimensions both historically and in response to contemporary environmental problems (p. 1).

> The objective here is to present a prismatic view of the potential and actual resources embedded in the world's religions for supporting sustainable practices toward the environment. An underlying assumption is that most religious traditions have developed attitudes of respect, reverence, and care for the natural world that brings forth life in its diverse

Box 4.16: Selection of Headings and Points for an Ecological Education (From LS, Chapter 6)

I.TOWARDS A NEW LIFESTYLE
1. Reduce "compulsive consumerism" (203).
2. Consider "the environmental footprint and the patterns of production" (206).
3. Be "always capable of going out of ourselves towards the other" (208).
II. EDUCATING FOR THE COVENANT BETWEEN HUMANITY AND THE ENVIRONMENT
4. "Restore the various levels of ecological equilibrium, establishing harmony within ourselves, with others, with nature and other living creatures, and with God"(210).
5. " Ecological education can take place in a variety of settings: at school, in families, in the media, in catechesis and elsewhere"(213).
III. ECOLOGICAL CONVERSION
6. "The ecological crisis is also a summons to profound interior conversion"(217).
7. "The ecological conversion needed to bring about lasting change is also a community conversion"(219).
IV. JOY AND PEACE
8."An integral ecology includes taking time to recover a serene harmony with creation, reflecting on our lifestyle and our ideals" (225).
V. CIVIC AND POLITICAL LOVE
9."Care for nature is part of a lifestyle which includes the capacity for living together and communion" (228).
10. "We must regain the conviction that we need one another, that we have a shared responsibility for others and the world"(229).

> forms. Furthermore, it is assumed that issues of social justice and environmental integrity need to be intricately linked for creating the conditions for a sustainable future (p.2-3).

At this point of noospheric development, humanity as a whole may represent a new theological category with a planetary role. Under the pressure of both a progressive planetary crowding and the necessity of moralisation to mitigate conflicts, humanity can find its own way for a new mystique of integration, harmony, and love that Teilhard de Chardin used to call "humanisation" (Caporali 2015), and that we now call "sustainable development". A spiritual evolution phase should ignite the entire process of life evolution on the Earth through a noospheric convergence brought about by both demographic pressure in a confined planetary space and a spirit of creative belongingness in a cosmic project. Science show us an inherently self-organising universe (Jantsch 1981), a dynamic creative system operating everywhere in the planet Earth in a sequential rise of complexity and consciousness (Teilhard de Chardin 1959), and a humanity invested by the task to guide the transition to a viable Earth community. There is the awareness to live "a new story, a story of cosmic and planetary dimensions" (Swimme 2003):

> We must concern ourselves, then, with discovering this macrophase of the human. We must learn to conceive of ourselves and our genetic powers within the total life process of this

> planet. Previous conceptions of the human have failed to reach this larger planetary dimension. Former conceptions might well have been adequate for certain earlier historical period, each with its own specific needs, but those situations have now disappeared (pp. 127-128).

To initiate this story in a correct way (Rees 2003, p.104), we should take note that:

1. our current environmental dilemma is due, in part, to a much distorted perception of reality;
2. our understanding of the environmental crisis is dangerously superficial, and the possibility of sustainable development based on the growth-oriented assumptions of neoclassical economics is illusory,
3. significant changes in sociocultural beliefs attitudes, and behaviour will be required before sustainable development can acquire any substantive meaning.

Under the influence of the dominant neoclassic economic categories, we consider the reality we live in as an immense planetary infrastructure that provides humanity with that flow of ecosystems services that human society perceives as valuable and without charge. The sustainable use of the planet requires that we maintain our ecological capital without imperils to its self-organisation. "'Environments' are out there and society/economy/community are 'in here'" (Luke 2001). The educational problem consists in inverting this epistemological stance and letting humanity perceive that we are a "smaller whole" of a "larger, planetary whole" in the "largest, cosmic whole". In this spirit, from now on this book will deal with agriculture, a human activity system that, since its origin, has accompanied the development of humanity as a trophic and sociocultural bridge with nature. A sustainable agriculture may be a model for exploring the right way of interacting with nature in partnership, to achieve mutual sustenance and benefits.

References

Alrøe et al (2006) Organic agriculture and ecological justice: ethics and practice. In: Halberg N et al (eds) Global development of organic agriculture: challenges and prospects. CAB International, pp 75–112

Anker P (2002) The context of ecosystem theory. Ecosystems 5:611–613

Bernstein BB (1981) Ecology and economics: complex Systems in Changing Environments. Ann Rev Ecol Syst 12:309–330

Bormann FH, Kellert SR (1991) Ecology, economics, ethics: the broken circle. Yale University Press, New Haven

Boyden S, Dovers S (1992) Natural-resource consumption and its environmental impacts in the Western world. Impacts of increasing per capita consumption. Ambio 21(1):63–69

Brauman KA et al (2007) The nature and value of ecosystem services: an overview highlighting hydrologic services. Ann Rev Environ Resour 32:67–98

Burns TP (1990) The power of the ecosystem concept and paradigm. In: Kawanabe H, Ohgushi T, Higashi M (eds) Ecology for tomorrow, Physiology and ecology Japan (special number), vol 27, pp 191–205

Callicott JB (2013) Thinking like a planet. The land ethic and the earth ethic. Oxford University Press, Oxford

Caporali F (2015) The ecological perspective of the phenomenon life. In: Cresti V, Galleni L (eds) Teilhard de Chardin and astrobiology (Italian and English version). Edizioni Erasmo, Livorno, pp 93–102
Carpenter SR et al (2009) Science for managing ecosystem services beyond the millennium assessment. PNAS 106(5):1305–1312
Cherret JM (1990) The contribution of ecology to our understanding of the natural world: a review of some key ideas. In: Kawanabe H, Ohgushi, T, Higashi M (eds), ecology for tomorrow. Physiol Ecol Japan (Special Number) 27:1–16
Cobb JB Jr, Castuera I (2015) For our common home. Process-relational responses to *Laudato si'*. Process Century Press, Anoka
Costanza R, Daly HE (1987) Toward an ecological economics. Ecol Model 38:1–7
Costanza R, Daly HE (1992) Natural capital and sustainable development. Conserv Biol 6(1):37–46
Costanza R et al (1991) Goals, agenda, and policy recommendations for ecological economics. In: Costanza R, Daly EE, Bartholomew JA (eds) Ecological economics. The science and management of sustainability. Columbia University Press, New York, pp 1–20
Costanza R et al (1997) The values of the world's ecosystem services and natural capital. Nature 387:253–260
Cowles H (1911) The causes of vegetative cycles. Bot Gaz 51(3):161–183
Cropper ML, Oates WE (1992) Environmental economics: a survey. J Econ Lit 30:675–740
Curry P (2011) Ecological ethics. An Introduction. Polity Press, Cambridge
Daily GC (1997) Nature's services. Societal dependence on natural ecosystems. Island Press, Washington, DC
Daily GC, Matson PA (2008) Ecosystem services: from theory to implementation. Proc Natl Acad Sci USA 105:9455–9456
Dale MB (1970) System analysis and ecology. Ecology 51:2–16
Daly HE (1974) The economics of the steady-state. Am Econ Rev 64(2):15–21
Daly HE (1991) Steady-state economics, 2nd edn. Island Press, Washington, DC
Daly HE, Cobb JB Jr (1994) For the common good. Beacon Press, Boston
Edwards D (2006) Ecology at the hearth of faith. Orbis Book, New York
Ehrlich PR, Mooney HA (1983) Extinction, substitution, and ecosystem services. Bioscience 33(4):248–254
Ehrlich PR et al (2012) Securing natural capital and expanding equity to rescale civilization. Nature 486:68–73
Evans FC (1956) Ecosystem as the basic unit in ecology. Science 103:1127–1129
Foley JA et al (2005) Global consequences of land use. Science 309:570–574
Georgescu-Roegen N (1971) The entropy law and the economic process. Harvard University Press, Cambridge
Giraldo of (2019) Political ecology of agriculture. Agroecology and post-development. Springer
Golley FB (1993) A history of the ecosystem concept in ecology. Yale University Press, New Haven
Gregersen NH (2008) Introduction. In: Pannenberg W (ed) The historicity of nature. Templeton Foundation Press, Pennsylvania
Hagens NJ (2020) Economics for the future – beyond the superorganism. Ecol Econ 169:106520
Hall CAS (1975) Look what' happening to our Earth. Bull At Sci 31(3):11–21
Hansen J et al (2005) Earth's energy imbalance: confirmation and implications. Science 308:1431–1435
IUCN/UNEP/WWF (1991) Caring for the Earth. Strategy for sustainable living, Gland
Jantsch E (1981) The self-organising universe. Pergamon Press, New York
Jax K et al (2013) Ecosystem services and ethics. Ecol Econ 93:260–268
Jennings B (2010) Ethical aspects of sustainability. Mind Nat 3(1):27–28
Kates RW et al (2001) Sustainability science. Science 292:641–642
Koltko-Rivera ME (2006) Rediscovering the later version of Maslow's hierarchy of needs: self-transcendence and opportunities for theory, research, and unification. Rev Gen Psych 10(4):302–317

Kosoy N, Corbera E (2010) Payments for ecosystem services as commodity fetishism. Ecol Econ 69:1228–1236
Lakoff G (1995) Metaphor, morality and politics, or, why conservatives have left liberals in the dust. Soc Res 62(2):177–212
Lakoff G (2010) Why it matters how we frame the environment. Environ Commun 4(1):70–81
Leopold A (1949) A Sand County almanac and sketches here and there. Oxford University Press, Oxford
Levin SA (1998) Ecosystems and the biosphere as complex adaptive systems. Ecosystems 1:431–436
Lindemann R (1942) The trophic-dynamic aspect of ecology. Ecology 23(4):399–418
Luke TW (2001) Education, environment and sustainability: what are the issues, where to intervene, what must be done? Educ Philos Theory 33(2):187–202
MAB (Man and Biosphere) (1971) International co-ordinating Council of the Programme on man and biosphere. Final report. UNESCO, Paris
Maslow A (1968) Toward a psychology of being, 2nd edn. Van Nostrand, Princeton
McGrath AE (2009) A fine-tuned universe. Westminster John Knox Press, Louisville
MEA (Millennium Ecosystem Assessment) (2005) Ecosystems and human well-being: biodiversity synthesis. World Resources Institute, Washington, DC
Meadows DH et al (1972) Limits to growth. A report of The Club of Rome's project on the predicament of mankind. Universe Books, New York
National Research Council (2002) The Drama of the commons. National Academy Press, Washington, DC
Odum PE (1969) The strategy of ecosystem development. Science 164:262–270
Pace P (1996). From Belgrade to Bradford- 20 years of environmental education. In: Filho, W.L, Murphy Z, O'Loan K (eds) A sourcebook for environmental education. The Parthenon Publishing Group, London, pp. 1–24
Palumbi SR (2001) Humans as the world's greatest evolutionary force. Science 293:186–1790
Pannenberg W (2008) The historicity of nature. Templeton Foundation Press, Pennsylvania
Pearce D (1987) Foundations of an ecological economics. Ecol Model 38:9–18
Peterson K (2012) Ecosystem services, nonhuman agencies, and diffuse dependence. Environ Philos 9(2):1–19
Peterson MJ et al (2010) Obscuring ecosystem function with application of the ecosystem services concept. Conserv Biol 24(1):113–119
Pickett STA, Cadenasso ML (2001) The ecosystem as a multidimensional concept: meaning, model, and metaphor. Ecosystems 5:1–10
Plumwood V (2002) Environmental culture. The ecological crisis of reason. Routledge, London
Pope Francis (2015) Laudato si'. Encyclical letter on Care for our Common Home. www.vatican.va
Pope John Paull II (1990) Peace with God the Creator, peace with all of creation. ST Paul Publications
Pope Pius XI (1931). Quadragesimo Anno, Encyclical letter, www.vaticaan.va
Popper K (1959) The logic of scientific discovery. Routledge
Porter GJ, Brown JW (1991) Global environmental politics. Westview Press, San Francisco
Quéré CL et al (2018) Global carbon budget. Earth Syst Sci Data 10:2141–2194. https://doi.org/10.5194/essd-10-2141-2018
Rapport DJ (2003) Regaining healthy ecosystems: the supreme challenge of our age. In: Rapport DJ et al (eds) Managing for healthy ecosystems. CRC Press, pp 5–10
Rees WE (2003) Sustainable development and the ecosphere. Concepts and principles. In: Fabel A, John DS (eds) Teilhard in the 21st century. The emerging spirit of Earth. Orbis Book, New York, pp 103–123
Reiners WA (1986) Complementary models for ecosystems. Am Natl 127(1):59–73
Rockstrom J (2009) A safe operating space for humanity. Nature 46:472–475
Rockstrom J et al (2009) Planetary boundaries: exploring the safe operating space for humanity. Ecology and Society 14(2):32. [online]. http://www.ecologyandsociety.org/vol14/iss2/art32/

Rolston H III (2006) Intrinsic values on earth: nature and nations. In: ten Have HAMJ (ed) Environmental ethics and international policy. UNESCO, Paris, pp 47–67
Rowe JS (1961) The level-of-integration concept and ecology. Ecology 42:420–427
Rowe JS (1992) Viewpoint. Biological fallacy: life equals organisms. BioScience 42(6):394
Rowe JS (1996) Land classification and ecosystem classification. Environ Monit Assess 39:11–20
Rowe S (2002) Home place. Essays on ecology. NeWest Press, Edmonton
Ruether RR (2015) Pope Francis' encyclical on care for creation. In: Cobb JB, Castuera I (eds) For our common home. Process-relational responses to *Laudato si'*. Process Century Press, Anoka, pp 16–20
Sadowski SR (2017) The potential of religion in the promotion and implementation of the concept of sustainable development. Papers on Global Change 24:37–52
Schaffer J (2009) Theological foundations for environmental ethics. Georgetown University Press, Washington, DC
Schumacher EF (1973) Small is beautiful. A study of economics as if people mattered. Blond & Briggs, London
Smuts JC (1926) Holism and evolution. Macmillan, New York
Solow RM (1974) The economics of resources or the resources of economics. Am Econ Assoc 64(2):1–14
Sponsel L, White L Jr (2016) One catalyst in the historical development of spiritual ecology. In: LeVasseur T, Peterson A (eds) Religion and ecological crisis: the "Lynn White Thesis" at fifty. Routledge, London, pp 89–103
Starke L (1990) Signs of hope. Working towards our common future. Oxford University Press, Oxford
Steffen W et al (2004) Global change and the earth system. Springer, Berlin
Stibbe A (2015) Ecolinguistic. Language, ecology and the stories we live by. Routledge
Swimme B (2003) The new natural selection. In: Fabel A, John DS (eds) Teilhard in the 21st century. The emerging spirit of Earth. Orbis Book, New York, pp 127–136
Tansley AG (1935) The use and abuse of vegetational concepts and terms. Ecology 16(3):284–307
Tansley AG (1939) British ecology during the past quarter-century: the plant community and the ecosystem. J Ecol 27:513–530
Tansley AG (2002) The temporal genetic series as a means of approach to philosophy. Ecosystems 5:614–624
Teilhard de Chardin P (1959) The phenomenon of man. Harper & Row, New York
Thienemann A (1918) Lebensgemeinschaft und Lebensraum. Naturwiss. Wochenschrift, N.F. 17: 282-290, 297-303
Thompson PB (1996) Sustainability as a norm, Technè. J Soc Philos Technol 2(2):75–94
Tucker ME, Grim JA (2001) Introduction: the emerging alliance of world religions and ecology. Dædalus 130(4):1–22
Turner BL II et al (2007) The emergence of land change science for global environmental change and sustainability. Proc Natl Acad Sci 104:20666–20671
Ulanowicz RE (2000) Toward the measurement of ecological integrity. In: Pimentel D, Westra L, Noss RF (eds) Ecological integrity: integrating environment, conservation, and health. Island Press, Washington, DC, pp 99–133
UN (United Nations) (1972) Report of the United Nations Conference on the Human Environment, Stockholm, 5–16 June 1972
UN (1992a) Agenda 21. Report of the United Nations Conference on Sustainable Development, Rio de Janeiro, 3–14 June 1992
UN (1992b) Rio declaration of environment and development, UN General Assembly, 1992
van Huyssteen JWV (ed) (2003) Encyclopedia of science and religion. Macmillan Reference USA
van Wensveen L (2000) Dirty virtues. The emergence of ecological virtue ethics. Humanity Books, New York
Wackernagel M, Rees WE (1995) Our ecological footprint: reducing human impact on the earth. New Society Publishers, Philadelphia

Wackernagel M, Rees WE (1997) Perceptual and structural barriers to investing in natural capital: economics from an ecological footprint perspective. Ecol Econ 20:3–24
WCDE (1987) Our common future: report of the world commission on environment and development. Oxford University Press, Oxford, WCDE
Weber M (2009) Die protestantische Ethik und der Geist des Kapitalismus. Anaconda Verlag, GmbH, Koln
Westman WE (1977) How much are nature's services worth? Science 197:960–964
Westra L (2003) The Ethics of ecological integrity and ecosystem health: the interface. In: Rapport D et al (eds) Managing for healthy ecosystems. CRC Press, pp 31–40
White L (1967) The historical roots of our ecological crisis. Science 155:1203–1207
Whitney E (2013) The Lynn White thesis. Reception and legacy. Environ Ethics 35(3):313–331
Wick R (2010) Consumption embedded in culture and language: implications for finding sustainability. Sustain Sci Pract Pol 6(2):38–48

Chapter 5
Agriculture, Ethics, and Sustainable Development

What we do need is an ethic that recognizes the need for agriculture to be conducted in a manner that makes a decent life for humans possible on this planet while, at the same time, retaining the ecological dynamics that sustain all life on the planet

(Kirschenmann F. 2004, p. 169)

We already recognised the systemic character of ethics as an intricate crossroads of values originating from facts. Agriculture is just one evolutionary fact that concretely expresses the relationship between man and nature in terms of food provision and land use change. Agriculture is an intentional alteration of natural ecosystems that undergo dramatic changes in their biotic and abiotic components, both above and below the soil surface. Agriculture is a human activity system in action since 10,000 years in the old continents and it is still encroaching on natural biomes in several parts of tropical and subtropical areas of new and old continents. Agriculture has traditionally meant a food resource under man's control (Caporali 2000) and, as such, it has been defined the primary activity for progressive human settlements and civilisation. In this historical frame of reference, agriculture has an implicit ethical character (Sanford 2011) in that it is a medium between the system of human values and the system of human needs (Fig. 5.1).

Human needs are the driving forces behind every human activity system. They operate as *attractors* for the other system components that undergo continuous disturbance as soon as the human needs change, quantitatively or qualitatively. In the history of human kind, human diet and human brain have co-evolved through reinforcing patterns and correspondent environmental or health impacts due to the predominant kind of civilisation. As humans, we have thrived in most of the world ecosystems with diets ranging from almost animal food among populations of the Arctic to almost plant food (tubers and cereal grains) among populations in the highlands of Andes (Fig. 5.2).

F. Caporali, *Ethics and Sustainable Agriculture*,
https://doi.org/10.1007/978-3-030-76683-2_5

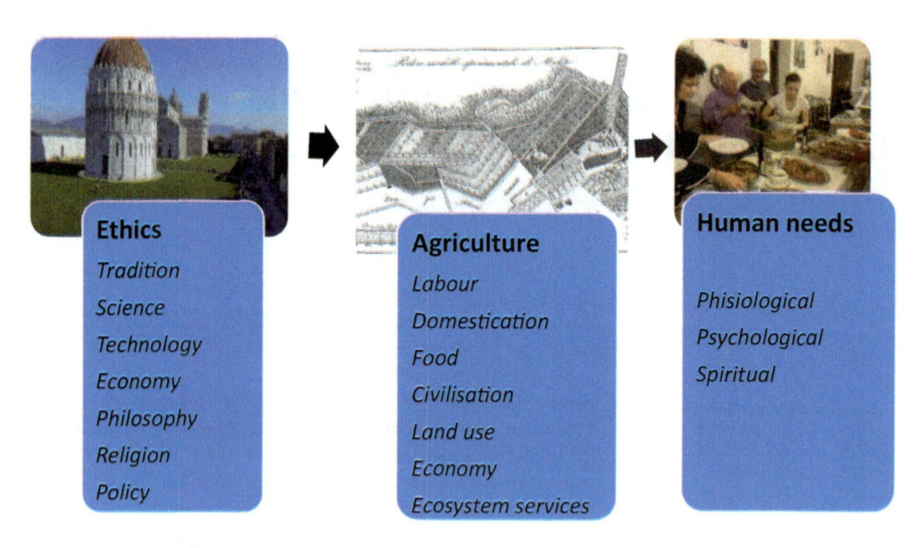

Fig. 5.1 Agriculture as a medium for meeting human needs and ethical values

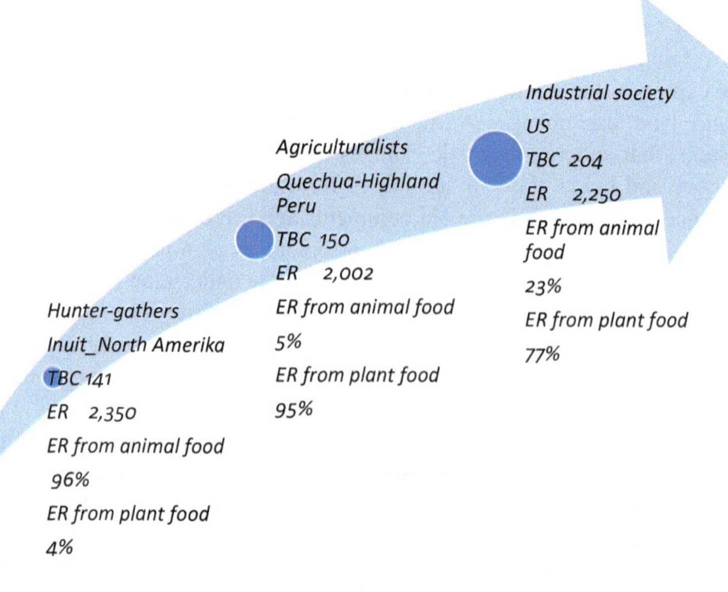

Fig. 5.2 Total blood cholesterol (TBC, mg dL^{-1}), energetic requirements (ER, Kcal day^{-1}) and diet components (%) in different kinds of civilisation. (After Leonard 2003, modified)

Dietary change played a determinant role in human evolution, as Leonard (2003) explains:

> Diet and brain expansion probably interacted synergistically: bigger brains produced more complex social behaviour, which led to further shifts in foraging tactics and improved diet, which in turn fostered additional brain evolution [...] Innovations such as cooking, agriculture, and even aspects of modern food technology can all be considered tactics for boosting the quality of the human diet.

Due to the reduction of physical activity, the challenge our modern society now faces is balancing the calories we consume with the calories we burn, how the cholesterol index reveals.

Beyond meeting the human physiological needs, agriculture has potential to meet both psychological and spiritual needs. All that is dependent on the favourable conditions that society as a whole is able to provide not only to farmers, but also to all stakeholders involved in agriculture as a complex, socio-ecological system.

Agriculture has also the potential to implement some or all the range of values that ethics displays. As to this topic, it is worth to quote the case of a recent paper asking a crucial question: "Do you see what I see? Examining the epistemic barriers to sustainable agriculture" (Carolan 2006). The paper examines those aspects connected of food production that are not readily revealed by direct perception. In doing that, it investigates how the tension between the "visible" and the "nonvisible" plays out in the debate between sustainable and conventional agriculture:

> Epistemology – that is, the study of how and what we know – cannot be abstracted from the site of practise and the socio-material relations from which it emerges. With this in mind, we can begin to understand that epistemic barriers are thoroughly performative, and thus social, in their nature; they are constituted through social interactions and organizing abstractions that shape how and what we know, and thus what we "see". (Carolan 2006)

The question of clarifying the importance of an epistemological approach within an ethical perspective merits a minimum of historical investigation. Philip T. Shepard (1985) arouse the challenge of how to mitigate moral conflicts in agriculture, starting with "to bring into the foreground much of what has been background". According to his opinion, it was necessary a re-examination of what was taken for granted as "normal" in the course of farming, extension, research, and policy-making. In practise, he made a comparison between two concurrent ethics in agriculture, as reported in Table 5.1.

Shepard shed light on a conflictual clash between two contrasting paradigms that persists unchanged until now:

> Agriculture today finds itself at the focus of numerous moral controversies, for example: whether increases in productivity morally justify a decreased quality of rural life and concentration of rural economic and political power in the hands of a few large agribusiness; to what extent, if any, it is morally permissible that short term increases in production jeopardize long run sustainability; whether researchers bear some moral responsibility for the social consequences of their work, and hence, whether they have an obligation to anticipate such consequences in selecting objectives and designing research. (Shepard 1985)

Table 5.1 Moral conflicts in agriculture: *productivity ethic* vs. *pluralistic ethic* as contrasting paradigms

Productivity ethic	Pluralistic ethic
Material prosperity	Spiritual growth
Competition	Cooperation
Free enterprise	Community
Scientific control	Empathic participation
Economic values	Systemic values
Indicators: yield per acre; net return for the whole farm enterprise.	Indicators: socioeconomic, environmental and cultural performances.

After Shepard (1985), modified

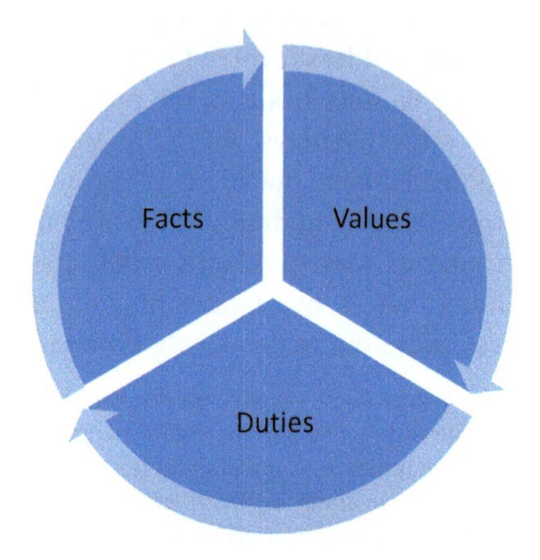

Fig. 5.3 Components of a virtuous circle of a pluralistic ethic

He recognised that moral conflict is part of an evolutionary process, where paradigmatic differences can be instrumental for making progress, because comparison and confrontation generate dynamic syntheses of value systems as they undergo evolution. As to moral issues in agriculture, coping with them requires scientists to work within a larger intellectual framework, including "metaphysics and value theory, history and social science, to see how facts and value are co-constructed out of a common cultural heritage" and come to shape the way to progress. All that under three basic assumptions: (a) "values are prospective and prescriptive rather than predictive", (b) "human values are systemic", and c) "facts and values depend on each other". From an ethical perspective, facts, values, and duties should be consequential, determining balanced judgement and legitimate action (Fig. 5.3).

Recognition of facts, value and duties as an interconnected sequence for framing behaviour has also a methodological meaning in order to acquire knowledge and finalise action in any kind of human activity system. Transferring this procedure to agriculture is what I will do from now on.

5.1 Agroecosystem Epistemology and Ontology

Ecology is a science originated by the human need to understand reality as both a whole and a composition of distinct parts. It is obvious that constructing a map of reality helps a lot in facilitating movements, interventions and desirable outcomes. The most remarkable achievement of ecology is just an epistemological innovation, the ecosystem concept, i.e. a transdisciplinary tool able to connect facts, values and duties in a frame of great coherence and meaning for design, management and control. This powerful instrument of knowledge is applicable to every kind of reality, including agriculture. In this regard, two assertions of John Harper sound explicitly truthful:

(a) "if ecology is the study of the relationship between organisms and their environment, the agriculture, and forest science are part of ecology- an applied sub-set of the science",
(b) "much of the development of ecology as a science in the post-descriptive phase will come from the study of man-managed ecosystems" (Harper 1982).

When applied to agriculture, the ecosystem concept defines an agricultural ecosystem or "agroecosystem", which is both a *real* ecosystem modified and used for agricultural purposes as well as a *model* that represent it (Caporali 2008, 2010). The agroecosystem concept is the epistemological tool, or unity of study and management, that grounds the science of Agroecology, or the science of ecology applied to agriculture (Caporali 2015). This cultural innovation has had recent development but has also ancient roots, as will be discussed later.

The recent development began with the cultural affirmation of the agroecosystem concept since the 1970s, when the first scientific papers, journals and books with the agroecosystem title appeared on the wave of the environmentalist movement that had started to shake the foundations of modern civilisation. A pioneering paper of this kind was "Agricultural ecosystems" by C.R.W. Spedding (1971), who explained how the general concept of agricultural ecosystems had gained widespread acceptance in scientific circles and why an understanding of the concept was indispensable for recognising the role of agriculture in the world. As to the epistemology of the concept, he asserted the occurrence of four pillars as shown in Fig. 5.4.

Spedding recognised that much agricultural research in the past had consisted of the application of scientific methods to *components* of agricultural systems and that eventually the agricultural systems themselves must be the subject of study:

(a) "man himself […], his intellect, his money, and his tools, are all important components of agricultural systems [that] may greatly influence agricultural systems, all of which have business and management components, as well as biological contributions";
(b) "agricultural systems are sub-systems of something larger […] and agriculture itself is one sub-system of the total activity of Man".

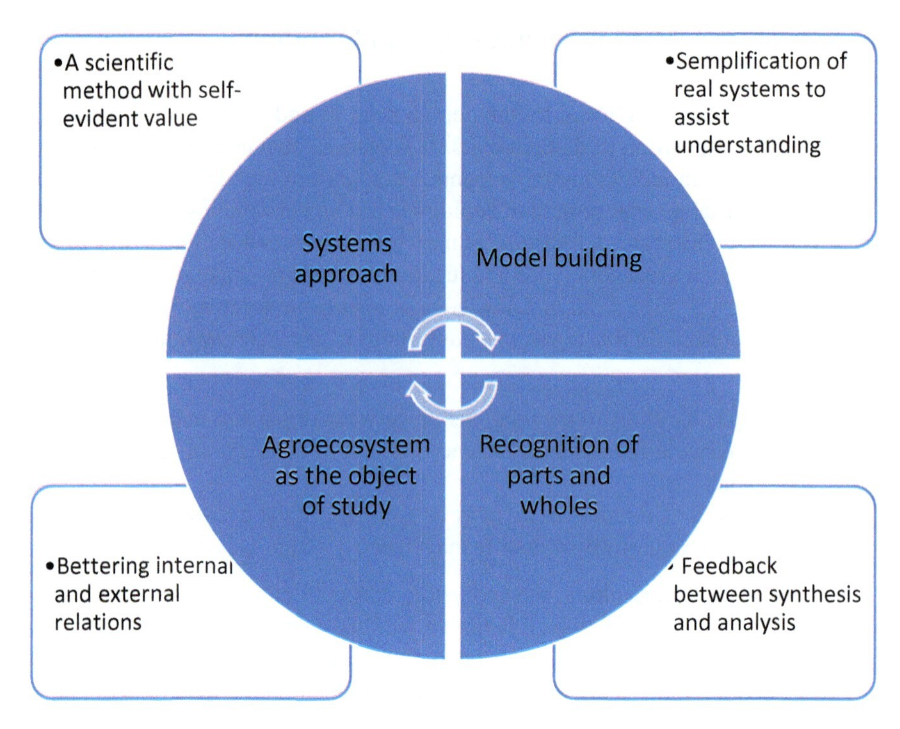

Fig. 5.4 Epistemological pillars of the agroecosystem concept. (Elaborated from Spedding 1971)

The sequence of the epistemic assumptions reported above shows clearly how the assimilation of the systems approach can be effective in revealing the *processual organisation* of agriculture as a socio-ecological system at every level of spatiotemporal scale, permitting connections between the local and the global, and the past, the present and the future. It is an all- around description revealing major agroecosystem properties, such as *hierarchy*, *emergence*, *communication,* and *control* (Checkland 1993; Caporali 2015). These properties make up the epistemological strength of the agroecosystem concept and ground its universal validity as a usual scientific methodology for surveying. The basic functional model illustrating the agroecosystem organisation derives from the ecosystem model, through the replacement of the natural components, plants and animals, with *crops* and *livestock*, as components introduced purposefully by the farmer (Fig. 5.5).

According to the agroecosystem concept, the processual organisation of agriculture functions as an input/output model mainly on the base of the information input of human nature that establishes not only the number and kind of basic agroecosystem components, but also the amount and quality of the inputs of energy, matter and capital that sustain those of natural origin. Design, management and performances depend enormously on human choices, both individual and social. Considering that "ethics is about choices" (Chrispeels and Mandoli 2003), *agriculture is one the most ethical enterprise of humanity*. Surprisingly, this connotation has not yet found recognition by society.

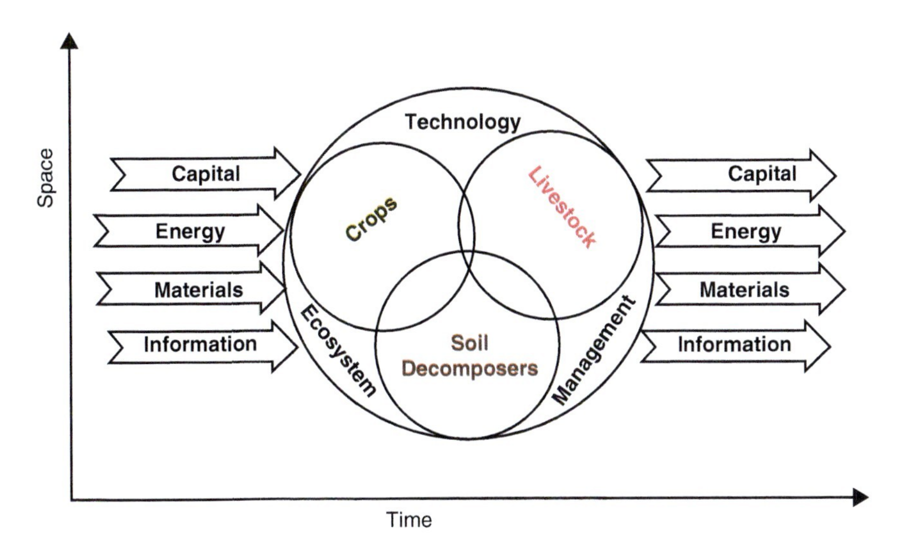

Fig. 5.5 Basic organisation model of a mixed agroecosystem. (After Caporali 2015)

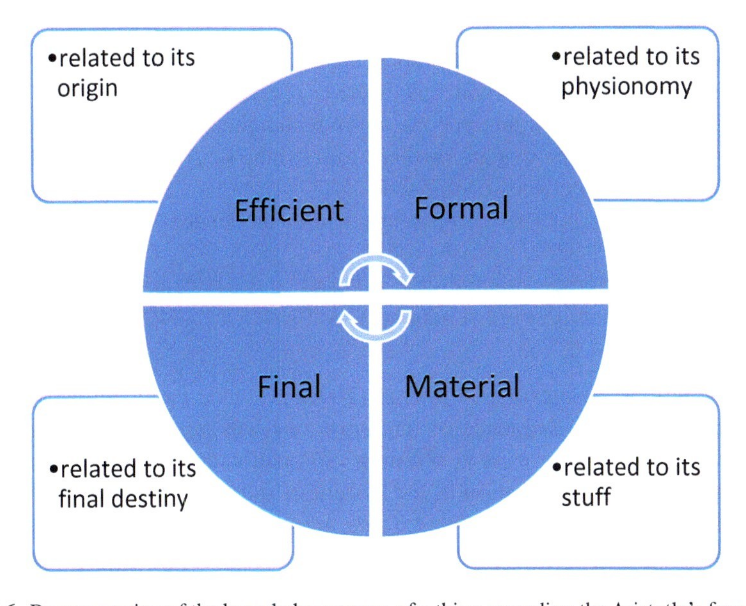

Fig. 5.6 Representation of the knowledge process of a thing according the Aristotle's four causes of knowledge

In a previous paper, Caporali (2008) has shown how the philosophical foundations of the systems paradigm has its roots in the Aristotle's *Physics*, specifically in the so- called four *causes* or principles of knowledge, i.e. *efficient, formal, material,* and *final causes.* Accordingly, a representation of the knowledge process of a thing, material or immaterial, appears in Fig. 5.6.

The analogy with the agroecosystem input/output model is self-evident, the *efficient* and *formal causes* being correspondent to the input of energy and information, respectively; the *material* to the stuff a thing is made of, and the *final causes* having to do with the output or future goals. In the case of the agroecosystem model, most of the causes are in the mind and the hands of the farmer, and therefore, they depend by his/her choices. Obviously, the process is recurrent and the final causes (effects or outputs) of a real agroecosystem operating in a context inform the project of the next agroecosystem model. Under this perspective, the agroecosystem is both a *real* ecosystem modified and used for agricultural purposes as well as a *model* that represents it. In this sense, agroecology, i.e. the agroecosystem science, exhibits a method that reflects a content, whereby epistemology and ontology coincides (Caporali 2008, 2010). This means that agroecology as a science has a high level of credibility, measured by the correspondence between its model of analysis and the reality analysed. In other words, the complementarity between analysis and synthesis that agroecology exhibits legitimates its value as a science appropriate for informing ethics.

As to the architecture through which conceiving and implementing the agroecosystem model, Spedding (1975) recommended the elements of conceptualisation listed below:

(a) the *purpose* for which the system is being carried;
(b) the *boundary* that defines which is inside or outside the system;
(c) the *context* or the external environment in which the system operates;
(d) the *main components* that are involved to form the system;
(e) the *relationships* between components;
(f) the *resources* or internal components within the system that are used in its functioning;
(g) the *inputs* or external resources that are used by the system;
(h) the *outputs* (main, desired products or performances and by-products, useful but incidental)

Following these suggestions, Caporali (2008) provided an iconographic sketch in which the temporal sequence of the agroecosystem model construction takes place. It is a four-step process concerning: (a) spatiotemporal scale framing; (b) boundary fixation, i.e. allocation of the system of interest into the spatiotemporal scale frame and delimiting an internal and an external context; (c) input/output exchanges; (d) final agroecosystem structure and functioning. The agroecosystem model has become the processual representation of agriculture that has paved the way to the modern science of agroecology.

Agroecology as a science found its concrete establishment in 1974 with the publication of "Agro-Ecosystems", a scientific journal of the international editor Elsevier, which celebrates with its title the cultural innovation based on the acceptance of the agroecosystem concept as the unit of study of agriculture. In that journal,[1] contributions coming from different disciplinary areas of agriculture have progressively accumulated to form a body of science that fundamentally has reflected and strengthened the four pillars of agro-ecological epistemology (Caporali 2010):

1. the agroecosystem concept as an input/output model, representing both the basic epistemological tool and the basic object of study in agroecology;
2. the representation of agriculture as a hierarchy of systems;
3. the representation of the farm system as a decision-making unity;
4. the representation of agriculture as a human activity-system.

All these pillars constitute a knowledge body, which defines agroecology as a transdisciplinary science and reveals its potential for permeating the value system of ethics. The agroecosystem model shows the universal pattern of organisation which operates at the level of both the whole planet earth and every its parts, including humanity as a powerful component that drives input and harvests output. Indeed, the agroecosystem model even goes beyond agriculture itself if, metaphorically, we assume the whole planet as the land to cultivate,[2] as the bible warns:

> "The Lord God took the man and put him in the garden of Eden to work it and keep it. (Genesis 2.15)

This means to take care of land in order to let it to produce and maintain its productivity as long as possible. To do that, it becomes indispensable to learn how to govern the basic processes of *energetics, matter cycling, biodiversity,* and *information* that interact both locally and globally for maintaining life and productivity.

5.1.1 Energetics

Energy is the driving force of natural and man -made ecosystems, including agriculture. Energetics concerns the study of energy sources, storages, flows, and conversions that have paved the way to the evolution of biosphere and civilisation (Smil 2008). Energy is both an input and output of every kind of ecosystems at every level of spatiotemporal organisation. In the sequence of energy transfer within the ecosystem components, measuring the transfer ratio between them, or *efficiency*, is what refers to as eco-energetics (Caporali 2010). While ecologists and agronomists have early started this kind of analysis in both cultivated fields (Transeau 1926) and natural ecosystem (Lindeman 1942), today the eco-energetics focus has moved principally toward the planetary scale (Tomlinson et al. 2014; Schramski et al.

[1] Currently, the journal is published under the title "Agriculture, Ecosystems & Environment".

[2] The metaphor of Earth as "land to cultivate" is etymologically justified in that "agro" comes from the Latin *ager*, that means "field", but also "land" in general (a farm or a whole region, for example, *Ager Gallicus* was the today's French land).

2015). As warned by Odum (1989), hierarchical theory offers a promising basis for integrating the local into the global provided respect of the following condition:

> Because ecosystems are thermodynamically far-from-equilibrium open systems where processes at lower levels are constrained by those at higher levels, what is called a "top-down" or "outside-to-inside" approach is suggested in which externals are considered first, then the internals [...] Then, energy, material, and organism inputs and outputs, and major functional processes (primary production, for example) of the system as a whole are examined.

Integrating the local into the global is just the task of agriculture, that should be organised respecting the limitations imposed by climate (external condition) and soil (internal condition) to be more efficient in using the complex of natural resources, reducing at the same time economic costs and environmental impact. However, there exists a socio-economic barrier to let this condition become operational:

> Applied scientists have been slow to use the holistic top-down approach to environmental problems. One reason is that "piece-meal" or "quick-fix" approaches often work well in the short-term of economic and political worlds [...] Science in general has become so reductionist that society is victimized by a "tyranny of small technologies" that arise from increasing specialization and the preoccupation with laboratory study. The open systems of real-world environments cannot be enclosed in glass tubes or laboratory walls. (Odum 1989)

A global perspective of energetics at planetary scale is today alarming. Schramski et al. (2015) propose a new appropriate metaphor for the planet earth under the form of "chemical battery", with a "cathode" charging (photosynthesis) and a discharging force (humanity) which radiates heat toward the chemical equilibrium of deep space ("anode"). This metaphor is particularly suited to the thermo-dynamic response that the whole planet is offering to the devastating pressure of humanity dominion. Considering an evolutionary frame, earth is "a chemical battery" with a "trickle-charge of photosynthesis" that allowed the accumulation of billions of tons of living biomass stored in forests and other natural ecosystems, and in vast reserves of fossil fuels. In the last few hundred years, humanity entered the era of industrialisation and extracted every kind of solar derived energy in the form of biomass, fossil fuels and humus from soil with the following outcome:

> This rapid discharge of the earth's store of organic energy fuels the human domination of the biosphere, including conversion of natural habitats to agricultural fields and the resulting loss of native species, emission of carbon dioxide and the resulting climate and sea level change. (Schramski et al. 2015)

Data on current production and historical human consumption of living biomass of the planet earth, where all energy available in the trophic web resides, appear in Table 5.2.

Data on Table 5.2 account for a continuous and progressive use of energetic natural capital stored in living biomass by humanity, amounting to a harvest rate of 45% in 2000 years (natural capital of living biomass almost halved!). If calculated on the flux rate of accumulation (PPN), the living biomass annual rate of human consumption (H) amounts currently to 75%. In other words, human consumption annually erodes ¾ of the increment of living biomass that should sustain the planet

Table 5.2 Energetics of production and human consumption of the living biomass of the planet earth as a battery (all data in ZJ = joules × 10^{21} or years, when appropriate)

Current production		Historical stock and human consumption	
1.Living biomass (B)	19	1. Living biomass at the birth of Christ	35
2. Annual primary production (NPP)	2	2. Living biomass in 1900	23
3. B/NPP (turnover time, years)	9.5	3. Living biomass in 2000 (B)	19
4. PPN/B (turnover ratio, year^{-1})	0.11	4. Human annual harvest (H)	1.5
		5. B/H (consumption time, years)	12.7
		6. H/B (turnover ratio, year^{-1})	0.08
B and NPP maintain biodiversity and regulate climate and biogeochemical cycling		Releases of carbon dioxide and heat undermine sustainability	

After Schramski et al. (2015), modified

life- support system. Explained in terms of land use change (Hooke et al. 2012), living biomass depletion means deforestation, desertification, agroecosystem expansion, urbanisation, and emergence of secondary effects such as pollution and unsustainable forestry and fisheries. On the base of these considerations, Schramski et al. (2015) add the following comments:

> As we burn organic chemical energy, we generate work to grow our population and economy. In the process, the high-quality chemical energy is transformed into heat and lost from the planet by radiation into outer space. The flow of energy from cathode to anode is moving the planet rapidly and irrevocably closer to the sterile chemical equilibrium of space.

Since the law of thermodynamics governing the neg-entropy of photosynthesis and the entropy of human- driven discharge of the earth's battery are universal and absolute, Schramski et al. (2015) prospect the following worrisome conclusion:

> The earth is shifting back toward the inhospitable equilibrium of outer space with fundamental ramifications for the biosphere and humanity. Because there is no substitute or replacement energy for living biomass, the remaining distance from equilibrium that will be required to support human life is unknown.

The thermodynamic interpretation of the biosphere energetics suggests that great part of material organisation of the economic 'production' is in fact 'consumption', which contributes to a constant increase in global net entropy or disorder. Therefore, sustainable development is "development that minimizes resource use and the increase in global entropy" (Rees 1990). This rule should be applied in the design and management of agroecosystems at every level of organisation.

Numerous studies of energetics of crops, farms, regions and states, took place since the 1970s and got published mainly in journals specialised in agroecology, such as "Agroecosystems" and "Agricultural systems", but also in more generalist scientific journals, such as "Science". In those studies, an energy accounting was set up, attributing an energy value to each meaningful agroecosystem component or process involved in the agricultural transformations. Those studies gradually let scholars and people appreciate agriculture as a sequence of nested agroecosystems, which are open to each other and reciprocally subjected to constraints from both the

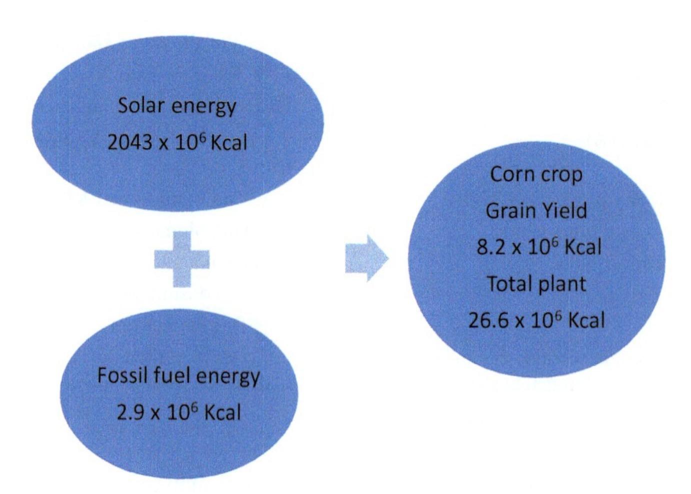

Fig. 5.7 Energetics of 1-acre corn field (USA, in 1970) during the growing season. (After Pimentel et al. 1973, modified)

top and the bottom of the hierarchy. A notable example is the paper of Pimentel et al. (1973), dealing with food production and the energy crisis, which shows the evolution of USA agriculture and food system during the "green revolution", through an account of the energetics of corn production. A synthetic representation of the energy sources involved and their contribution to corn grain production appears in Fig. 5.7.

Each corn field feeds on the direct flow of solar energy for photosynthesis and on the indirect flow of solar energy stored in fossil fuels for field management through machinery, agrochemicals, irrigation, electricity, and so on. Direct solar energy is plentiful (704 times the fossil fuel energy), for free, diffuse and clean, whilst fossil fuel energy is finite, concentrated, costly, and polluting. During the growing season, corn crop accumulates as plant biomass 11% of the fossil fuel energy flow with an efficiency of 1.26%. Even more interesting, Pimentel et al. (1973) compared corn crop performances in the period 1945–1970, just the time of "green revolution", with new corn hybrids, genetically selected for responding to increasing inputs of agrochemicals and irrigation. Table 5.3 reports some selected data of the input components of energetics accounting for corn crop production in USA, one of the most important leading crop of "green revolution" in the world.

Corn typifies the transition from traditional to conventional farming, where technological inputs derived from cheap, fossil fuels energy replaced an agronomic management of organic resources. In two decades, nitrogen fertilisation skyrocketed, boosting corn yield by more than double and leading the list of contributors to energy spending with a share of 32.5%. Industrial synthesis of nitrogen fertilisers bears a very high energy expenditure. In the form of N-NO_3, nitrogen is easily leachable and in the form of N-NOx volatiles in the atmosphere, contributing to both water eutrophication or pollution, and GHG emissions. All that implies the necessity for control of cycles of nutrients and agrochemicals in agroecosystems in order to promote their internal use and to reduce their external impact.

Table 5.3 Energy input composition of 1-acre of corn crop in USA (1950–1970)

Energy inputs		1950 (Data in the second column are %)		1970 (Data in the second column are %)	
Biological inputs					
Labour	(kcal × 10^3)	9.8	0.8	4.9	0.2
Seeds	(" ")	40.4	3.3	63.0	2.2
Technological inputs					
Machinery	(" ")	250	20.7	420	14.6
Gasoline	(" ")	616	51.0	797	27.5
Nitrogen	(" ")	126	10.4	941	32.5
Electricity	(" ")	54	4.5	310	10.7
Other inputs (P-K fertilisers, agrochemicals, irrigation, etc.)		110	9.3	361	12.4
Total input	**(Kcal × 10^6)**	1.2	100	2.9	100
Output (Corn grain yield)	**(" ")**	3.8		8.2	
Output/total input		3.2		2.8	

After Pimentel et al. (1973), modified

5.1.2 *Matter Cycling*

From the perspective of matter cycling, agriculture is a way of organising trophic cycles in favour of human beings. It is a change of the natural trends that entail a fair distribution of energy and matter to all members of living community. In a way, agriculturalists force nature to follow "unprecedented" patterns that contrast with an internal law, exactly a "natural law", which brings benefits at large for biodiversity maintenance and regeneration as a whole, and not only for a single species. Latin agricultural literature largely grounds its basis on the principle that "Omnibus Justissima Tellus" (Earth is very just in regard to all) (Caporali 2015), warning agriculturalists that agriculture should be 'docile', trying to mould nature with the spirit of accommodating its patterns instead of upsetting them.

A trophic cycle is the means through which energy flow, associated with matter, diffuses and sustains all members of a trophic community, where each single component captures, transforms and provides its share of energy/matter. We call it a 'cycle' because each chemical element that constitutes matter follows repetitively a circular path entering and leaving some form of biological entities or some form of physical habitat like soil, water and air. The global Carbon cycle describes appropriately how living beings take part in spreading energy-matter all over the world through the life-web continuum (Houghton 2007). All members of a living community live in a communion of energy-matter sharing, whereby each one performs its role of interconnectedness as individual and species. This 'drama' of trophic dependence, or mutual 'grazing', provides regeneration and creation of new forms of life in continuous development. Energy, matter and biodiversity are different perspectives of one unique cosmic event. Astonishingly, an uninterrupted string of matter, since its origin in stellar metabolism, appears to unite all the past, present and future events in an "undivided wholeness" (Davis 1990, p. 112), an unfinished cosmic tapestry made up of uncountable nodes of energy, matter and information

(Deacon 2012). An informed agriculture should not neglect the frame in which it is enveloped and respect its principle of orderly interconnectedness. Agriculture being a human organised part of a larger natural economy bears the responsibility of organising agroecosystems that satisfy the basic synergistic order that works for a lasting outcome, a common good, the health of both the whole biosphere and its ecosystem components. Agriculture has to play a crucial role in furthering evolution with the task of letting humans to become full earthlings in the way suggested by Thomas Berry (2003), i.e. by living a new story of cosmic and planetary development:

> Within this story, a structure of knowledge can be established with its human significance from the physics of the universe and chemistry through geology and biology to anthropology, and so on to an understanding of the entire range of human endeavour from language, literature, art, history, and religion to medicine and law, to psychology and sociology, to economic and commerce, and so to all those studies whereby human beings fulfil their role in the Earth process. In all these studies and in all these functions, the basic values depend on conformity with the Earth process. To harm the Earth is to harm the human; to ruin the Earth is to destroy humankind.

To organise agriculture is not only a technical and economic question, but also involves the value of establishing fair membership within a cosmic context of life. Agriculture is a human activity system with many components, where climate, soil, plants, animals and human management interact locally in a given context that has both an internal and an external environment, whereby numerous connections intervene. The spatiotemporal scale of an agroecosystem existence is a field crossed by currents of physical, biological and socioeconomic drivers, which transfer matter within and outside a web of internal and external components. Trophic chains of grazing, detritus, and biological control are typical of internal organisation, while market transfers of both input and output concern external organisation. Physical currents, such as water runoff, erosion, leaching, volatilisation, etc. are only partially controlled, but can be monitored or estimated, in a way to provide structural or management measures of prevention.

A well-organised report on cycling of mineral nutrients in agricultural ecosystems appeared in a 1977 issue of the journal "Agro-Ecosystems" (Frissel 1977), as the outcome of the first symposium on the topic, cosponsored by the International Association for Ecology and Elsevier Scientific publishing Company, and held in Amsterdam in 1976. That report provides important methodological suggestions, such as:

(a) the idea of 'control' of the drivers of the nutrient transfer process, which is a life supporting service for our planet and the fertility of both land and water;
(b) the recognition of water as the usual transport medium of most nutrients for both soil organisms and crops;
(c) the choice of "the farm level" as the unit for agroecosystem study, because of its easily recognisable boundaries and the probable availability of data concerning nutrient movements;
(d) the choice of a model for nutrient cycling with three components (pools) –plant, livestock and soil within the farm level.

As to the principle of 'control' as a method to assess the movements of nutrients, it is a choice that recognises the importance of the cybernetic nature of ecosystem development under the perspective of the trophic-dynamic aspect (Patten 1959; Patten and Odum 1981). According to the theory of ecological succession, organic evolution proceeds in the direction of increasing information to gain better conditions of long-term stability or sustainability. Since humanity is part of the organic evolution of the planet earth, it is perfectly coherent to adopt a rational method to verify an equation of agroecosystem control, such as:

$$\text{Control} = \text{inputs choice} + \text{transformation options} + \text{output balance}$$

To better frame the task of control, it is important to consider the hierarchical structure of agroecosystem representation. Figure 5.8 shows how the nested hierarchy of agroecosystems should appropriately inform the decision-making process at both institutional and individual levels, starting from the basic and more constraining conditions, those at the environment level, that are in large part unmodifiable.

Decisions made at the farm level should conform hierarchically to environmental constraints, and to socioeconomic and political constraints if appropriate, whereby this last decisional step between institutional and individual components should emerge from a procedure of public confrontation and shared consensus. Control is a recurrent, cybernetic step to verify if the whole agroecosystem and its parts function well according to the expectations. Therefore, control implies to verify the correspondence between the three basic components of the agroecosystem organisation, i.e. inputs, internal transformations and outputs. In practice, the correspondence emerges from investigation on nutrient transfers within the system and transfers across the system boundaries. The former ones concern the internal transfers between the crop pool, the soil pool and the livestock pool through grazing and detritus chains in an integrated web of synergistic productive community. The latter ones consist of inputs (or supplies), such as fertilisers or feed, and outputs (or

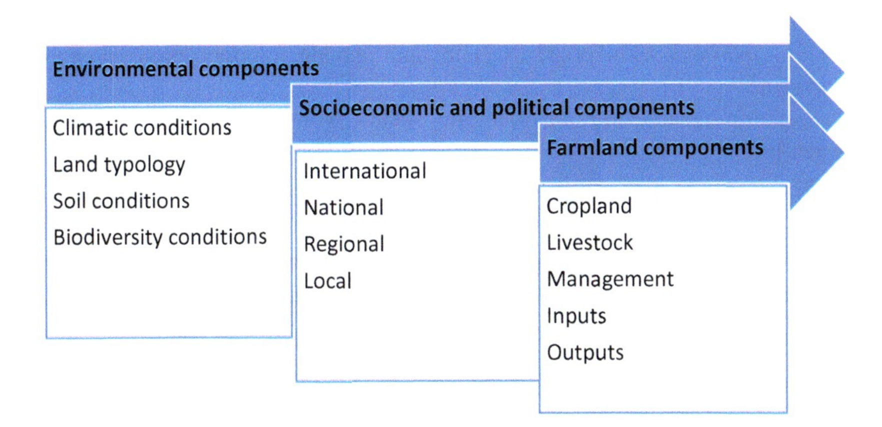

Fig. 5.8 Hierarchical assemblage of agroecosystem components for control assessment

removals), such as crop or livestock products, or leachable or volatile losses. In order to be sustainable, the best recommended practices are those that maximise nutrient retention in internal pools to potentiate soil fertility, and crop and livestock productivity for marketable products, while reducing nutrient losses that promote environmental dis-functions. In agroecosystems, soil fertility is a key factor for ensuring crop productivity by retaining (in organic form) and delivering (in mineral form) nutrients as appropriate, according to the seasonal rhythms of microbial mineralisation and plant uptake. In a Mediterranean environment, where the seasonal regime is very marked for both temperature and rainfall pattern, sustainability of crop production is a question of harmonizing crop growing with microbial activity in a way to sustain microorganisms with appropriate plant residues and plant nutrition with released nutrients by microorganisms. Adequate crop rotations, intercropping systems and green manures can modulate this plant/soil synergy under the farmer's control for the benefit of both production and reduced environmental impact (Caporali and Campiglia 2001; Caporali 2015).

Monitoring nitrogen dynamics in plant, soil and water constitute an interesting thread for following transfers between abiotic and biotic pools within and between different agroecosystems. Nitrogen is the most important nutrient for cereal crops that feed now not only most part of humans but also livestock and, in some share, new human mechanical 'slaves', such as cars, with ethanol production. Nitrogen cycle involves the following steps:

1. *removal* from atmosphere through:
    (a) physical fixation and transport by water precipitations to soil as NH_4-N or NO_3-N;
    (b) biological fixation by soil bacteria free or symbionts of legume species;
    (c) industrial fixation in fertilizers and subsequent application to crops;
2. *soil microbial transformations* in mineral or organic compounds available either for crop uptake and soil particle absorption, as well as for losses to air through volatilization and denitrification and losses to water through runoff, erosion and leaching;
3. *feeding livestock* with forage or grains and leaving livestock body as excreta and urine;
4. *undergoing composting* for manuring fields with other losses to air and water bodies.

The overall magnitude of anthropogenic relative to natural sources of fixed nitrogen (210 Tg N year^{-1} anthropogenic and 203 Tg N year^{-1} natural) is so large it has doubled the global cycling of nitrogen over the last century (Fowler et al. 2017). Nitrogen organic and mineral forms are easily detectable in water suspension or solution and monitored through routine sampling and chemical analysis of water bodies all over the world. Early studies at the scale of watershed level in both mountain and hilly areas of Central Italy (Nannipieri et al. 1985) provided data on nitrogen inputs in rainfall, nitrogen fixation in permanent meadows, nitrogen fertiliser applications in farms and nitrogen outputs as measured in river streamflow. Nitrogen

Table 5.4 Nitrogen balances by watershed (kg ha $^{-1}$ year^{-1})

Inputs and outputs	RENO watershed	ERA watershed
	4090 ha, 1640-609 m ab.sl.	9630 ha, 603-98 m ab.sl.
	Forest 97%, pasture, 3%	Forest 36%, arable land,64% (1/3 meadows)
	Annual precipitation 1800 m	Annual precipitation 840 mm
Precipitation	15	22
Biological fixation (permanent meadows)	4	18
Fertilisers	0	100
Total inputs	19	140
Outputs in streamflow	59	16
Total inputs minus outputs in streamflow (biomass uptake or release)	–40	124
Linear equations between nitrogen losses(Y) and stream flows (X)	NO_3-N Y = 0.360 + 1.181 X r = 0.99 NH_4-N Y = 0.023 + 0.074 X r = 0.95 Organic-N Y = −1.437 + 2.519 X r = 0.96	NO_3-N Y = −1.184 + 4.360 X r = 0.98 NH_4-N Y = 0.030 + 0.072 X r = 0.85 Organic-N Y = −0.494 + 2.007 X r = 0.93
Annual range of NO_3-N in stream flows (ppm)	1.0 – 2.0	2.5 – 6.3

After Nannipieri et al. (1985), modified
Nitrogen losses and flows in linear equations are as g s^{-1} and m^3 s^{-1}, respectively

balance of two watersheds, one almost totally forested and the other with 64% of arable land, is shown in Table 5.4. At level of land use, two factors prevail in shaping the nitrogen balance, one is the climate, in particular the rainfall regime, and the other is the human disturbance due to forestry or agricultural practices. The rainfall regime in a Mediterranean climate is characterised by higher precipitations from autumn to spring and lower or absent precipitations during summer time. In mountain areas, rainfall is usually higher in both quantity and intensity. Runoff and soil erosion that affect nitrogen cycle, as well the cycle of the other nutrients, are strictly dependent on quantity and intensity of rainfall during autumn-winter times. Streamflow from the forested watershed was about six times that from the agricultural watershed in hilly areas.

From both watersheds practically all monthly N losses, calculated as the product of the average flow and the respective average N concentration, occur during autumn and winter and are more dependent upon total stream flow than upon concentration and land use. From the forested watershed yearly organic-N and NO_3-N losses account for the same proportion of total N losses. From the agricultural watershed yearly NO_3-N losses, owing to higher concentrations, are nearly double the organic-N losses. Organic-N and inorganic-N losses amounted to 59 kg ha^{-1} year^{-1} in the forested watershed. Much of the losses was in organic form (30 kg ha^{-1}) and probably derived from the litter layer of the predominant coppice woodlands. A wide range of nitrate losses from disturbed forests has been reported by Vitousek et al. (1979), but such losses seem to occur in a limited way in forested watersheds of Central Italy. The higher nitrate concentrations observed in the stream flow of the agricultural watershed, especially in autumn-winter seasons, depends on the factors that control the release of nitrate from the soil reserves. A common agricultural practice in the Mediterranean area is to plough the soil during summer and leave it bare until mid-autumn, when sowing of winter cereals and nitrogen fertilisation are carried out. On bare soils, with increasing humidity and still mild temperature, microbial mineralisation intensifies. Therefore, the rise in streamflow nitrate concentration in autumn and winter, when rainfall is particular heavy, may be due to mineralisation of soil organic matter, nitrogen fertilisation, and scanty crop uptake.

Nutrient imbalances in agricultural development have been signalled in different world regions (Vitousek et al. 2009). Input of nitrogen and phosphorus are essential for high crop yields, but losses to water and air can diminish both environmental quality and human well-being. Harvested crops remove nitrogen, phosphorus and other chemical elements from agricultural soils, while sustainable agricultural productivity requires their replacement, whether through biological processes such as legume planting or addition of organic wastes from livestock or application of fertilisers. Since the advent of "green revolution", fertiliser application has more than doubled the input of nitrogen and phosphorus to terrestrial biosphere, with a series of environmental consequences, such as degradation of water quality, eutrophication of both internal and costal marine ecosystems, increasing of photochemical smog and greenhouse effect by nitrous oxide. Simultaneously, food production in some parts of the world is nitrogen-deficient, highlighting inequities in the distribution of nitrogen containing fertilisers. Optimizing the need for a key human resource while minimizing its negative consequences requires an integrated interdisciplinary approach and the development of both strategies and incentives to promote the adoption of nutrient-conserving practices and processes (Galloway et al. 2008; Vitousek et al. 2009).

Special cases of matter cycles concern the use of synthetic agrochemicals in both crop and livestock husbandry. In these cases, the cycle starts with industrial production of an agrochemical, for instance a pesticide or an antibiotic, its application to crops or livestock, its potential diffusion into the environment, its transformations along the feed or food chain, its probable return to human body through diet, air or water. These 'cycles' fall under full human responsibility, even if the cycle is often out of human control. The 'environmental predicament' aroused by Rachel Carson in her "Silent Spring" was just originated from the unexpected effects on flora and fauna caused by the use of persistent pesticides. The 'environmental predicament' is exactly the complementary face of the 'human predicament', in that human health and environmental health constitute two faces of a same coin, the health of the living planet earth. The use of synthetic agrochemicals in agriculture is issue of continuous controversy between the multinational corporations that produce them and the ecologists in general, be their politicians, researchers, farmers or ordinary citizens, who press for banning them.

Dichlorodipheniltrichloroethane (DDT) is a potent insecticide used worldwide for agricultural and public health purposes from the 1940s to 1970s, when concerns for bioaccumulation up the food chains and its toxic effects on wildlife and humans led to restrictions and prohibitions on its use. Since the 1990s, international negotiations started to control the use of DDT and other persistent organic pollutants (POPs) in the world. In the framework of United Nations Environment Programme, countries joined together and negotiated a treaty known as the Stockholm Convention on POPs (22 may 2001) for banning their use. The Convention includes a limited exemption for the use of DDT to control mosquitoes that transmit the microbe that causes *malaria* - a disease that still kills millions of people worldwide. In September 2006, the World Health Organization (WHO) declared its support for the indoor use of DDT in African countries where *malaria* remains a major health problem, citing that benefits for health outweigh environmental risks. The WHO position is consistent with the Stockholm Convention on POPs, which bans DDT for all uses except for *malaria* control. The continued need for DDT for disease vector control, which is subject to evaluation by the Conference of the Parties during its regular meetings held every 2 years, was confirmed in 2015.Global trends in the production and use of DDT are provided by van den Berg et al. (2017).

The case of DDT is concrete exemplar of matter cycling induced by man and performed by nature processes of translocation, degradation, storage, and feedback on an extensive spatiotemporal scale. Stemmler and Lammel (2009) provided a model for shaping the DDT global fate in the period 1970–1990 on the base of FAO's data concerning agricultural usage and cropland distribution. The chemical DDT is characterised by a low vapour pressure and water solubility and a medium lipophilicity. Degradation is slow in soils and marine sediments and is set to zero in seawater. DDT removal from the model is by degradation in soil, represented as a first-order process (4.05×10^{-9} s^{-1} at 298 K) and assumed to double per 10 K temperature increase. A simple sketch of DDT cycling is in Fig. 5.9.

After application to the crop system (arrow 1), and due to the DDT persistence (residence time 14.9 years), DDT volatiles in part to atmosphere, from which can also be redeposited (arrow 2), depending on weather conditions, as humid or dry fallout. Atmosphere operates as a large exchange sink of DDT volatilisation for both receiving and depositing DDT from the crop system and the ocean (arrow 3). According to the model, the DDT volatilisation flux from the global ocean to the atmosphere was 2472 kt $year^{-1}$ in 1970 but only 301 kt $year^{-1}$ in 1990. In this last year of DDT agricultural applications, 292 kt were deposited from the atmosphere to the ocean and 41 kt were exported from the ocean surface layer (i.e. <90 m) to the deeper levels. The phenomenon of "outgassing of DDT", i.e. the net release of dissolved DDT from ocean to atmosphere, shows regional connotations. It started in the Atlantic ocean in 1970s, while started 20 years later in the Pacific ocean. The delay was due to both the different peaks of DDT applications to crops and the upwind position of the application areas with regard to the oceanic regions. In conclusion, the results of this DDT cycling model points out that until the 1970s the

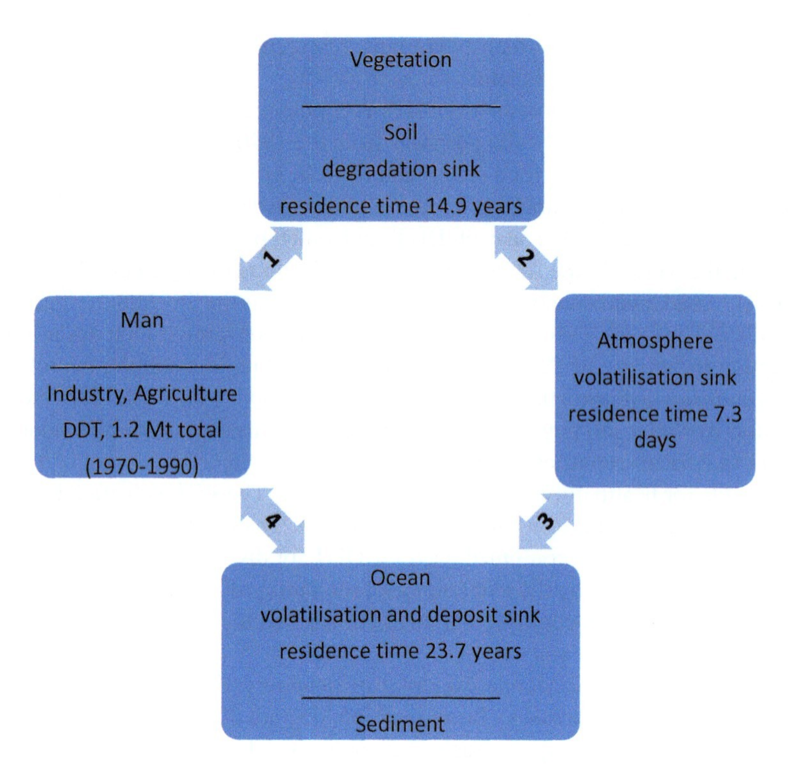

Fig. 5.9 Main steps of DDT cycling from man to man. (After Stemmler and Lammel 2009, modified)

ocean acted as a global sink of DDT due to the massive applications on crops and to the subsequent DDT volatilisation in atmosphere, which in turn dissolved most of DDT into the ocean. From the end of 1970 onwards, the net DDT air-sea flux was reversed and the ocean became a source of volatile DDT to be dissolved in atmosphere, which in turn became a vector of DDT deposition on land. To some extent, even atmospheric concentration over the European continent may have been affected by DDT returned from the ocean via long-range transfer in atmosphere.

Another well-known way of DDT return to man is through food chains, both from land, via crops and livestock (arrow 1), and sea, via fishery products (arrow 4). Due to its lipophilicity, DDT dissolves in vegetable and animal fats and bioaccumulates along the food chain from soil to man, originating adverse effects on wildlife and humans (McKinlay et al. 2007, Eskenazi et al. 2009, Vall et al. 2014). In humans, DDT and its metabolite DDE (dichlorodiphenildichloroethylene) can undergo transgenerational transfer through maternal milk and cause epigenetic effects and endocrine disturbances (Kortenkamp 2007; Kabasenche and Skinner 2014).

A recent paper of Vall et al. (2014) reports on a study of the prenatal and postnatal exposure to DDT by breast milk in Canary Islands. This study found detectable levels of DDT in 34 (47.2%) of 72 breast milk samples obtained from a population of lactating women from Tenerife (Canary Islands). The presence of DDT in breast milk correlated with both the average consumption of vegetables and a frequent intake of poultry meat. However, DDT levels documented in this study were considerably lower than the levels found in human breast milk from other countries, such as China, South-Africa and other in the Mediterranean area. The variability observed in DDT concentration between different countries could be explained in part because some of them still produce and/or use DDT. Anyway, in many cases, providing the necessary nutrients for the correct development of the infant, human milk is also a source of lipophilic environmental pollutants. This study concludes with the recognition that, despite DDT banning in Spain since 1977, Canary Islands population is still exposed to this insecticide. Nevertheless, it is emphasized that with one isolated exception, concentrations were under the acceptable criteria by WHO, suggesting that infants in the island are exposed to small quantities of DDT.

A final comment regards ethical considerations for intergenerational environmental justice advanced by a paper of Kabasenche and Skinner (2014) concerning the DDT effects on epigenetic inheritance of diseases. A variety of environmental factors that include toxicants, such as DDT, nutrition and stress have been shown to induce epigenetic transgenerational inheritance of disease (Anway et al. 2005; Kortenkamp 2007; Diamanti-Kandarakis et al. 2009). According to Kabasenche and Skinner (2014):

> Epigenetic transgenerational inheritance requires the germline (sperm or egg) transmission of epigenetic information that alters disease or phenotype, in the absence of direct environmental exposures. Transgenerational phenomenon have been demonstrated in humans, rodents, worms, flies, and plants. Therefore, even though you have never had a direct exposure, your ancestors' environmental exposures may influence your disease development. Environmentally induced epigenetic transgenerational inheritance of disease is a factor in disease etiology that needs to be considered in environmental policy.

Recognition of ethical rights to future generations should take in account the following points:

1. *Consent/Respect for Autonomy*: members of future generations cannot consent to risks and harms imposed by earlier generations.
2. *Non-maleficence*: members of future generations are harmed, via health deficits associated with epigenetics, due to exposure of ancestors to DDT (and other toxicants).
3. *Justice*: members of future generations bear a disproportionate balance of risks and harms, whereas members of the current generation, when DDT is being used, enjoy disproportionate benefits. (Kabasenche and Skinner 2014)

On the base of the recent empirical findings showing that DDT is likely to cause intergenerational harm, policies that determine its ongoing use should be re-framed incorporating these new ethical concerns.

5.1.3 Biodiversity

Agriculture is an organisation of resources, both natural and man-made, under human control and responsibility. Biodiversity is both a main component of and a synthetic expression for a systemic pattern of life, meant today as ecosystem, where abiotic and biotic components interact functionally to maintain a living community. Biodiversity components are single organisms, different species and different local and regional ecosystems that all together constitute the whole biosphere. Agroecosystems are special type of ecosystems where the human component has a dominant role and has replaced other natural biodiversity components, like native plants and animals, with domesticated plants (crops) and animals (livestock). These operations result in new ecosystems, i.e. agroecosystems, where biodiversity components can be distinguished in two sub-components, *planned biodiversity* and *associated biodiversity* (Vandermeer and Perfecto 1995; Altieri 1999). *Planned biodiversity* depends on farmer's choice, while *associated biodiversity* intervenes spontaneously as coloniser of the ecological niches freed by farmer's disturbances. The last one consists of native microorganisms, plants and animals that survive human disturbances or even thrive on them, such as *weeds, pathogens* and *insects* harmful to crops. The representation of a mixed agroecosystem components that appears in Fig. 5.10 shows how biodiversity, in all its biological forms, constitutes the heartbeat of the whole agroecosystem body, emerging from its roots in the soil

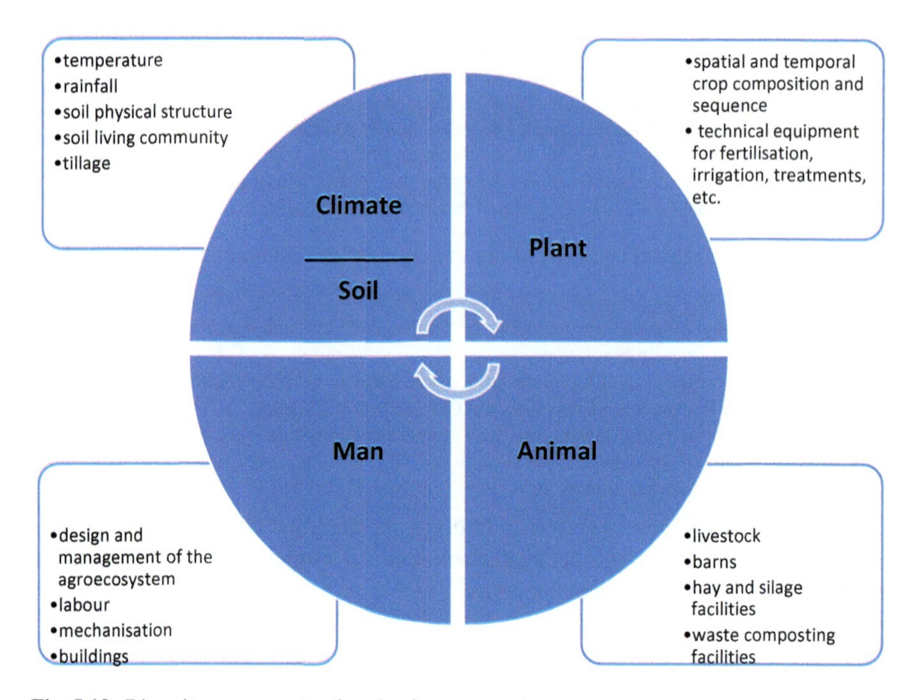

Fig. 5.10 Diversity components of a mixed agroecosystem

and permeating successively the interface soil/atmosphere where crops develop sustaining livestock and human needs, both material and immaterial. This complex structure operates under farmer's cybernetic control, labour, equipment, and resources. Farmers reciprocally exchange materials and services with both society, through the market, and the environmental context where the farm is located, through the natural process of energy-matter and information transfer.

This kind of agroecosystem organisation is valid as a model that holds at each hierarchical level, from the field, the farm, to landscape levels, be they local, regional, national or international. Olson and Francis (1995) include in the descendent hierarchy of agroecosystems even the plot and micro-plot levels that can be of interest for research to agronomists and microbiologists, respectively. With each hierarchical level corresponds a process of decision-making concerning purpose, design, management, monitoring, and agroecosystem performance evaluation, that is a matter of interest for the stakeholders involved (farmers, researchers, advisors, traders, politicians, consumers, etc.). The farm level of organisation is the most important for farmers that make a living or business with agriculture, or for consumers that are interested in knowing how the food they eat was made. Instead, the landscape level is more interesting for planners, architects, engineers, and politicians who take care and responsibility for environmental management and enhancement of agriculture multifunctional role.

The current predominant interpretation of agriculture as a business enterprise has distorted its traditional role of sustaining the civil society while conserving the environment quality throughout millennia. In making decisions to mitigate the current human predicament, it is essential to understand the role that biodiversity plays in agriculture, at both farm level and landscape levels. Biodiversity operates at every hierarchical level with the same principle of integration of available resources (solar energy, water, soil and nutrients) through a system of different biological forms (microorganisms, plants and animals) that carry out complementary roles in maintaining the system, which they belong to, self-regenerating, self-regulating, and self-productive. This ecosystems capacity of self-organisation, also named autopoiesis (Varela et al. 1974), was so efficient to generate man and its culture. According to Perry et al. (1989), this property of self-organisation reminds to a curious metaphorical comment derived from the physicist Paul Davis (1990, p. 48):

> The idea of a system of particles generating themselves in a self-consistent loop of explanation is reminiscent of the story of the boy who fell into a bog and hauled himself out by pulling on his own bootstrapps, so physicists call such modes of explanation "bootstrapping".

Following this metaphor, a farmer should be good at "bootstrapping", i.e. able to organise his/her farm resources in such a way to allow the maximum of the agroecosystem productivity by its self-organisation. From a productive perspective, the agroecosystem key-property is soil fertility. In order to explain soil fertility from an agro-ecological perspective, it is very useful to represent the soil as the place where the other components converge, i.e. the agroecosystem crossroads. Soil organic matter (SOC) is the soil component that brings memory of the past events of life

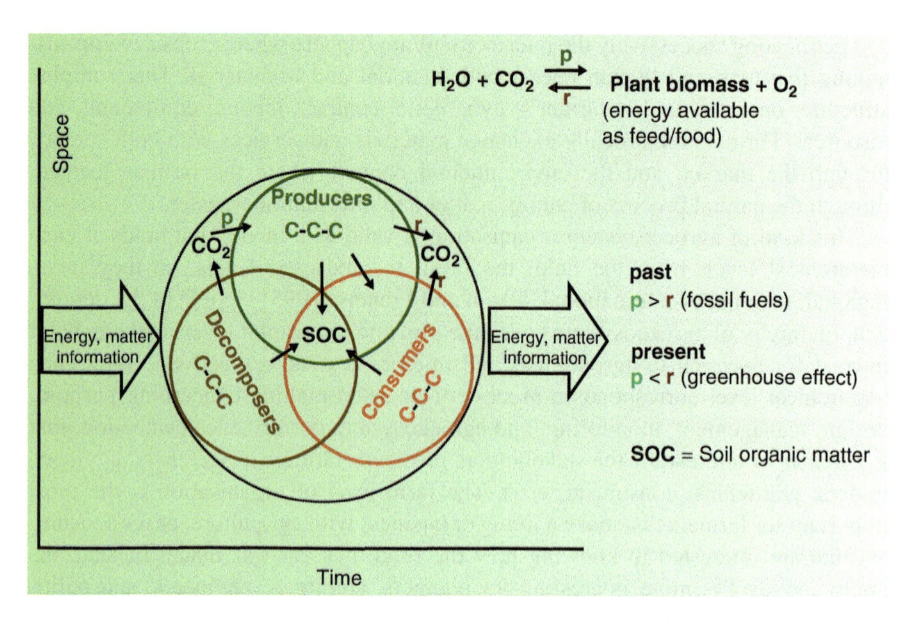

Fig. 5.11 Soil organic matter (SOC) as memory of ecosystem life development

diversity deposited as litter of the grazing chain and source of energy-matter for the detritus chain (Fig. 5.11).

According to soil scientists like Amundson and Jenny (1997), soil is an ecosystem component that represents "a state factor model" expressed in general mathematical form by the following equation:

$$\text{Soil and ecosystem properties} = f\begin{pmatrix} \textit{climate, organisms, topography,} \\ \textit{parent material, time, humans} \end{pmatrix}$$

Soil appears as a "process structure" of terrestrial ecosystems that has memory of the past and potential for the future. As a process structure in agroecosystems, soil undergoes human interventions at farm level (tillage, cropping sequence, fertilisation, irrigation etc.). Those interventions greatly modify physical, chemical and biological characteristics, or soil fertility, that finds its synthetic indicator in SOC, i.e. the soil carbon reserve in the arable layer which reflects the long-term balance between C input and loss rates (Amundson 2001). SOC is the soil component constituted prevalently of *humus*, the complex of organic substances elaborated by soil microorganisms, which binds together the soil particles in aggregates able to hold water, air and nutrients for the benefits of plant growing. Amundson (2001) points out that "soils are now a focal point of scientific interest because of the large amount of organic C stored in them and the relatively rapid turnover rates". He presents a schematic diagram of the global C cycle in order to put in evidence the soil contribution (Fig. 5.12) and the implication for agriculture.

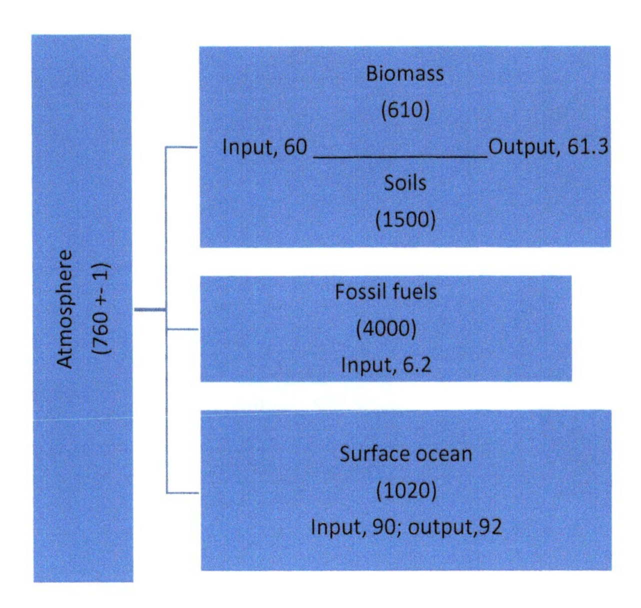

Fig. 5.12 Global C cycle estimation (Gt): pool share (in parentheses) and annual fluxes. (After Amundson 2001, modified)

C stored in soils is about double than C in atmosphere. Annually in the land pool, soils emissions are slightly less than the plant biomass uptake. C stored in land reservoir (biomass + soils) roughly equals half of C stored in fossil fuels. Most of emissions from fossil fuels remains in the atmosphere, a part dissolves in ocean surface and a part is fixed by plant biomass. The average C atom in atmospheric CO_2 takes roughly 12 years to pass through SOC somewhere in the world, on the base of the stock/exchange ratio. The global C cycle involves all kind of biodiversity and accounts for the whole history of life on the planet Earth if you remind that fossil fuels represent the result of a process of transformation of organic remains across geological times. Considering that agricultural soils derive from conversion of native ecosystems, mostly forests and prairies, it is important to be aware of an ecosystem-based distribution of the global soil C pools and fluxes, as reported in Fig. 5.13.

Soil C balance is strongly affected by climate, through both Carbon input of biomass to soil and C output by microbial respiration and losses by leaching, runoff and erosion. C input to soil depends on the ecosystem productivity that increases with the best combination of humidity and temperature, the former depending on quantity and seasonal distribution of precipitation and the latter from seasonal temperature trends that affect both plant C fixation and microbial respiration. Long seasonal periods of plant growth favour the development of both complex plant systems and litter accumulation to soil, but also a rapid microbial consumption that can be slowed down only by soil water saturation, as in the case of wet tropical forests that show the maximum C stock in the soil but also the shortest Carbon

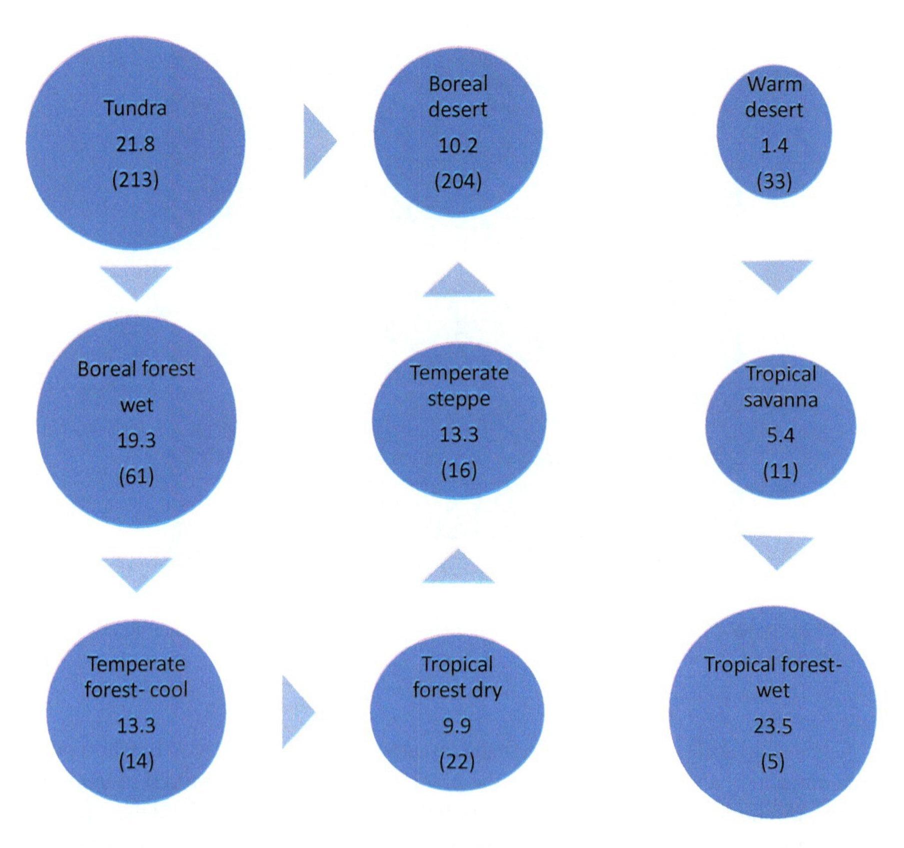

Fig. 5.13 Ecosystem-based distribution of global soil C pools (kg m^{-2}), and residence time (years, in parentheses) in selected life zones. (After Amundson 2001, modified)

residence time. Temperate steppes and temperate forests are the most balanced climatic life zones prone for traditional cultivation, due to their medium soil Carbon density and residence time that are good indicators of soil fertility, environmental biodiversity richness and resilience towards human disturbances. Cultivation of soil through tillage entails inevitably C losses due to increased aeration and microbial mineralisation, with a globally averaged reduction of initial soil C content by about 30% (Amundson 2001). The rate of C loss is a function of climate, increasing from higher to lower latitudes in a way that a viable agriculture can exists, in the absence of appropriate fertilisation, for only a few years in tropical deforested soils.

The same pattern of anabolic/catabolic organisation performed by a synergetic integration of abiotic and biotic ecosystem components operates at both macro-level and micro-level scale. Each great biome or regional ecosystem and each rhizosphere biome or root hair ecosystem operate according to the same cooperative pattern that allows life to penetrate and thrive on any available spot on Earth. This tension to affirm life through biological forms in a spatiotemporal continuum worldwide is a strategic form of colonisation that the biologist Edward O. Wilson (1993) has recognised as a meaningful component for framing his "biophilia hypothesis", i.e. "the innately emotional affiliation of human beings to other living organisms". Looking at how a soil/plant integrated system develops according to an aboveground-belowground community ecology (Ohgushi et al. 2018) should be of great help in moulding an attitude to farming while recognising the value of biodiversity potential. Under a climate favourable for cultivation, SOC is the most important indicator of soil fertility in the arable layer. In practice, SOC expresses the soil capacity to store solar energy in an organic form that ensures the building up of an environment physically structured, chemically enriched, biologically diversified, and astonishingly vital to maintain itself for centuries, if not disturbed. According to van Breemen (1993), soils can be regarded as "biotic constructs favouring net primary productivity". This character explains why pristine grassland ecosystems and forests ecosystems have traditionally undergone conversion to agricultural ecosystems. In agricultural conversion, only the soil fertility is of interest to farmers and society for exploitation, whilst selected crops replace the other above ground system components. Looking at comparisons between forest and grassland soil for their potential fertility under conditions of constant environment (Minnesota and Illinois, USA), interesting data on organic carbon concentration and C/N ratio are reported by Jenny (1994, pp. 158–159). Total organic carbon and, accordingly, organic matter is more abundant in the prairie than in the forest soil, especially in the first 60 cm of soil depth. The same holds true for total nitrogen. The carbon-nitrogen ratio is wider for the forest, a feature that seems to be generally characteristic of forest soils, denoting a lower level of soil fertility in comparison with prairie soils. Jenny's final comments are as follows:

> In conclusion, it may be said that the Illinois data are corroborated by those from Minnesota insofar as they both reveal a higher degree of translocation of soil materials under timber than under prairie. The Illinois soil types have had a longer period of formation, and therefore the profile characteristics are more distinct than in the Minnesota series. To the soil surveyor, the most striking difference between the two vegetational types is the darker colour of the prairie profile, which is the result of a higher content of organic matter.

Grassland ecosystems have had the primacy in creating a big reservoir of energy, nutrients and water holding capacity to be the most appropriate soils, or natural capital, to invest in the agricultural enterprise. As to the evolution of human society, they have constituted a natural, profitable heredity for agriculture, which is second only to fossil fuels for industry. To justify their recognised biodiversity value, hold the following citations:

> At the small scale, below 10 m^2, grasslands are probably the most species-rich plant communities in the world. Knowledge of the structure and functioning of the aboveground and belowground floral and faunal communities is essential for the correct management for maintenance and preservation of these globally important ecosystems. (Sanders 1995, p. 225)

> Soils contain more (known and unknown) species diversity than other terrestrial habitat. Biodiversity is therefore strongly an underground phenomenon. (Kuyper and Giller 2011)

> New insights from studies on aboveground-belowground interactions should be used to improve our predictions of the effects of human-induced environmental changes on biodiversity and ecosystem properties and to enhance the efficiency of human interventions in restoration and conservation efforts. (Wardle et al. 2004)

An excellent description of soil formation under different climate and vegetational cover is provided by the former USSR school of pedology set up by Dokuchaiev and extended by Vil'yam's (1968) to agricultural applications. Russian soil scientists describe steppe as a sequence of four climatic zones: forest-steppe, meadow-steppe, dry- steppe and desert-steppe. According to Moore 1982), the four steppe types have in common (a) a climate typically continental with winter temperature abundantly below freezing point and very hot summer with prolonged period of drought, and (b) a steppe soil known as *chernozem* or *chestnut* soil types.[3] The steppe type is similar to the grassland-dominated vegetation in North America (USA and Canada) known as prairies. They differ for the geographical distribution, in that there is a north-south zonation in the former USSR while an east-west zonation is present in North Amerika. Vi l'yams explains the transition from forest-steppe, meadow-steppe to dry-steppe with *chernozem* soils as a phenomenon of natural evolution and community adaptation starting from an emerging condition of coniferous forest thinning on a *podzol* soil.[4] The main actors of change are three types of grasses (*rhizomatous*, *spars-bunch* and *dense-bunch*) that colonise clearings originated by natural forest thinning and initiate the process of sod soil formation typical of meadow-steppe. Grasses colonise the interface soil/atmosphere with different tillering patterns and a root development that allows a diffuse penetration of roots across the soil profile and a synergic interaction within the soil living community. The fungal decomposition of the coniferous forest litter releases a flow of mineral nutrients that is met by the widespread dense network of hairy roots permeating the first superficial layer of soil colonised by rhizomatous grasses. Unlike rhizomatous grasses, the roots of sparse-bunch grasses develop to a depth of 30 cm and more, whilst the dense-bunch grasses carry scarcely ramified roots with a mat of mycelium filaments of either endotrophic or ectotrophic mycorrhiza or both for nutrient supply instead of hairy roots. The lack of mineral nitrogen compounds in the decomposition products of forest litter justifies the presence of *Leguminosae* in the flora of thinning forests. Their ability to fix nitrogen thanks to the symbiosis with *rhizobium* bacteria allows, after their decay, the release of mineral nitrogen compounds for grasses uptake. The deep-reaching perennial roots of legumes, often

[3] Chestnut Soils.—Brown or grayish-brown soils of the short-grass region. Considerable organic matter, neutral or alkaline reaction, lime horizon near the surface, indications of columnar structure, little profile development. Also known as dark-brown soils (definitions according to Jenny 1994).

Chernozem Soils.—Rich in organic matter to a considerable depth. Neutral to slightly alkaline or acid reaction. Pronounced lime horizon a few feet below the surface, columnar structure, essentially AC profile, traces of horizon development (definitions according to Jenny 1994).

[4] Podsol Soils.—Characterized by pronounced *A*, *B*, and *C* horizons. Surface rich in organic matter, followed below by a white o rash-gray leached horizon that is above the brown zone of accumulation of aluminum and iron. Feebly developed podsols are often designated as podsolized soils (definitions according to Jenny 1994).

more than 2 m long, do not compete with the grasses for mineral nutrients. The intense assimilation of calcium through the roots of legumes permits, after their decay, its distribution across the soil profile and its absorption by humus that firmly cements soil crumbs. The soil acquires a strong crumb structure that enhances physical properties, such as permeability, capillarity and water retention, approaching the ideal condition for resilience and fertility. This ideal condition becomes real when environmental limiting conditions for forest development, such as rainfall deficit, together with more appropriate edaphic conditions due to the presence of clay and/or calcium arise. When this happen, the dry-steppe apex of biodiversity is achieved with the generation of “rich chernozems”:

> On the whole, the sod period of soil formation under meadow steppe on Permian moraines proceeds in the same manner as on calcareous moraines, giving rise to the rich clayely chernozems of the eastern and southern parts of the USSR. These chernozems markedly differ from other chernozem varieties in their perfect, deep (100–120 cm) structure and high natural fertility. There is one horizon: the sod horizon, containing 10–20% organic matter (Vil’yams 1968, p. 181).

Grasslands in general have a high decomposition rate in that up to 90% of primary production enters the decomposer subsystems (Gibson 2009). This shows at once that the maintenance of the soil heterotrophic community has an elevated energetic cost, but also that this community is rather stable. The decomposer community includes a large varieties of organisms that are classified according to their size: *microorganisms* (bacteria and fungi), with populations in temperate grasslands of the order of 10×10^6 bacteria g^{-1} soil and 3000 m of fungal hyphae g^{-1} soil; invertebrate animals, classified as *microfauna* (nematode and protozoa), *mesofauna* (Collembola, Acari, Isoptera, Diplura, Protura, Diptera, etc.) and *macrofauna* (earthworms, Diplopoda, Isopoda, Coleoptera, Mollusca, Orthoptera, Dermaptera, etc.). Faunal population counts are of the order of 500×10^6 microfauna m^{-2}; 37,000 mesofuana m^{-2}, and 1400 macrofauna m^{-2} (Swift et al. 1979). Abundance and diversity of species has been also classified under the aspect of functional role carried out within community for essential ecosystem processes. Recognised key functional groups of soil biota include: *microsymbionts*, such as N-fixing organisms and mycorrhiza; *decomposers* or *saprothrophs*, such as cellulose and lignin degraders; *elemental trasformers*, such as nitrifiers and denitrifiers; soil *ecosystems engineers* or *bioturbators*, such as earthworms, ants and termites; *soil borne pest and diseases*, such as plant-parasitic nematodes and root-rots; *microregulators*, such as grazers, predators, parasites (Barrios 2007, Kardol et al. 2018). All these functional roles contribute to perform ecological services for the maintenance of the whole soil community, modifying physically, chemically and biologically the soil context of life. For instance soil bioturbators, such as earthworms, mixing soil layers by feeding on substrates and depositing faeces, can perform multiple functional roles of support and regulation influencing (a) soil porosity and soil glomerular structure, that facilitate the movements of water through the soil profile, capillarity formation, water retention and plant nutrition; and (b) soil nitrogen availability, plant growth and the resource acquisitive species that are the strongest competitors for N (Kardol et al. 2018). Grassland biodiversity is also important as a repository of genetic resources in that most of cereal crops are grasses, their ancestors arouse in grasslands, and modern improvement of cereal and forage cultivars through breeding draws upon grassland genetic resources (Gibson 2009).

5.1.4 Information

The conceptualisation of information is not an easy question (Schroeder 1973). As a broad definition, it is plausible to assume that information is the elementary unit of knowledge in the decision making process, where knowledge is an ordered package of information. Information is necessary in order to think, know, understand, decide, and act. According to well-known cyberneticist Gregory Bateson (1972, p. 459), information is "any difference that makes a difference", whereby the broader source of information is the environment meant as the context of life. At level of a single organism, we can distinguish an external environment that we appreciate by senses, and an internal environment, made up by the complex community of body cells that constitutes at the same time our structure and phylogenetic memory. Our body- mind unity is able to integrate the information coming from the two environments during the learning process and to make decisions. Information allows linkages among different levels of knowledge organisation, from the cellular, individual, social, ecological, to the cosmic one. Each individual is born in a context, a physical place and a culture, and it is from this context that he/she obtains the elements for his/her own physical and cultural development. The social community has set up special institutions for processing and transferring knowledge, such as schools and universities.

The definition of agriculture as a human activity system reminds us that only human choices determine what kind of system agriculture is. Choices emerge at individual and institutional levels; at local, regional, national, and international levels; at different professional levels (farmers, traders, consumers, researchers, politicians, etc.), whereby the agro-food system appears as the most pervasive human activity system in modern society (Caporali 2010, 2015). Systems thinking methodology as proposed by Checkland (1993) is particularly appropriate for generally identifying system components and fitting them into a mnemonic form representing the whole system, such as the acronym CATWOE, where:

> C = *customers* (beneficiaries or victims); A = *actors*; T = *transformation process*; W = *Weltanschauung* = to give *meaning* to the system; O = owners; E = *environmental constraints* of the system. (Checkland 1993, p. 18)

As to agriculture, *customers* are all those who buy agricultural products; *actors* are the *stakeholders* involved; *transformation process* is the whole process of *land use management*, including input and output; *Weltanschauung = sustainability or business-as-usual*?;*owners* are people from *farmers* to *corporations*; *environmental constraints* are biophysical and socioeconomic characters, both local and global, of the system context.

In agriculture CATWOE, the question mark is about making the right or wrong decision on *Weltanschauung* for establishing agriculture goals, which is not only a matter for farmers but concerns all people and the entire living community.

To make good decisions is an ethical challenge and touches the responsibility of each stakeholder, taking in account that stakeholder responsibility increases with higher level of social organisation. The first step to make good decisions is to disseminate information among stakeholders in order to promote participatory process of social learning at local level. That methodological step is part of the process of constructing knowledge with a bottom up approach for shared consensus, designing common projects, creating a sense of belongingness and identity, providing funds for research and professional training, and lobbying politicians to make innovative legislation at international, national and local levels.

Ecology is a science that generates a culture for sustainable development, which is the main goal for the present and future generations. Ecology teaches that sustainability is a feature and an asset of the biosphere. It is up to humanity to recognise ecological sustainability as an inspiring principle and a goal for all human activity systems. Agriculture is one of the basic system of human activity, the most ancient and long lasting, that needs to be re-oriented through the paradigm of sustainability. The following Chap. 6 deals with the topic of *sustainable agriculture* according to the principles of *agroecology* and the specific concept of *ecological intensification*.

References

Altieri MA (1999) The ecological role of biodiversity in agroecosystems. Agric Ecosyst Env 74:19–31

Amundson R (2001) The carbon budget in soils. Annu Rev Earth Planet Sci 29:535–562

Amundson R, Jenny H (1997) On a state factor model of ecosystems. Bioscience 47(8):536–543

Anway MD et al (2005) Epigenetic transgenerational actions of endocrine disruptors and male fertility. Science 308:1466–1469

Barrios E (2007) Soil biota, ecosystem services and land productivity. Ecol Econ 64:269–285

Bateson G (1972) Steps to an ecology of mind: collected essays in anthropology, psychiatry, evolution, and epistemology. Jason Aronson Inc., Northvale

Berry T (2003) The new story. In: Fabel A, John DS (eds) Teilhard in the 21st century. The emerging spirit of Earth. Orbis Book, New York, pp 77–88

Caporali F (2000) Ecosystems controlled by man. In: Frontiers of life, vol 4. Academic, New York, pp 519–533

Caporali F (2008) Ecological agriculture: human and social context. In: Clini C, Musu I, Gullino ML (eds) Sustainable development and environmental management. Springer, Dordrecht, pp 415–429

Caporali F (2010) Agroecology as a transdisciplinary science for a sustainable agriculture. In: Lichtfouse E (ed) Biodiversity, biofuels, agroforestry and conservation agriculture. Springer, Dordrecht, pp 1–71

Caporali F (2015) History and development of agroecology and theory of agroecosystems. In: Monteduro et al (eds) Law and agroecology. A transdisciplinary dialogue. Springer, Berlin, pp 3–29

Caporali F, Campiglia E (2001) Increasing sustainability in Mediterranean cropping systems with self-reseeding annual legumes. In: Gliessman SR (ed) Agroecosystem sustainability. Developing practical strategies. CRC Press, Boca Raton, pp 15–27
Carolan MS (2006) Do you see what I see? Examining the epistemic barriers to sustainable agriculture. Rural Sociol 71(2):232–260
Checkland P (1993) Systems thinking, systems practice. Wiley, Chichester
Chrispeels MJ, Mandoli DF (2003) Agricultural ethics. Plant Physiol 132:4–9
Davis P (1990) God and the new physics. Penguin Books
Deacon TW (2012) Incomplete nature. How mind emerged from matter. W.W Norton & Company, New York
Diamanti-Kandarakis E et al (2009) Endocrine-disrupting chemicals: an Endocrine Society scientific statement. Endocr Rev 30(4):293–342
Eskenazi B et al (2009) The Pine River statement: human health consequences of DDT use. Environ Health Perspect 117(9):1359–1367
Fowler D et al (2017) The global nitrogen cycle in the twenty-first century. Philos Trans R Soc B 368:20130164. https://doi.org/10.1098/rstb.2013.164
Frissel MJ (1977) Cycling of mineral nutrients in agricultural ecosystems. Agro-Ecosystems 4:7–16
Galloway JN et al (2008) Transformation of the nitrogen cycle: recent trends, questions, and potential solutions. Science 320:889–892
Gibson DG (2009) Grasses & grassland ecology. Oxford University Press, Oxford
Harper JL (1982) After description. In: Newman EI (ed) The plant community as a working mechanism, British Ecological Society. Special publication, N° 1, pp 11–25
Hooke RL et al (2012) Land transformation by humans: a review. GSA Today 22(12):4–10
Houghton RA (2007) Balancing the global carbon budget. Annu Rev Earth Planet Sci 35:313–347
Jenny H (1994) Factors of soil formation: a system of quantitative pedology. Dover publications, New York, first edition 1941
Kabasenche WP, Skinner MC (2014) DDT, epigenetic harm, and transgenerational environment justice. Environ Health 13:62. http://www.ehjournal.net/content/13/1/62
Kardol P et al (2018) Soil biota as drivers of plant community assembly. In: Ohgushi T, Wurst S, Johnson SN (eds) Aboveground-belowground community ecology. Springer, Cham, pp 293–318
Kirschenmann F (2004) Ecological morality: a new ethic for agriculture. In: Rickerl D, Francis C (eds) Agroecosystems analysis. American Society of Agronomy, Madison, pp 167–176
Kortenkamp A (2007) Ten years of mixing cocktails: a review of combination effects of endocrine-disrupting chemicals. Environ Health Perspect 115(supplement 1):98–105
Kuyper TW, Giller KE (2011) Biodiversity and ecosystem functioning below-ground. In: Lenné JM, Wood D (eds) Agrobiodiversity management for food security. CAB International, Wallingford, pp 134–149
Leonard WR (2003) Food for thought. Sci Am 13(2):62–71
Lindeman R (1942) The trophic-dynamic aspect of ecology. Ecology 23(4):399–418
McKinlay R et al (2007) Endocrine disrupting pesticides: implications for risk assessment. Environ Int 34:168–183
Moore DM (1982) Green planet. The story of plant life on earth. Cambridge University Press, Cambridge
Nannipieri P et al (1985) The effect of land use on the nitrogen biogeochemical cycle in Central Italy. In: Caldwell DE, Brierley JA, Brierley CI (eds) Planetary ecology. Van Nostrand Reinhold Company, New York, pp 453–460
Odum PE (1989) Input management of production systems. Science 177:177–182
Ohgushi T et al (2018) Aboveground-belowground community ecology. Ecological studies 234. Springer, Cham
Olson RK, Francis CA (1995) A hierarchical framework for evaluating diversity in agroecosystems. In: Olson RK, Francis CA, Kafka S (eds) Exploring the role of diversity in sustainable agriculture. American Society of Agronomy, Madison, pp 5–34

Patten BC (1959) An introduction to the cybernetic of the ecosystem: the trophic-dynamic aspect. Ecology 40(2):221–231
Patten BC, Odum EP (1981) The cybernetic nature of ecosystem. Am Nat 118:886–895
Perry DA et al (1989) Bootstrapping in ecosystems. Bioscience 39(4):230–237
Pimentel D et al (1973) Food production and the energy crisis. Science 182:443–449
Rees WE (1990) The ecology of sustainable development. Ecologist 20(1):18–23
Sanders IR (1995) Grassland ecology. In: Encyclopedia of environmental biology, vol II. Academic, San Diego, pp 225–235
Sanford AW (2011) Ethics, narratives, and agriculture: transforming agricultural practice through ecological imagination. J Agric Environ Ethics 24:283–303
Schroeder MJ (1973) The difference that makes a difference for conceptualization of information. Proceedings 2017 1:221. https://doi.org/10.3390/IS4SI-2017-04043
Shepard PT (1985) Moral conflict in agriculture: conquest or moral coevolution? In: Edens TC et al (eds) Sustainable agriculture & integrated farming systems. Conference proceeding 1984. Michigan State University Press, East Lansing, pp 245–255
Schramski JR et al (2015) Human domination of the biosphere: rapid discharge of the earth-space battery foretells the future of humankind. Proc Natl Acad Sci USA 112(31):9511–9517
Smil V (2008) Energy in nature and society. General energetics of complex systems. The MIT Press, Cambridge
Spedding CRW (1971) Agricultural ecosystems. Outlook Agric 6(6):242–247
Spedding CRW (1975) The biology of agricultural systems. Academic Press, London
Stemmler I, Lammel G (2009) Cycling of DDT in the global environment 1950–2002. World ocean returns the pollutant. Geophys Res Lett 36:L24602. https://doi.org/10.1029/2009GL041340
Swift MJ et al (1979) Decomposition in terrestrial ecosystems. Blackwell, Oxford
Tomlinson S et al (2014) Applications and implications of ecological energetics. TREE 29(5):280–290
Transeau EG (1926) The accumulation of energy by plants. Ohio J Sci 26(1):1–11
Van Breemen N (1993) Soils as biotic constructs favouring net primary productivity. Geoderma 57:183–211
van den Berg H et al (2017) Global trends in the production and use of DDT for control of malaria and other vector-borne diseases. Malar J 16:401. https://doi.org/10.1186/s12936-017-2050-2
Vall O et al (2014) Prenatal and postnatal exposure to DDT by breast milk analysis in Canary Islands. PLoS One 9(1):e83831. https://doi.org/10.1371/journal.pone.0083831
Vandermeer J, Perfecto I (1995) Breakfast of biodiversity: the truth about rainforest destruction. Food First Books, Oakland
Varela FJ et al (1974) Autopoiesis: the organization of living systems, its characterization and a model. Biosystems 5:187–196
Vil'yams RV (1968) Basic soil science for agriculture. Israel Program for Scientific Translation Ltd. S. Monson, Jerusalem
Vitousek PM et al (1979) Nitrate losses from disturbed ecosystems. Science 204:469–474
Vitousek PM et al (2009) Nutrient imbalances in agricultural development. Science 324:1519–1520
Wardle DA et al (2004) Ecological linkages between aboveground and belowground biota. Science 303:1629–1633
Wilson EO (1993) Biophilia and the conservation ethics. In: Keller SR, Wilson EO (eds) Biophilia hypothesis. Island Press, Washington, DC, pp 31–40

Chapter 6
Sustainable Agriculture Through Ecological Intensification

Fidelity to fields, watershed, crops, livestock, farmers, and guests is crucial for the maintenance of a healthy and flourishing life

(Wirzba 2019, p. 19).

The idea of sustainable agriculture emerges from facts, inspires value judgements and requires adaptation measures – it is an ethical challenge. All that can be summarised in Fig. 6.1 as a synthesis of the insights made by Richard B. Norgaard in the early 1980s (Norgaard 1984a, b, 1988).

Facts concern the modernisation of agriculture that has become unsustainable due to the dependence on fossil fuels energy, the adverse environmental effects of agrochemicals and the negative social impact on traditional rural communities. Values stem from a change in understanding agricultural reality through a systems paradigm that recognises relations and new potential for strengthening them. The challenge consists in letting these values become operational in society worldwide.

In a general framework and for practical goals, agriculture ought to mean 'a good of public utility' that delivers ecological services (or life-support services) in favour of man and the living community as well. To meet these expectations, agroecosystem design and management should ground on the promotion of major ecological drivers that operate in nature, such as solar energy, material recycling and biodiversity.

In particular, agriculture being a site-specific biodiversity assemblage under human control and responsibility should possess general structural and functional elements to express the agroecosystem properties as shown in Fig. 6.2.

Agricultural biodiversity is the underpinning of more resilient agroecosystems and it is essential to cope with the current impacts of climate change, yield stability, farmer income, rural community vitality, land use conflicts, health challenges and social justice. Many of the benefits deriving from agricultural biodiversity pertain to different agroecosystem levels of organisation, and cut across political divisions,

F. Caporali, *Ethics and Sustainable Agriculture*,
https://doi.org/10.1007/978-3-030-76683-2_6

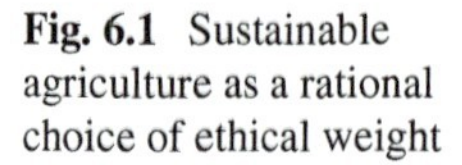
Fig. 6.1 Sustainable agriculture as a rational choice of ethical weight

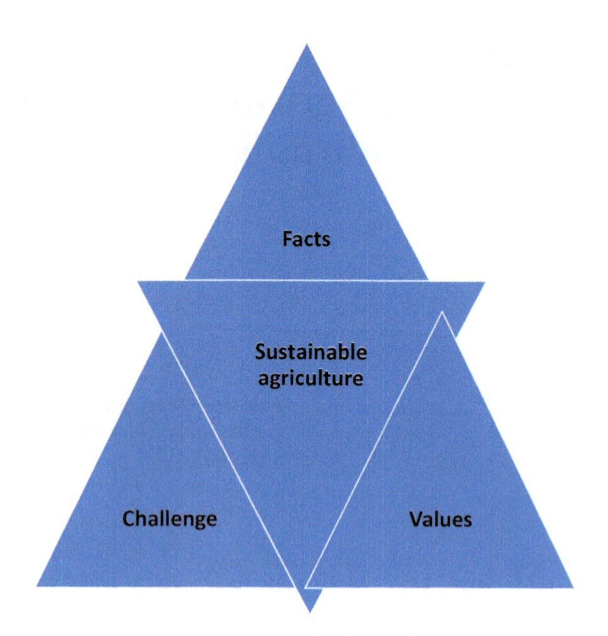

requiring a cross-sectoral approach to reassess the role of agriculture in society in a perspective of sustainable development (Frison et al. 2011). Site-specificity is a recognised indispensable attribute of sustainable agriculture (Caporali 2010), whereby the local dimension of enquiry, adaptation and management are stepping-stones to pave the way for success. After that, an outlook on the international and planetary context can help make decisions. In any case, the concept of "a coevolutionary model of agricultural development" advanced early by Norgaard (1984a, b) can help frame the systems view underlying the relation man/nature that agriculture embodies:

> Coevolution in biology refers to an evolutionary process based on reciprocal responses of two closely interacting specie [...] The concept can be broadened to encompass any feedback processes between two evolving systems. For agricultural social and ecological systems, man's activity modify the ecosystem while the ecosystem's responses provide cause for individual action and social organisation. Thus, agricultural development can be viewed as a coevolutionary process between a sociosystem and an ecosystem that, fortuitously or by design, benefits man (Norgaard 1984a, p. 528).

A sketch of agriculture as a coevolutionary socioecological system is presented in Fig. 6.3.

Farmers and rural environment constitute the system's internal context, the organisation of which takes form through the structural elements of fields, ditches, crops, hedges, internal roads, buildings, stables, barns, etc. The other system components constitute the external context that contribute to its organisation providing inputs and/or acquiring outputs through market or information exchanges. This kind of representation unveils the fact that the entire system can be also qualified as an *ecological food system* (Francis et al. 2003) involving consumers, transformers,

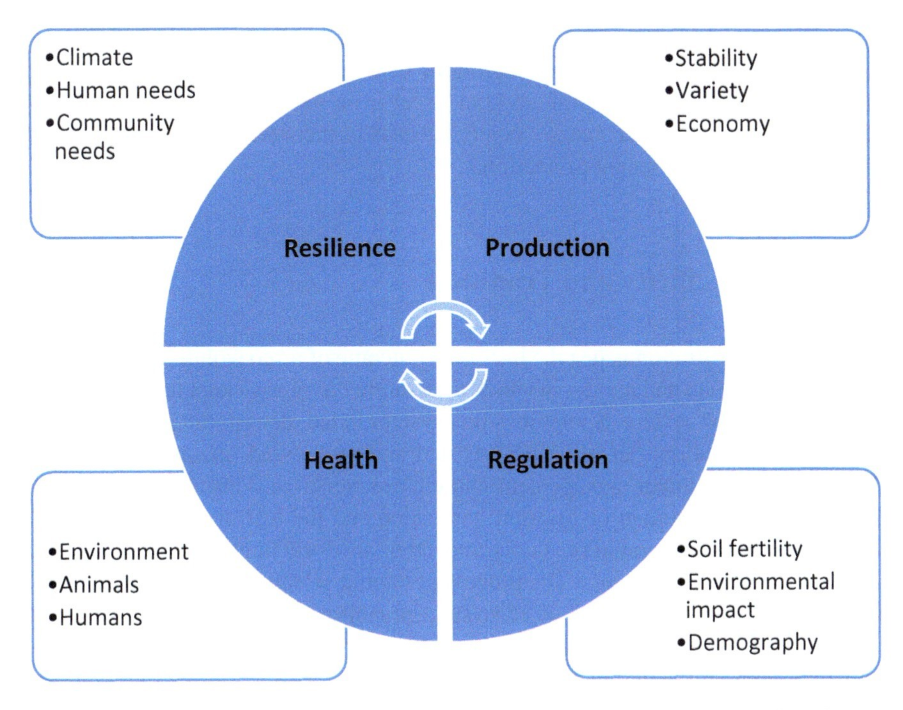

Fig. 6.2 Properties of agriculture as a 'good of public utility' that delivers ecological services

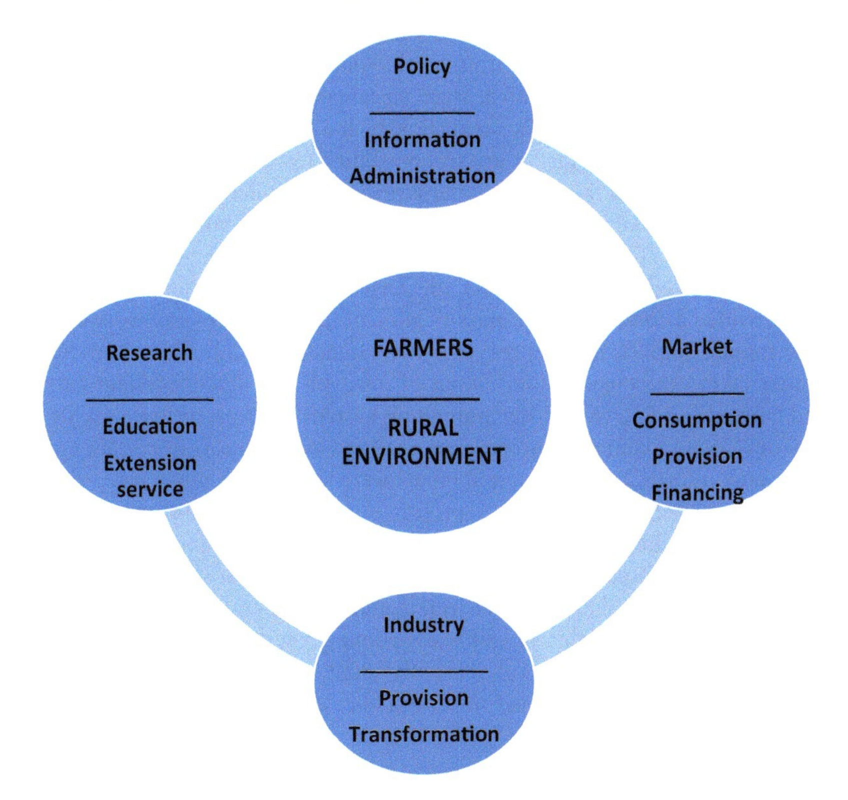

Fig. 6.3 Main components of agriculture as a coevolutionary socioecological system

advisers, researchers, traders and private as well as public institutions. All those stakeholders should develop their feeling of belongingness to agriculture and the food system, just with the ultimate goal to let it be sustainable. From a systems perspective and an ethical stance, society entrusts farmers with a double task, food production and environment protection.

6.1 Site-Specificity and Tradition

What we call civilisation is just the historical outcome of a "coevolutionary development" between man and nature, in which agriculture has always kept its primary role of ecological food system. Sometimes this system failed, determining the end of its promoting civilisation, but often has resisted and progressed, maintaining a good balance between human management and environmental sustenance. As any other biological system, human civilisation has undergone the test of natural selection, which goes beyond the rapid predatory tension of ancestral human origin, and at the end determines fall or success. The process of natural selection is a universal invariant that operates at any levels of ecosystem organisation, although with different speeds, processes at small scale being rapid and processes at large scale being slow. Humanity has been acting at planetary scale since decades and a few limits to the predatory habit of humanity have already been overcome (Rockstrom et al. 2009). Human civilisation is now a unique driving force at planetary level and needs to find a right balance between environmental global resources and global human management. Site-specificity has growing from local to planetary and requires an adaptation effort extended from local to global, since sustainability is a property of the entire system, and not of an individual system (Bakshi and Fiksel 2003). As to agriculture, the institutional hierarchy that bears responsibility for this epochal change spans from the family level to the international level, in an uninterrupted, coherent chain that entails agreement on goals, share of means, and responsibility of action.

A systems view of "good" agriculture, which operates as a general framework, is necessary and needs recognition. At the onset of agroecosystem study, Frissel (1977) provided a comprehensive classification of farm agroecosystems worldwide, valuing inputs, outputs, internal recycling, and quantity and quality of production, this latter taking in consideration output of nitrogen. An agroecosystem category named "mixed farming or self-sustaining unit system" or "mixed livestock farms", as they developed mainly in Europe, was qualified as "the most important agricultural system on earth", in that external inputs were reduced because crop productivity, livestock transformations and appropriate nutrient recycling were organised internally in a circular fashion realising agroecosystem 'bootstrapping' while reducing environmental outputs. That picture is very similar to that suggested today as a model of multifunctional agricultural with a circular economy (IPES FOOD 2016). Frissel (1977) added that this model was operating on comparatively small plots supporting only one family or another usually small, social group. In this model, the cultivation techniques evolved and improved over a long period of time, showing much more sophistication than one might think. Currently, FAO recommends a policy for the sustainable intensification of smallholder crop production and for family farming (FAO 2011; FAO and IFAD 2019). On this basis, it is possible to propose a sketch of a mixed agroecosystem model that functions mostly on the organisation of natural resources and processes (Fig. 6.4).

What kind of 'sophistication' a mixed agroecosystem shows is a matter of interest for discussion in anthropology, history, agronomy, engineering, ecology, policy and ethics. The aboveground system components, crops and livestock, represent the main rings of the man-organised grazing chain. To develop this chain in a balanced and stable manner has taken not only centuries but several millennia of years spent for the process of domestication of both wild plant and animals, which is not yet finished. Crops and livestock are the biological "workers" for accumulating energy-matter on the fields and transforming it into animal proteins, respectively. Swift and Anderson (1994) define crops and livestock jointly as *the productive biota*, in that it produces food, fibre or other products for consumption, use or sale. This biota is a free choice by the farmer and is the main determinant of the biodiversity and complexity of the agroecosystem. The belowground components, soil and manure, represent the main rings of the detritus chains, where the soil functions of physical support and fertility regulation need to be continuously restored by organic manure applications. Soil is the place where organic residues from crops and livestock are in great part converted by soil biota to mineral nutrients available for both crops and microorganism uptake, and in small part to humus for long-term effects on the physical, chemical and biological soil conditions. Soil biota is defined by Swift and Anderson as *the resource biota*, which contributes positively to the productivity of the system even if does not generate a product for the farmer. Crops and soil components define together the aboveground/belowground subsystem, through which the solar energy flux penetrates, affecting the crop medium that extends its aboveground organs in the atmosphere and its belowground organs into the soil. After harvest, crop residues (roots, stables, etc.) remain both on and into the soil as a reservoir of organic energy-matter available for soil biota decomposition. In the history of agriculture, livestock has played different roles; originally, as work co-provider for ploughing fields and commodity transfer; currently and mostly, as provider of marketable commodities (milk, meat, wool, etc.). The livestock role as provider of manure is still important for all kind of traditional and organic agriculture worldwide. All agroecosystem components manifest openness, i.e. mutual exchanges of energy-matter, a thread connecting and alimenting all parts in a structural and functional unity under human organisation. While solar energy is the agroecosystem physical driver, human mind is the agroecosystem informational driver. Each agroecosystem component is also a component of a larger living community, which envelops it in a spatially and temporally historical context that comes from the past and moves to the future. Agroecosystem components exchange energy, matter and information within this larger, vital, cosmic community we call environment.

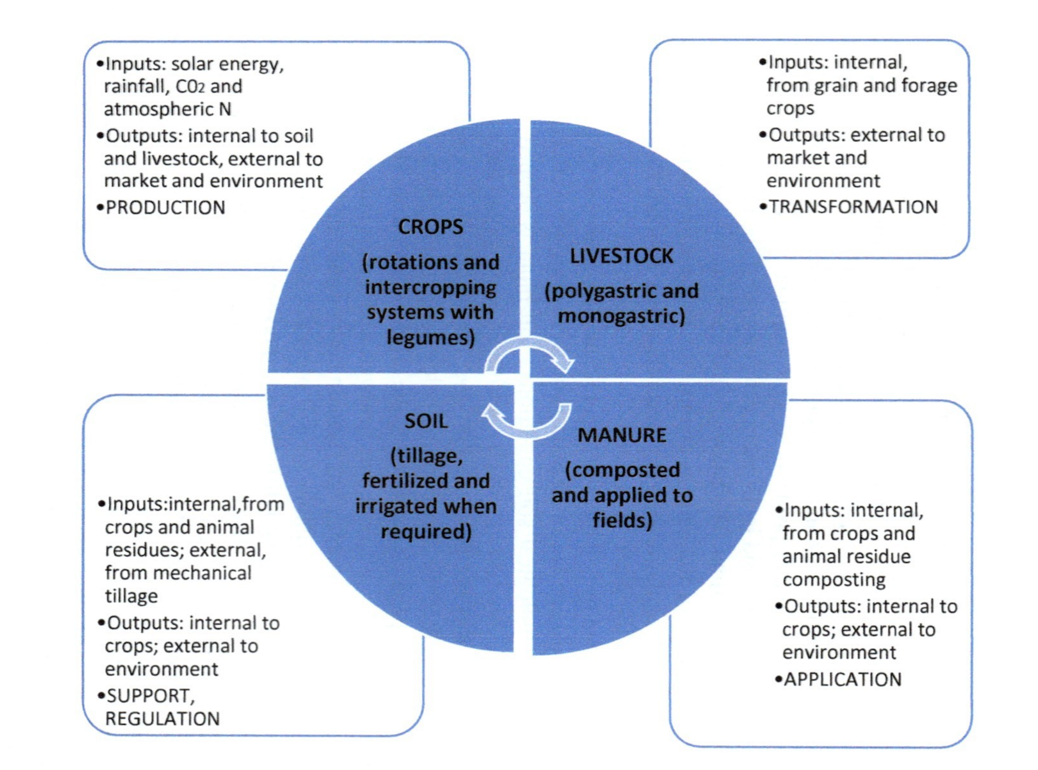

Fig. 6.4 Basic components and processes of a mixed agroecosystem grounded on natural resources

6.1.1 Domestication and Adaptation

Plant and animal domestication is the most important development in the past 13,000 years of human history (Diamond 2002). Both the biological components of agroecosystems under control of man, crops and livestock, derive from natural ancestors selected by humans through a process we call domestication. According to the classic definition of Harlan (1995), "the processes of domestication are evolutionary in nature [...] Domestication involves genetic changes in populations tending to confer increased fitness for human-made habitats and away from fitness for wild habitats". The consequence of domestication is that fully domesticated plants and animals are completely dependent on human care for survival.

Harlan adds some important consideration as to the etymology of domestication, the name deriving from Latin *domus*, which means house or dwelling, whereby *domestic* plants and animals are reared to serve the master (*dominus*, in Latin). Etymologically, dominion on crops and livestock appears completely justified. The trend to domesticate not only plants and animals, but also the whole nature has increased with human population and technological power, in a way that now the entire biosphere is a *domus extensa,* an enlarged house, where humans, as *domini* (masters, lords) can exercise their power. Changing pristine ecosystems to agroecosystems is an act of domestication that involves problems of quantity and quality of land use conversion – an ethical contract could be necessary between man and the rest of nature.

A more recent definition (Stetter et al. 2017) defines "domestication as the process of adaptation to agro-ecological environments and human preferences by anthropogenic selection". It is now clearer the co-evolutionary character of the process that involves reciprocal adaptation of both the 'domesticator' (human being) and the 'domesticate' (plant and animal) under the site-specificity of climatic and edaphic conditions. Therefore, current perspectives on the future of domestication studies entail "reconsidering plant and animal domestication within an integrated evolutionary and cultural framework" (Larson et al. 2014).

The first act of domestication was choosing the suitable species for human needs and the second one changing their original traits to favour a domestic partnership. Most likely, annual plants were selected before perennial and arboreal species to become crops because of their short cycle of life that allows faster performance monitoring and trial resetting. For cereal crops, the improvements of natural selection for optimising harvest are listed in Table 6.1.

The above reported characters, common to most herbaceous crops, constitute what has been named "domestication syndrome" (Gepts 2004):

> In selecting plants to fulfil their needs for food, feed, and fiber, humans have – perhaps inadvertently – selected crops that, while they do extremely well in cultivated fields, are unable to grow and reproduce successfully for more than a few seasons in natural environments, away from the care of humans who provide adequate seed beds and reducing competition from weeds.

Table 6.1 Traits of cereal crops modified through domestication (after Harlan 1995; Gepts 2004, modified)

Original traits	Modified traits
Branching and tillering habits	Evolution of forms with single stalks (maize, sorghum) and reduced tillering capacity (wheat, barley, oats, rice)
Uneven ripening	The routine of harvesting and planting has developed plant architecture with more even ripening
Shattering (fragmented seed dispersal)	Abscission layers suppressed and abscission delayed until harvest; lack of seed dispersal at maturity
Appendages as seed planting device	Reduction of appendages due to artificial planting
Presence of vestigial flower structures	Recovery of fertility in reduced or vestigial flower structure (more seed production)
Seed dormancy	Regulation of dormancy until next planting
Reduced seed size	Seeds of domesticated races are larger, produce more vigorous seedlings and can germinate from greater depths
Low harvest index (20–30%), ratio of the harvest part (grain) to the total aboveground biomass	Contemporary advanced cultivars show a harvest index of 60% or more

Domestication of fruit trees, like Olive (*Olea europea* L.), Grape (*Vitis vinifera* L.), Date (*Phoenicx dactylifera* L.), and Fig (*Ficus carica* L.), dates back to the fourth millennium B.C., as known from well-preserved carbonised olive stones discovered in Palestine in close association with cereal grains, dates, figs and pulses (Zohary and Spiegel-Roy 1975). All this plants were, and still are, major agricultural crops of the Near East and the Mediterranean Basin. Their pattern of domestication has produce the effects reported in Box 6.1.

In the case of classic fruit trees, the vegetative manipulation of the reproduction process through uninterrupted propagation of vegetative parts has practically generated a drastic, cultural interference in the process of natural evolution, such as a kind of 'genome fossilisation'. The successful of fig cultivars, such as "Dottato",

Box 6.1 Major Steps in the Domestication of Fruit Trees (After Zohary and Spiegel-Roy 1975, Modified)

Characters of ancient ancestors

1. Wild forms are cross-pollinated and wild populations manifest wide variation, maintaining a high level of heterozygosity.
2. All wild forms reproduce from seed; sexual reproduction prevails.
3. Wild forms are often dioecious, i.e. fruit formation depends on the presence of both male and female individuals.
4. Fruits numerous, small and slightly palatable.

Characters of domesticates

1. Fruit tree domestication is a shift from sexual reproduction to vegetative propagation.
2. The cultivated forms are clones maintained and multiplied by vegetative manipulations like cuttings (grape, fig), basal knobs (olive), transplanting offshoots (date), and graftings, much later.
3. Cultivated clones persisted hundreds or even thousands of years.
4. The climatic requirements of the cultivars closely resemble those of their wild relatives.
5. Classic fruits have not been pushed much beyond the climatic requirements of their wild ancestors.
6. Domestication has brought about a genetic shift from dioecism to hermaphroditism.
7. Pollination in dioecious fruit trees replaced by parthenocarpy in some cases (common fig and Corinth type grapes).
8. Fruits numerous, large and highly palatable.

currently cultivated in the Southern part of Italy and largely exported as dry-fig, has likely propagated uninterruptedly since the Roman time, as appears by historical reports since Latin age. Even iconographical representations, such as in Pompei frescos and in Orvieto dome medieval sculptures, celebrate this renowned fig appreciation. Fig 'cultivar Dottato' could be a current exemplar of 'fossil genome' that witnesses how a co-evolutionary process between man and nature has happened, maintaining a sustainable balance between site-specificity (Mediterranean climatic and edaphic conditions), a plant resource (fig), and a sapient human management. Fig's genome is so much adapted to the climate, edaphic and biological conditions of Mediterranean area that a fig plant usually does not need irrigation, fertilisation and pesticide treatments, when grown in mixed poly-cultural systems.

Historically, plant domestication marks both the transition from a civilisation of hunter-gatherers to a civilisation of farmers and the contemporary transition of socio-political organisation from tribes to States. Flannery (1972) describes this transition as a cultural evolution of civilisation promoted by agriculture. Making a summary of theories on the origins of the State, he proposes seven "mechanisms of State formation": population growth, warfare, irrigation, trade, symbiosis between contrasting peoples or environment, cooperation and competition, and the integrative power of religions or great art styles. Most of them, except warfare, have direct connections with agriculture as represented in Fig. 6.5.

Agriculture has been a historical innovation that has changed human civilisation starting with mutual domestication of plant, animals and humans, first in local environments and then in the global environment of the whole biosphere. Agriculture is a practical expression of cultural evolution that has invested the whole planet Earth

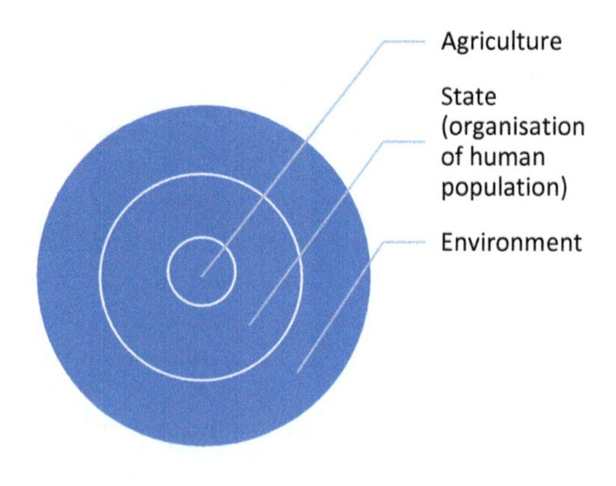

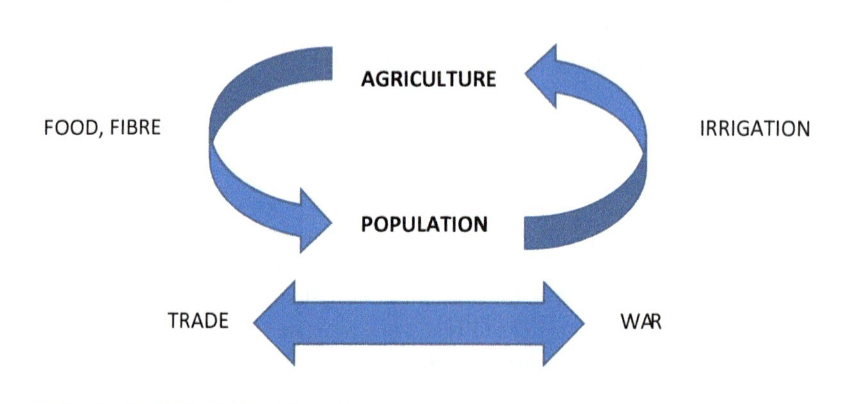

Fig. 6.5 Site-specificity, domestication, agriculture and State organisation (trade and war) as components of a co-evolutionary planetary event. (Adapted from Flannery 1972)

showing how "cultural and genetic evolution can interact one another and influence both transmission and selection" (Creanza et al. 2017). Through the process of adaptation, the most successful combinations of crops, livestock and human capacity of organisation (cultural, technological, economic, social and political) have spread in most part of liveable Earth in that "like genes, cultural traits can be more or less adaptive depending on the environment and spread accordingly" (Creanza et al. 2017). Diamond (2002) traces back a very convincing history of domestication original centres and their diffusion worldwide starting with the assumption that:

> Domestic plants and their wild ancestors evolved as consequences of wild plant selected, gathered and brought back to camp by hunter-gatherers, while the roots of animal domestication included the ubiquitous tendency of all people to try to tame or manage wild animals.

As homelands of agriculture are regarded "those regions to which the most numerous and most valuable domesticable wild plants and animal species were native". Today there are at least nine areas in the world that are recognised as homelands of agriculture (Diamond 2002). For instance, the well-renowned Fertile Crescent of southwest Asia was home to wild wheat, barley, peas, sheep, goats, cows and pigs, beginning the world's first farmer and herder homeland since around 8500 BC. Starting with agriculture, a chain of development civilisation steps followed with more food, population, trade, mining, technology, empires, and professional armies. From these tools of conquest, a new civilisation arouse with a subsequent spreading of genes and culture diffusing west into Europe and North Africa and east into western and central Asia. Interestingly, agriculture expansion occurred more rapidly along east-west axes than along north-south axes:

> The reason is obvious: locations at the same latitude share identical day-lenghts and seasonalities, often share similar climates, habitat and diseases, and hence require less evolutionary change or adaptation of domesticates, technologies and cultures than do locations at different latitudes (Diamond 2002).

The Eurasia's east-west axis of civilisation development became one of the main ultimate reasons why Eurasian people conquered people of New-Continents (America and Oceania) and not vice versa. Unfortunately, the simultaneous expansion of the new consortium of life (climate, soil, plant, livestock and man) has involved also the expansion of epidemic infectious diseases that can sustain themselves only in large dense populations, whereby they are often termed 'crowd diseases' (Diamond 2002).

Recently, a publication sponsored by FAO under the editing guide of Koohafkan and Altieri (2011) has highlighted the great meaning of recognised "globally important agricultural heritage systems" (GIAHS) defined as follows:

> Remarkable land use systems and landscape which are rich in globally significant biological diversity evolving from the co-adaptation of a community with its environment and its needs and aspirations for sustainable development (p. 1).

These systems are exemplars derived from the intergenerational investment of local people in moulding, cultivating, living, enjoying, defending and maintaining their effective 'homeland', which condenses culture in a piece of land:

> Through a remarkable process of co-evolution of Humankind and Nature, GIAHS have emerged over centuries of cultural and biological interactions and synergies, representing the accumulated experiences of rural peoples (p. 1).

The achievements of human ingenuity and labour represented in GIAHS are listed in Table 6.2.

Altogether, GIAHS represent practical solutions of sustainable development cumulated by experiential learning of extended generations of farmers that have grounded a rural civilisation for each piece of local land with its diverse endowment of physical and biological resources. These systems are effective vital 'monuments' where nature and culture intermingle successfully in landscapes that are a sum of values (Fig. 6.6) not yet completely understood by modern society.

Table 6.2 Types of GIAHS and their main functions (after Koohafkan and Altieri 2011, modified)

Types of GIAHS	Drivers and functions
1. Mountain rice terrace agroecosystems 2. Multiple cropping/ polyculture farming systems 3. Understory farming systems 4. Nomadic and semi-nomadic pastoral systems 5. Ancient irrigation, soil and water management systems 6. Complex multi-layered home gardens 7. Below sea level systems 8. Tribal agricultural heritage systems 9. High-value crop and spice systems 10. Hunting-gathering systems	Managed by an estimated 1.4 billion people, mostly family farmers, peasants and indigenous communities Repository of ancestral and local varieties of plant and animal races, with low external inputs, capital or modern agricultural technologies They produce between 30% and 50% of the domestic food consumed in developing countries They contribute substantially to food security at local, national and regional level They are exemplars of traditional systems that have stood the test of time testifying to resilient and successful strategies, representing models of sustainability

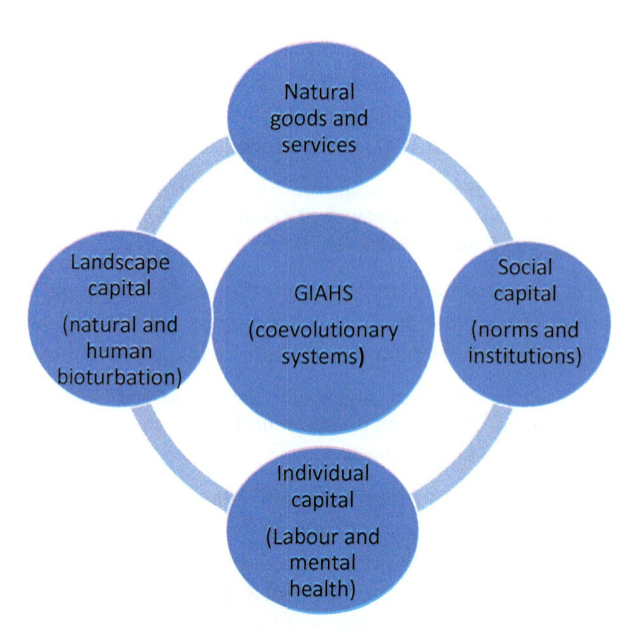

Fig. 6.6 'Facts and values' interconnection in GIAHS. (After Koohafkan and Altieri 2011, modified)

6.1.2 Contrasting Dismantling Tendency

The alarming fact is that GIAHS are rapidly shrinking under the pressure of a predatory economy that pretends to make money replacing well-balanced socio-ecological systems for a local sustainable economy with unbalanced mono-cultural agroecosystems for the benefit of the international market drivers. A paradoxical situation emerges where, on one hand, rural civilisations are threatened to be destroyed when, on the other hand, there is an increasing call for forms of farming requiring the same characters that GIAHS show (Fig. 6.7).

Behind that apparent paradox, there should be some hidden constraints, some "distorsions" as Murdoch (1990) calls them, which move the balance needle toward the minus instead of the plus in Fig. 6.7. Murdoch argues that the major constraints on food production are not physical, biological, agronomic, or even technical, but structural distorsions that derive from the current maldistribution of productive resources. There are distorsions within the agricultural system itself and distorsions in the relations between agriculture and the rest of economy. Their combination generates a slow dismantling of the monumental heritages of past generations in favour of fast accumulation of capital that urges people to abandon their homeland to concentrate in industrialised and urbanised environments in search of less and less likely work and safe conditions for making a life. In this process, land property becomes more and more concentrated in a few hands, or better, under the wheels or the wings of increasingly sophisticated machines representing a new generation of mechanical slavers that replaces the previous generation of animal and human labour. The environmental consequences of this epochal change constitute the core of what has been defined 'the human predicament'. What are the benefits for human society and the planetary living community for accepting such a never signed agreement? How should we stop this tendency and reverse it to avoid a desertified Earth and a human collective suicide? Looking more carefully at Murdoch's argument, the current structural distorsions within the agricultural system could be represented as shown in Fig. 6.8.

The basic distorsion of the current agricultural systems is the uneven distribution of land that brings about a cascade of events. Land is a primary good and its property determines a gradient of social power historically consolidated, whereby more land property equals more power. This power legitimates bank trust and credit facilities, which allows purchasing machines and technological means of production. Machines in large part replace animal and costly human labour, and eventually larger properties dismiss smaller properties causing former farmers' exodus and violent urbanisation. On the front of production, "large farmers use little labour per acre and have low yields. Small farmers have much higher yields in spite of their inadequate technology" (Murdoch 1990). This happens because small farmers have little land and therefore they try to maximize yields per acre, with a great investment of their own labour. Curiously, the current powers of human society (political, economic, social and cultural) seem to prefer more the labour of machines than the labour of man, even if the labour of man can make the soil more productive and more protective.

Fig. 6.7 The challenge to recognise, defend and expand GIAHS. (After Koohafkan and Altieri 2011, modified)

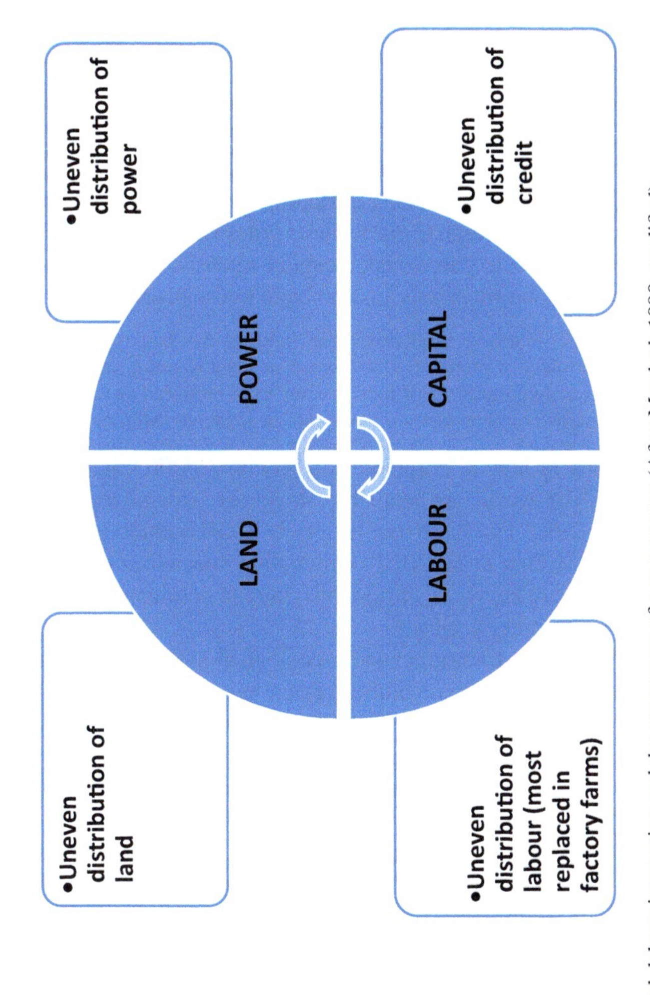

Fig. 6.8 Land, capital, labour interaction and the emergence of a new power. (After Murdoch 1990, modified)

The second kind of structural distorsion concerns the relation between agriculture and the rest of economy, which Murdoch (1990) explains in the international context as an adaptation of the previous "colonial dual economy", where:

> The colonial powers suppressed existing indigenous food-producing agriculture and cottage industry, which jointly formed the foci from which integrated economies could have developed. They created economies geared instead to the export of raw materials and the import of manufactured goods. In postcolonial times, the economic and political hegemony of Europe and the United States maintained these structures intact. During this historical process, a local economic and political elite was entrenched or created that was associated with the administration of the colony, with the export enclave, with the new trade, and with urban centers of economic and political power. It was to this urban and outward-oriented class that power was transferred in recent postcolonial history (p. 14).

Until now, the food producing subsistence farming in less developed countries serves as a repository of both cheap labour and cheap land to buy (*land grabbing*) for feeding commercial food-producing agriculture and cheap raw materials to be exported, via oligopolistic foreign firms, towards industrialised nations. Increasingly growing cities are home to large-scale, capital-intensive, and import-intensive industry, while rural economy is only instrumental to urban economy:

> The small-scale, labor-intensive rural and urban industry has largely been denied the resources and opportunity of growth [...] A low urban wage is still a necessity, so cheap urban food remains a high priority. This food is likely to be produced on large commercial farms, with the exploited rural population supplying cheap labour (Murdoch 1990, p. 17).

Updated accounts of how a 'corporate food regime' (CFR)[1] has developed and expanded all over the world are available in recent papers, reports and books (McMichael 2009; IPES-FOOD 2016; Heinrich Boll Foundation 2017, Mooney 2018, Gonzales de Molina et al. 2020). Production specialisation, intensification and commodification are key words for identifying the dominant agro-food processual drivers in contemporary society.

The hidden strength of traditional socio-ecological systems that have survived the industrial transition of 'green revolution' can now emerge as agroecology, which is both a science and a set of practices (Altieri and Toledo 2011; Caporali 2015a), recognises and promotes an innovative model of sustainable development based on highly knowledge-intensive agroecosystems. When regarded under the aspect of efficient use of local resources, traditional agricultural systems unveil a technological approach based on diversity, synergy, recycling and integration, implemented by social processes that value community involvement, social learning and participative decision-making. In practise, they are exemplars to imitate of a "triple dimensions of the agroecological revolution, namely, cognitive, technological and social" (Altieri and Toledo 2011). On the wave of this newly acquired awareness, a transnational social movement defending small farming and peasant life, *La Vıa Campesina*, emerged first in Latin America, and then at a global scale, during the 1980s and the early 1990s. Particular emphasis is now given to *La Vıa Campesina*'s fight to gain

[1] CFR refers to a food regime where a few large companies control not only most of the market of seed, agrochemicals and farm machinery, but also the distribution and retail sale of food.

legitimacy for the *food sovereignty* paradigm,[2] to its internal structure, and to the ways in which a shared peasant identity is a key glue for creating a true peasant internationalism (Martinez-Torres and Rosset 2010). Among the analytical tools used by this international peasant movement is the comparison between the energy efficiency of traditional small farm agriculture and modern industrial agriculture. A paper of Martinez-Alier (2011) traces back the history of agricultural energetics, and then looks at the use of the concept of EROI (energy return on energy input) by *La Via Campesina* when it claims that 'industrial agriculture is no longer a producer of energy but a consumer of energy', and that 'peasant agriculture cools down the Earth'. A summary of the proposal of *La Via Campesina* to contrast a policy in favour of industrialised agriculture as reported by Martinez-Alier (2011) is shown in Box 6.2. Now, *La Via Campesina International* (*LVC*) focuses on "the construction of peasant to peasant processes (PtPPs) as a dispositive for agroecological scaling up and transformation, the mobilization of a peasant political project and the building of a historical and political subject within the universe of organizations linked to LVC" (Val et al. 2019). The meaning of agroecological scaling up concerns the following vision:

> reinforces autonomy, biocultural diversity, spirituality, and conviviality. It situates agroecology as one key element of a broader societal transformations that challenge capitalism, colonialism, standardization, industrialization, patriarchy, and other forms of injustice (Ferguson et al. 2019).

According to De Schutter (2012), "rather than treating smallholder farmers as beneficiaries of aid, they should be seen as experts with knowledge that is

Box 6.2 *La Via Campesina's* Proposals for a Sustainable Agriculture (After Martinez-Alier 2011, Modified)

BASIC ASSUMPTION

Industrial agriculture is one of the main drivers of climate change, carrying food around the world and imposing monocultures and mechanization and the use of agrochemicals while destroying biodiversity and its ability to capture carbon and 'transforming agriculture from a producer of energy into an energy consumer'.

FALSE SOLUTIONS

La Via Campesina asserts that false solutions are being promoted in the face of climate change, such as agrofuels from monocultures (including tree plantations), which are undermining food sovereignty.

[2] The most widely used definition of food sovereignty is the "Declaration of Nyéléni" by delegates of a transnational network of farmer-led-organisations from 80 countries at a forum in Mali: "Food sovereignty is the right of peoples to healthy and culturally appropriate food produced through ecologically sound and sustainable methods, and their right to define their own food and agriculture systems. It puts those who produce, distribute and consume food at the heart of food systems and policies rather than the demands of markets and corporations" (Nyéléni 2007).

PROPOSALS

The solutions that La Via Campesina puts forward are:

1. small scale agriculture, which is labour intensive, uses little fossil fuel energy and can actually help stop the effects of climate change;
2. a genuine agrarian reform to strengthen peasant agriculture;
3. promoting food production as the primary land use;
4. considering food as a basic human right that should not be treated mainly as a commodity;
5. supporting local food production because it avoids unnecessary transport;
6. reducing patterns of production and consumption that promote wastes.

These proposals have strong empirical and theoretical footing in the study of energy flows in agriculture by academics over many decades.

complementary to formalized expertise" so that "participation can ensure that policies and programmes are truly responsive to the needs of vulnerable groups".

A recent document of FAO (2018a) recognises the relevance of agroecology scaling up for performing the sustainable development goals established in the 2030 Agenda. FAO's document claims that

> scaling up agroecology matches the transformative ambitions of the 2030 Agenda and will support countries to meet their commitments. Transitions require innovations in policies, rural institutions and partnerships, as well as in the production, marketing and consumption of nutritious food, leading to sustainability and equity throughout the entire food and agricultural system. Scaling up agroecology requires overcoming key challenges while harnessing emerging opportunities.

Challenges to scale up agroecology as listed by FAO's document (FAO 2018a) are listed below:

(i) Research, education and extension systems do not sufficiently respond to the needs of agroecology to effectively transform food and agricultural systems
(ii) Lack of awareness of agroecology among policy makers
(iii) Agroecology transitions require an enabling environment
(iv) Political and economic support needs to prioritize sustainable approaches
(v) Current market systems are not responding to agroecological approaches
(vi) Lack of coordinated action and collaboration in policy and governance

To bring agroecology to scale and transform food and agriculture systems requires a process of political adaptation that implies setting up steps for an enabling social context that can favour a spread of local sustainable agroecosystems as reported in Box 6.3.

Box 6.3 Public Policy Grounds for Self-Sustainable Agroecosystem Scaling Up (Adapted from Gonzales de Molina et al. 2020)

ENABLING ENVIRONMENT

1. Guarantee access to land and other natural resources
2. Raise the income of family farmers by assessing the ecological services they provide and by means of shorter distribution food chains
3. Promote sociotechnical innovation for family farms
4. Enhance the use of indigenous genetic material adapted to climate and soil conditions

LOCAL AGROECOSYSTEMS

1. Reduce dependence on external inputs
2. Build shorter and more equitable distribution chains
3. Promote more sustainable food consumption
4. Promote higher level of agricultural resilience to socio-economic and climatic disturbances

6.1.3 Terracing for Ecosystem Domestication

Land topography is a commonality shared in all continents, where hilly and mountain areas may constitute the most part of terrestrial land. Humans have developed settlements early in these areas as 'bioturbators' or ecological engineers, bringing about substantial interventions for modifying a critical natural character like the slope gradient which affects movements of both physical and biological components. A terrace can be easily built collecting local stones and setting them up as a cross-slope barrier to contain soil and filter water. Terraces have been successful man-made constructions, widespread in many parts of the world (Wei et al. 2016; Varotto et al. 2019), for mitigating the slope gradient in view of multiple ends (Fig. 6.9), involving a radical change of the pristine ecosystems in their physical, biological and cultural assets. Terraces can be identified as primordial efforts in order to 'domesticate' not species but whole natural ecosystems.

According to the principles of restoration ecology (Bradshaw 1996), the primary functions of ancient terraces were to store soil and regulate the hydrogeological cycle with the use of local available materials:

> It should be axiomatic that they should be used wherever possible, firstly because they cost nothing in themselves (although they may cost something to initiate); secondly, they are likely to be self-sustaining because they originate from within nature (although they may need nurturing in some situations); and thirdly, they can be used on a large scale.

While the distribution of terraces varied across continents, most often terracing practices were found in regions where agricultural civilization firstly developed. The earliest practices of terracing were recorded in Palestine and Yemen about

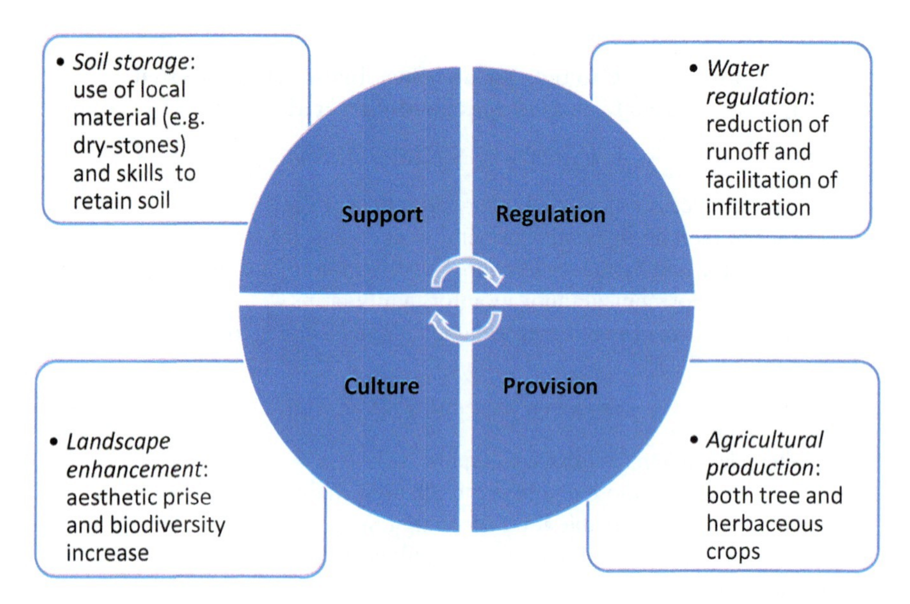

Fig. 6.9 Terracing multiple ends according to the theory of ecosystem services (MEA 2005)

5000 years ago, then spread to the drier regions of the Mediterranean. While massive terracing practices in the Mediterranean region mainly began from the late fourteenth century during the Renaissance period, older terracing practices recorded in the Alpine Region, the Maya Lowlands, the Middle East and sub-Mediterranean areas of Europe, date back to the Iron Age or even earlier (Wei et al. 2016). As to terracing classifications, multiple concepts can be used. Terraces in the Mediterranean region and in Central and South America (e.g., Mexico, Colombia, Ecuador, Peru and Chile), for example, have mostly been constructed using dry-stone walls, while terraces in North America, Vietnam, Thailand and NW China are mostly built of soil. As to agricultural use, terraces in the Asian humid regions are mainly used for rice cultivation, while terraces in Europe are used for grapevines and olive trees. In both of the semi-arid regions (e.g., western Kansas and Nebraska) and humid regions (e.g., Indiana and Kentucky) of North America, parallel terraces, bench terraces, contour terraces and parallel-tile-outlet terraces (discharging runoff through subsurface drains) were mostly used for corn, soybean and wheat cultivation (Wei et al. 2016). According to agronomic criteria, terracing is a soil and water conservation measure together with other agronomic practices of contour tillage and contour planting that conserve soil and moisture better than up and downslope practices (Barrow 1999; Chapagain and Raizada 2017). From this technical point of view, terracing is a preliminary soil conservation measure for adopting other virtuous farming practices and introducing important productivity drivers like irrigation. From an ecological point of view, terracing is the form of land use that better than any other connects and integrates the local physical resources with the human ingenuity in order to build up a sustainable agroecosystem, which delivers all four categories of ecosystems services, including the cultural ones.

A special case of cultural heritage, which originated from *ante litteram* restoration ecology studies, has been the birth of academic studies in the field of agriculture in Italy. Academic studies in agriculture have established in Italy since 1844, before the setting up of the Italian State, with an academic 3-year degree course in "Agriculture and Animal Husbandry" implemented at the Agrarian Institute of the University of Pisa, then belonging to the Great-Duchy of Tuscany. That curriculum was part of an academic activity based on research carried out in two experimental farms of the Agrarian Institute. The founder of this successful initiative (the first academic agriculture curriculum in Europe) was marques Cosimo Ridolfi, a well-known agronomist in Europe with an active membership of "Accademia dei Georgofili" in Florenz. Ridolfi was such a passionate pedagogist with expertise in *ante litteram* 'restoration ecology' or 'reclamation ecology' to have established an experimental farm of 16 ha (Fig. 6.10) in his property (*fattoria* "Meleto") – with a residential college for students where he taught how to reclaim degraded land by erosion to productive state-before the curriculum implementation in Pisa.

Ridolfi taught in Pisa only one academic year (1844–1845) and then left his teaching assignment to Pietro Cuppari who, according to Caporali (2015b), was a precursor of Agroecology in Italy.

When adopted at large on a territory, terraces confer a monumental landscape character that compares with natural scenery. For instance, the rice terraces of Ifugao in the Philippines are nationally considered as the "eighth wonder of the world" and were described as an "irreplaceable treasure" in the Philippines by presidential decree in 1973. Large parts of the terraces are included in the UNESCO World Heritage List and are among the Globally Important Agricultural Heritage Sites selected by the Food and Agriculture Organization of the United Nations (Castonguay et al. 2016). In Peru alone, indigenous farmers built more than 1 million ha of agricultural terraces. The magnificently sculpted terrace landscape of the

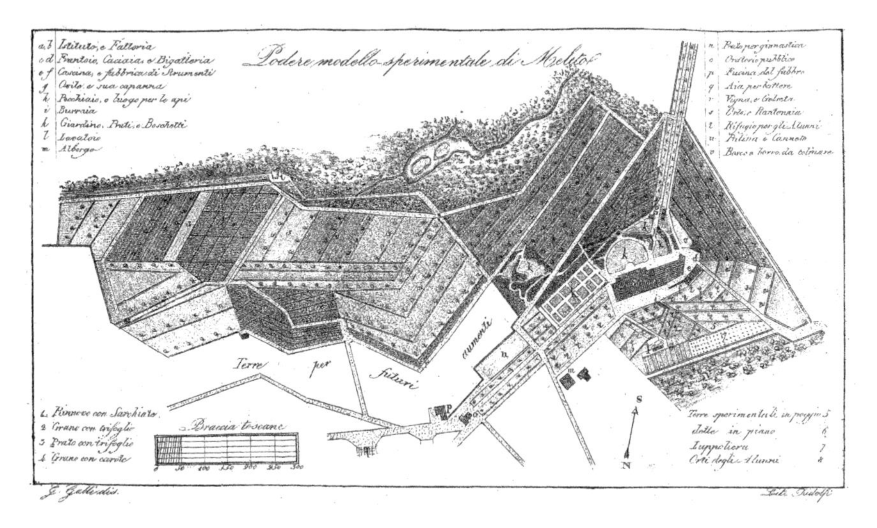

Fig. 6.10 Map of Cosimo Ridolfi's experimental farm with fields in terraces (Ridolfi 1838)

Colca Valley is a triumph of Andean engineering. The valley contains more than 14,000 ha of fields two-thirds of them are walled, irrigated valley-side terraces (Treacy 1987). Terraces reflect not only a technical character, but transmit a message of land care that conforms to the goals of restoration ecology as expressed by Diamond (1987) as follows:

> The goal of restoration should be recreate a natural community, or to recreate a self-sustaining community, or perhaps to preserve a community for posterity in essentially its natural state […] It is a choice based on values, and it is only one of the many possible choices. We know that different people with different values would make different choices about the same site (p. 331).

It is evident that a concept of restoration ecology in a framework of sustainable development encompasses man not only as the driver of restoration but also as its keeper, generation after generation. According to Aldo Leopold (1949), restoration is merely a first step of sustainable development:

> A starting point in the long and laborious job of building a permanent and mutually beneficial relationships between civilised men a civilised landscape (cited in Jordan III et al. 1987, p. 7).

This stance is still valid (LaFevor 2014) and needs recognition as a point of common interest for strengthening the attitude to take care of anthropogenic heritage with poly-functional roles:

> Agroecosystems are largely humanized environments, created, maintained, and managed by and for human use. As such, they require the continued attention of cultivators if they are to produce the domesticated plants and landforms that characterize them, regardless of whether the initial intent of restoration is to produce food or to mitigate environmental degradation.

In all Mediterranean basins, during the past centuries, the need for cultivable and well-exposed areas determined the extensive anthropogenic terracing of large parts of hillslopes and mountain areas. Most of the historical terraces are of the bench type with dry-stone walls and require maintenance by hand. Terraced areas exist all over Italy, from the Alps to the Apennines and in the cost on the sea, in both the hilly and mountainous areas, representing distinguishing elements of cultural identity. Contour terraces and regular terraces remained in use until the second post-war period, as long as sharecropping contracts guaranteed their constant maintenance (Tarolli et al. 2014). Recently, rural depopulation in both hilly and mountain areas has caused farmland and terrace abandonment in Europe, with special reference to the Mediterranean region, as well documented by several studies (Garcia-Ruiz and Lana-Renault 2011; Van der Sluis et al. 2014; Socci et al. 2019). The cessation of maintenance of dry stone terraces due to the crisis of traditional agriculture can be cause of failure. During heavy rainfall events, such as those registered recently in the famous area of "Cinque Terre", hundreds of landslides, mud flows, and erosions have occurred (Agnoletti et al. 2019). The main feature of the Cinque Terre landscape is the presence of terraced cultivations of grapevines on steep slopes facing the sea (Fig. 6.11). The area represents a remarkable cultural landscape, is a National Park, and is included in the World Heritage List of the UNESCO.

Fig. 6.11 Dry-stone terraces of "Cinque Terre", Liguria (IT), with grapevines and natural vegetation encroachments

In order to create more understanding and appreciation for terraced landscapes and traditional agricultural systems, a wider communication of scientific studies and the building of local management capacities are necessary. The first International 'Terraced Landscape Conference' took place in the city of Menzi, in Honghe Hani and Yi Prefecture, Yunnan Province, south-west China, on 11–15 November 2010. The conference was innovative for bringing together, for the first time, most of the countries with terraced field landscapes, irrespective of their degree of development or the watering methods, and regardless of whether the terraced fields are the main food source for the local people or mainly kept as a heritage landscape. The most important contribution of this conference to the current debate on landscape conservation lies in the 'Hani Terrace Declaration' (Honghe declaration), which is the official document for terraced landscape conservation accepted by this international forum. After the conference, the 'International Terraced Landscapes Alliance' (ITLA) was established with the main objective of both gathering international resources for conservation and the sustainable use of terraced landscapes, and publishing internationally about key issues in conservation policies (Yehong et al. 2011).

Terracing is an anthropogenic element that modifies both physical and biological diversity, providing new habitats and ecological niches that altogether generate new ecotones. The final local effect is also aesthetically appreciable, as that in Fig. 6.12, which is relative to a terraced home garden of a family farm in a hilly area of central Italy.

Fig. 6.12 A terraced home garden at the scale of habitat availability in a hilly area of central Italy (Viterbo, Lazio Region)

The anthropogenic element of the dry-stone walls is completely absorbed under a cover of an exuberant natural vegetation that compares with the linear order of the crop canopy, which in turn is completely segmented in rows of different species and variety of summer and autumn maturing crops like, from the left to the right, potato, tomato, aubergine, bean, grapevine and olive. This biodiversity explosion, which is supporting, regulating, productive, and inspiring at the same time, derives from both ancient tradition and current human governance. Such an agricultural system can be sustainable, provided human care is assured, and this is a question of non-limiting social conditions that favour individual property, competence, skill, irrigation, biodiversity, functionality, stability, productivity and beauty. In that case, farmers "have a personal relation to the land – and not simply an economic relation to it" (Barbour 1993, p. 96) – which allows agriculture to be a way of life in a harmony-with-nature-development. This teaching comes from afar and it remained conserved to the benefits of future generations into the medieval monastic gardens where the art of maintaining horticultural and gardening understanding and traditions was a practise of life (Jagger 2015). To sustain their life, monks had to conform to the rule of self-subsistence, as expressed by St. Benedict who stated "whenever possible the monastery should be so laid out that everything essential, that it to say water, mills, garden, and workshops for the plying of the various crafts is found within the monastery walls" (as cited in Jagger 2015, p. 633). Manual labour in the home garden and contemplative meditation in the cloister were complementary activities for giving meaning to life according to the Bible statement – "the Lord took the man, and put him into the Garden of Eden to cultivate it and to keep it".

In general, research highlights the human dimension of terraces as human-engineered ecosystems that greatly depend on humans for their maintenance over time. Climate, vegetation dynamics, fire, market demands, production costs, etc., govern the terrace system equilibrium (Savo et al. 2014). Heritage interest in terraced landscape areas has grown in recent times leading to a sound valorisation process such as the UNESCO's recognition of dry-stone walling construction as Intangible Cultural Heritage of Humanity since 2018. This connects with the recent concept of biological cultural heritage (BCH), or biocultural heritage, which identifies domesticated landscapes resulting from long-term biological and social relationships. Moreover, also pastoral enclosures of dry-stones arise as traditional rural constructions that engaged local communities recover as biocultural value in terms of identity and positive conservation outcomes (Grove et al. 2020). Pastoral or horticultural enclosures emerging at some height on the soil can be able to affect positively microclimate in unfavourable environments, like islands in open seas and oceans, especially for their protective action against wind damages and evapotranspiration losses. For instance, UNESCO (2014) has identified the community of Pantelleria island as "… the true custodians of the traditional knowledge regarding the technique of cultivating the 'head trained bush vines' (vite ad alberello) …, and Pantelleria terraces and gardens have to be considered as whole corpus of traditional knowledge recognized" (Barbera et al. 2018). When farming is a way of living and the farmer live in and with his/her homeland, a symbiotic relationship between man and environment takes the form of a co-evolutionary agroecosystem where cultural and natural diversity (biotic and abiotic) intermingle with reciprocal sustenance.

In view of these challenges, the United Nations proclaimed the United Nations Decade of Family Farming (2019–2028) in December 2017, providing the international community with a Global Action Plan, which is an extraordinary opportunity to address family farming from a holistic perspective. The Global Action Plan represents a tangible result of an extensive and inclusive global consultation process involving a wide array of different partners around the world. The purpose of the Plan is to mobilize concrete, coordinated actions to overcome challenges family farmers face, strengthen their investment capacity, and thereby attain the potential benefits of their contributions to transform our societies and put in place long-term and sustainable solutions (FAO and IFAD 2019). It is a great chance for substantial transformations in current food systems that will contribute to achieving the 2030 Agenda Goals for Sustainable Development.

6.2 Sustainable Agriculture as a Turning Point of the Human Predicament

Sustainable agriculture is an asset of public utility that needs implementation and protection. While successful traditional agricultural systems have given proof of their resilience, modern industrial agricultural systems have not. Much of the facts of modern or *conventional* agriculture do not meet much of the values that people attach to it. Making decisions to change this tendency is an ethical step:

> The world that will exist in 100 and 1000 years will, unavoidably, be of human design, whether deliberate or haphazard. The principles that should guide this design must be based on science, much of it done only sketchily to date, and on ethics [...] A sustainable world will require an ethic that is ultimately as incorporated into culture and as long lasting as a constitutional bill of rights or as religious commandments (Tilman 2000).

Ian Barbour (1993), considering ethics in an age of technology and agriculture as a vast field of technological application, sheds lights on the drivers that have moulded in new fashion the face of agriculture worldwide (Fig. 6.13).

The framework of conventional farming in Fig. 6.13 is rather eloquent in explaining how the farmer's independence in organising his/her farm is impaired by an overarching hierarchy of allied powers, both public and private, that reduce

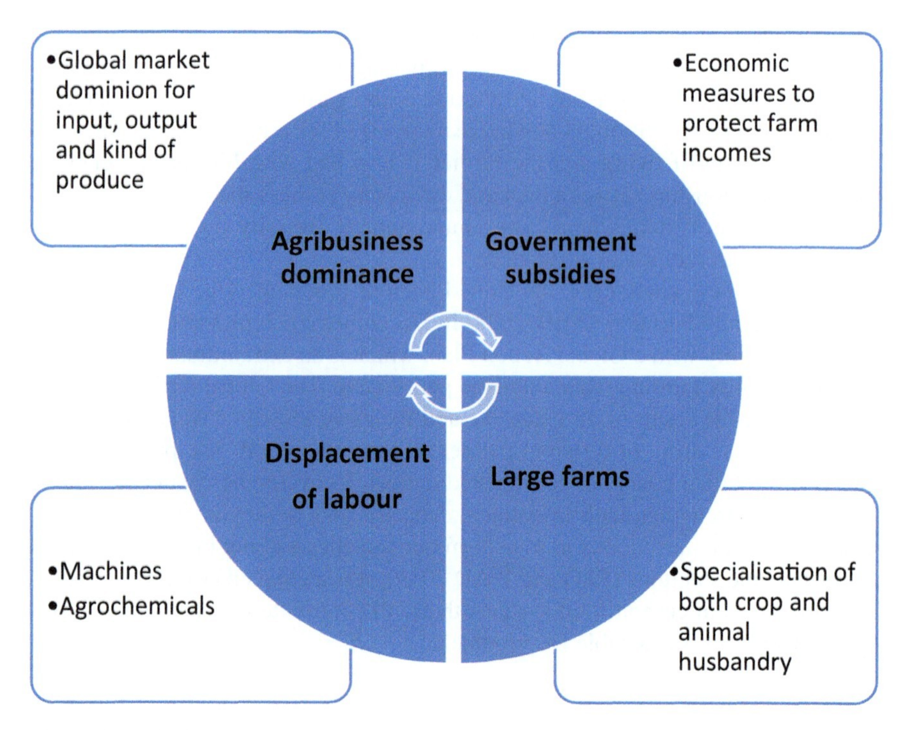

Fig. 6.13 Main drivers of industrialised agriculture according to Barbour (1993, pp. 92–93, modified)

agriculture to a simple ring of a longer buyer- and seller-chain under the agribusiness dominance. Agriculture being a simple component of a market chain completely separate from the spot of production, any relationship with the soil, the land, the natural biodiversity and the local society vanishes, although concretely and negatively affected. This framework has largely contributed to the processes of de-ruralisation of the countryside, urbanisation, unemployment, land consumption in flat, fertile areas, and environmental pollution. There is no surprise in claiming that this kind of development is unsustainable and a general inversion is necessary. The turning point should be an ethical choice in the interest of the living community. Agriculture, which concerns the organisation of natural resources for the benefit of humanity, should start the process. Agroecology, as the science for a sustainable agriculture, should legitimate principles, criteria, design and management for sustainable agroecosystems.

International institutions at the highest level have recently adopted resolutions in favour of sustainable development (UN 2015) that directly involve agriculture. Out of 17 Sustainable Development Goals of the 2030 Agenda for Sustainable Development, the first (*End poverty in all its forms everywhere*) and the second (*End hunger, achieve food security and improved nutrition and promote sustainable agriculture*) are clearly referred, directly or indirectly, to agriculture enhancement, especially in poor countries. In particular for the second goal, points 2.3 and 2.4 state, respectively:

> By 2030, double the agricultural productivity and incomes of small-scale food producers [...] through secure and equal access to land, other productive resources and inputs, knowledge, financial services, markets and opportunities for value addition and non-farm employment;
>
> By 2030, ensure sustainable food production systems and implement resilient agricultural practices that increase productivity and production, that help maintain ecosystems, that strengthen capacity for adaptation to climate change, extreme weather, drought, flooding and other disasters and that progressively improve land and soil quality.

Small scale-food producers are the main target of the UN programme, while the means for increasing productivity and maintaining the resilience of agroecosystems is a matter of agro-ecological concern. A recent survey of FAO (Lowder et al. 2016) on the number, size and distribution of farms, smallholder farms, and family farm worldwide, shows that there are more than 570 million farms worldwide, most of which are small and family-operated. Moreover, it shows that small farms (less than 2 ha) operate about 12% and family farms about 75% of the world's agricultural land. In low- and lower-middle-income countries of South Asia and Sub-Saharian Africa, about 70–80% of farms are smaller than 2 ha and operate about 30–40% of the land. More poverty and smaller farm property appear strongly joined. Recently, an explanatory document of the European Parliament (2017) alerts that agricultural land concentration is a topic with European relevance:

> Figures from 2010 show that in the 27 member EU, only 3% of farms already controlled 50% of the land used for farming purposes, while in contrast, in 2012, 80% of farms had the use of 12% of the farmland [...] As with the concentration of financial wealth, too high a concentration of agricultural land splits society, destabilises rural areas, threatens food

> safety and thus jeopardises the environmental and social objectives of Europe […] Ownership is the best way of securing a responsible relationship with the land and its sustainable management. It promotes a sense of belonging and thus encourages people to stay in rural areas (p. 14).

FAO (2018a) has issued a document destined to summarise the science and practice of Agroecology in 10 principles (to be analysed later on Sect. 6.3) that should help stakeholders to adopt a strategy of sustainable agriculture based on the five pillars reported in Fig. 6.14.

The construction of the five pillars for achieving the final goal of sustainable agriculture requires co-operation among the stakeholders of the agricultural system at any hierarchical levels, from the farm to the national and international levels, in that the agricultural system can function successfully only if its parts are organically connected. The farm level is the physical point where agriculture takes practical form under the flow of information, internal and external to the farm system, that the farmer receives and elaborates in his/her decision-making process that defines the farm organisation. The system model reported in Fig. 6.15 shows this process, where the main environmental components operate as inputs of information for the farmer's process of elaboration that produces decisional output and actions

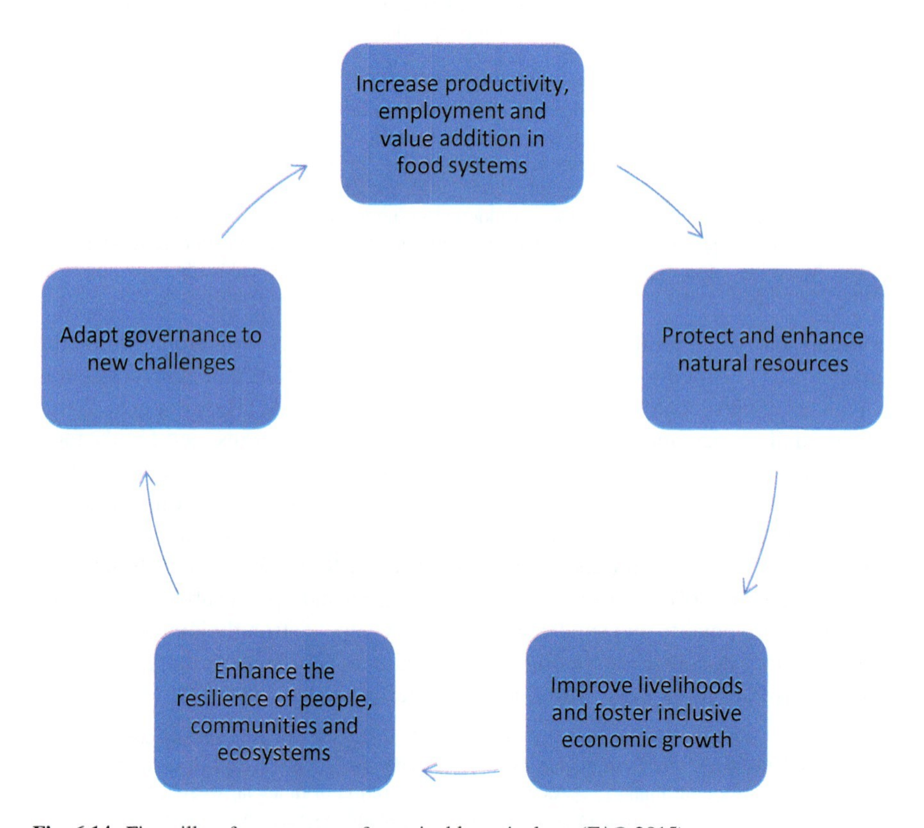

Fig. 6.14 Five pillars for a strategy of sustainable agriculture (FAO 2015)

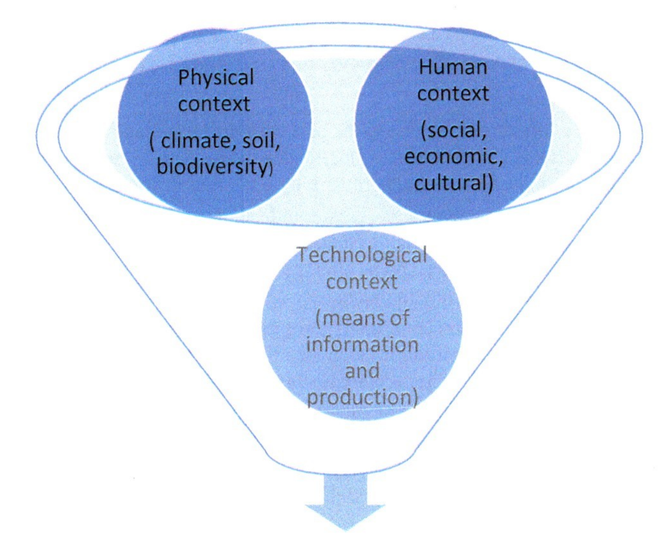

Fig. 6.15 The farmer's decision-making process according to a systems view

moulding structure and function of his/her farm. Decisions made as farming design and practices produce continuous variations in the environment components that subsequently act as inputs to remodel the farmer's knowledge. The individual decision-making process is therefore cyclic, dynamic, and continuously subjected to remodelling by new elements of knowledge and therefore hardly predictable (Caporali 2010).

Decision-making at the family farm level is the best candidate for implementing the guidelines for sustainable agriculture suggested by international institutions such as UN, FAO and EU. In Table 6.3,a summary of the appropriate conditions of owner-operated small or midsize family farms for implementing sustainable agriculture goals is reported.

On the base of the above reported arguments and judgements, Barbour's conclusions are as follows:

> In sum, small and midsize family farms do contribute to widely dispersed ownership and stronger rural communities, offering greater opportunities for what I have called participation and personal fulfilment [...] This suggests the need to correct current biases toward large farms by giving more help to midsize and owner-operated farms (p. 96).

The kinds of argument and judgement discussed in Table 6.3 reveal, in their succession, a thread of interconnection under an overarching hierarchy of components that unwinds across the farm level, the local community level, and the landscape level. This interconnection needs an organisation to be successful because it constitutes the rationale connecting the farmer's values with social values within environmental values (Fig. 6.16).

Table 6.3 Conditions facilitating sustainable agriculture goals in small or midsize family farms according to Barbour (1993, pp. 94–96, modified)

Kinds of argument	Judgements
Cultural and moral tradition	Family farms are a tie to the past with distinctive cultural identity. Thomas Jefferson suggested that small independent farmers are morally superior: virtuous, honest, and hardworking – the foundation of a democratic society
Widely disperse ownership	The owner-operator has greater independence and opportunity for participation in decisions than a corporate farm manager Political influence spread among many persons rather than concentrated in the hands of a few individuals or corporations
Stronger rural communities	The family farm community have greater stability and more and better services and institutions (schools, libraries, hospitals, etc.). Dispersal of small industries in rural areas strengthens rural life
Land stewardship	Farmers plan to pass on their farms to their children and take better care of their land than corporations. They diversify their operations, raising both livestock and crops with manure and rotations. They treat their animals better than large commercial feedlot operators

Today the question of reconnecting people with their larger environment is vital for ensuring sustainable development, and sustainable agriculture is a model to verify environmental pragmatism as currently meant:

> The observation that the human sphere is embedded at every point in the broader natural sphere, that each inevitably affects the other in ways that are often impossible to predict, and that values emerge in the ongoing transactions between humans and environments, for example, are all central concepts for the pragmatists – as for many contemporary philosophers of environment (Parker 1996, p. 21).

> In a pragmatic vision, the ultimate value is that "the being of any existent thing, human or non-human, is constituted in its relations with other things in a context of meaningful connections" (Parker 1996, p. 34).

The connections of each farmer (and each farm) within the larger context of local community and the environment in general should be *meaningful*, in the sense that the paradigm of sustainable development should be borne by both the farmer and the human institutions that cooperate with him/her in the vast area of agriculture as a human activity systems. This condition of coherence is often neglected because cultural barriers exist that impede to recognise the importance of establishing a synergic relation between the individual and the public institutions. Bellah et al. (1991) provide a deep investigation of American society in the attempt of prospecting how "the individual is realized only in and through community" and "how better institutions are essential if we are to lead better lives":

> Our present situation requires an unprecedented increase in the ability to attend to new possibilities, moral as well as technical, and to put the new technical possibilities in a moral context (p. 5).

Bellah et al. (1991) go on warning that

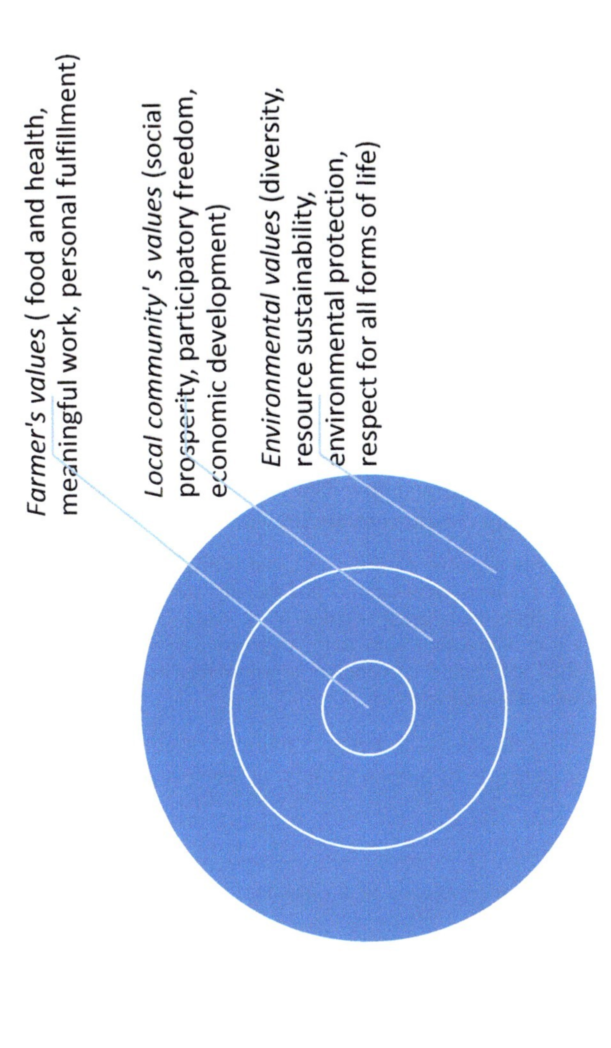

Fig. 6.16 Hierarchy of agroecosystem components and values at local level. (Adapted from Barbour 1993, p. 81)

(a) the culture of individualism makes the very idea of institutions inaccessible to people;
(b) it is hard for people to think of institutions as affording the necessary context within which a person can grow;
(c) if some of institutions have indeed grown out of control and beyond people comprehension the answer is to change them and not to imagine that people can escape them.

Bellah et al. (1991) assume that a "great society" is not enough, and that the challenge is "transforming the great society into the good society":

> The great classic criteria of a good society – peace, prosperity, freedom, justice – all depend today on a new experiment in democracy, a newly extended and enhanced set of democratic institutions, within which we citizens can better discern what we really want and what we ought to want to sustain a good life on this planet for ourselves and the generations to come (p. 9).

The theme of cooperation emerges also as an important principle of evolutionary dynamics, in that cooperation is necessary for evolution to construct new levels of organisation (Novak 2006). This conclusion stems from a mathematical formalistic comparison of five mechanisms for evolution as reported in Table 6.4. For each mechanism, a simple rule holds that specifies whether natural selection can lead to cooperation.

This comparison provides insights about the fundamental rules that specify whether cooperation can evolve and how their relative weight plays in the dynamics of evolution. The following conclusions are drawn:

> New levels of organization evolve when the competing units on the lower level begin to cooperate. Cooperation allows specialization and thereby promotes biological diversity. Cooperation is the secret behind the open-endless of the evolutionary process. Perhaps the most remarkable aspect of evolution is its ability to generate cooperation in a competitive world. Thus, we might add "natural cooperation" as a third fundamental principle of evolution beside mutation and natural selection.

Whether we will be successful in being convinced or not that cooperation is the decisive organising principle of human society, is the very challenge to sustainable development.

Table 6.4 Mechanisms for the evolution of cooperation and their operational patterns (after Novak 2006)

Mechanisms	Operational patterns
Kin selection (KS)	KS operates when the donor and the recipient of an altruistic act are genetic relatives
Direct reciprocity (DR)	DR requires repeated encounters between the same two individuals
Indirect reciprocity (IR)	IR is based on reputation: a helpful individual is more likely to receive help
Network reciprocity (NR)	NR means that clusters of co-operators outcompete defectors
Group selection (GS)	GS is the idea that competition is not only between individuals but also between groups

6.3 Agroecology as a Transdisciplinary Field of Cooperation

Agroecology is today a transdisciplinary field of enquiry that is capable of changing our common vision of both agriculture and society (Caporali 2010, 2015a; Gliessman 2015, 2018; Gallardo-López et al. 2018; Fernández González et al. 2020). A selection of publications and resources that shows agroecology as a "transformative vision and practice" is in Anderson and Anderson (2020). Its paradigmatic value reflects into a recent definition of agriculture, accepted at international level (IAASTD 2006), that derives just from agroecological knowledge, as shown by the under reported definitions of both agroecology and agriculture:[3]

Agroecology. The science of applying ecological concepts and principles to the design and management of sustainable agroecosystems. It includes the study of the ecological processes in farming systems and processes such as nutrient cycling, carbon cycling/sequestration, water cycling, food chains within and between trophic groups (microbes to top predators), lifecycles, herbivore/predator/prey/host interactions, pollination, etc. Agroecological functions are generally maximized when there is high species diversity/perennial forest-like habitats.

Agriculture. A linked, dynamic social-ecological system based on the extraction of biological products and services from an ecosystem, innovated and managed by people [...] It encompasses all stages of production, processing, distribution, marketing, retail, consumption and waste disposal.

The systems paradigm that defines agroecology has affected the current definition of agriculture. Turning back to Lakoff's concept of 'framing', agroecology has re-framed the concept of agriculture. FAO's choice of Agroecology as the science guiding theory and practice of sustainable agriculture results in the recent publication of the report "The 10 elements of agroecology guiding the transition to sustainable food and agricultural systems" (FAO 2018a). Figure 6.17 displays how the tentative separation of the 10 elements in two complementary parts of biophysical and socioeconomic components cover the vast transdisciplinary field of agroecology enquiry. Social sciences and environmental sciences find their final connection into agricultural sciences, defining altogether the new systemic science of agroecology (Tomich et al. 2011). The five elements of biophysical components express the logic of self-organisation underlying structure and function of a sustainable agroecosystem, where biodiversity is the most relevant driver for promoting synergies, efficiencies, recycling, and resilience as the ultimate capacity to maintain a dynamic stability even in adverse environmental conditions. These five elements govern the strategy of development of natural ecosystems without disturbance. In situations of *planned* biodiversity, such as those of agroecosystems, the challenge is to find appropriate combinations of both crops and animal husbandry that join together to satisfy productive and protective goals. Climate, soil, crops, and livestock are the

[3] Definitions drawn from the glossary of the IAASTD's Global Report "Agriculture at a Crossroads".

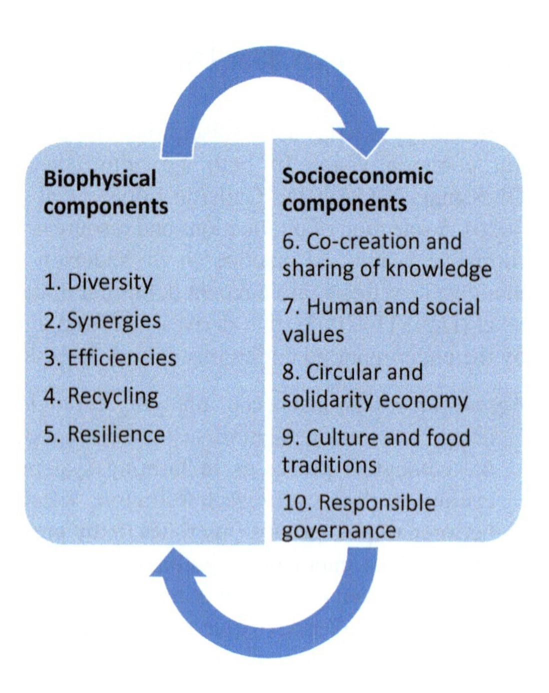

Fig. 6.17 Ten agroecology elements (or key words) that frame agroecosystem sustainability. (Adapted from FAO 2018b)

main diversity components that need to become cooperative under the farmer's control.

As to the five socioeconomic components, they reveal how wide is the range of elements to consider in making decisions on the organisation of a farm, the choice depending on external factors such as available technology, economic profitability, cultural values, environmental impact, social cohesion, and political incentives or limitations. Cooperation in the field of socioeconomic components is a slow process, and requires decades to develop, often affecting two or more generations of stakeholders. The 10 elements altogether contribute to both design and implementation of farm system organisation. Cooperation of both internal and external drivers converges toward a common goal, an extended sustainable agroecosystem from the local to the global scale (Fig. 6.18). The internal drivers operate as spontaneous processes and represent the 'work' of nature; the external drivers are human-dependent and should merely facilitate the 'work' of the internal drivers that govern the ecosystem services of support, regulation and production. Soil provides the biophysical functions of support and regulation for the terrestrial living community, including crops, livestock and human beings. In agricultural terms, soil means *fertility*, i.e. the natural capacity to sustain directly crop production, and indirectly, animal transformation and human sustenance. Soil is a limited resource, non-renewable in short-times, and easily impaired or lost if ill disturbed. Therefore, the first condition to ensure the development of sustainable agroecosystem is *protecting the soil*, against misuses for human settlements (Hellerstein et al. 2002; Akkari et al. 2019), and mismanagement from agricultural practices (Lal 2014).

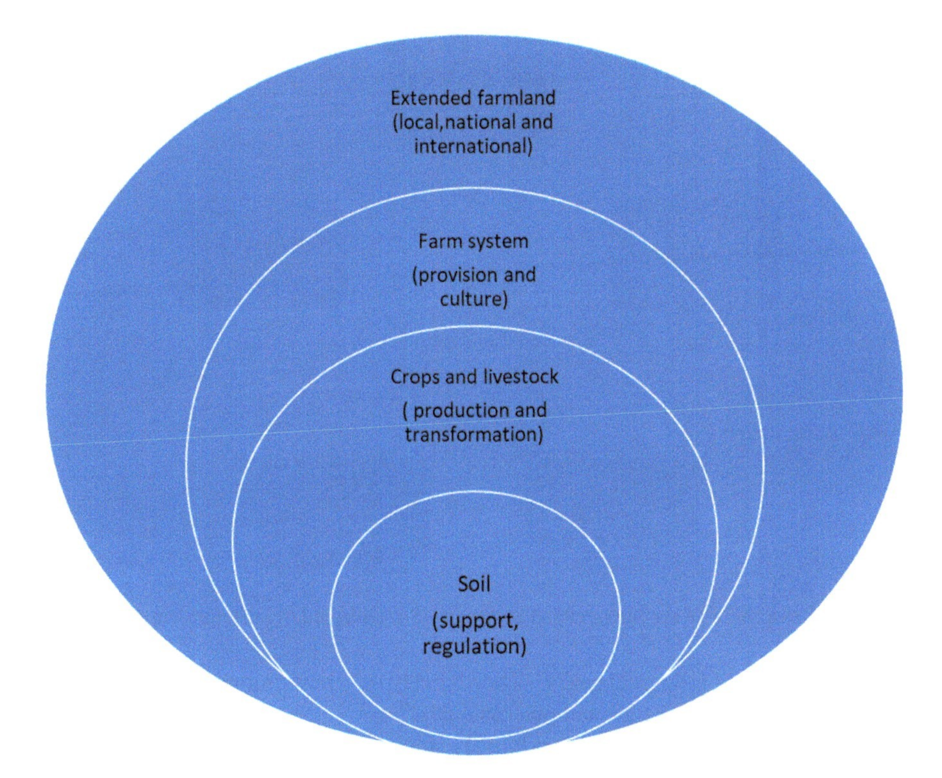

Fig. 6.18 The extended nature of agroecosystem sustainability and ecosystem services

Concerning the must of protecting the soil for agricultural use against alternative uses, Caporali (2015a) warns against the presence in Europe of contradictory experiences:

> The challenge of sustainability for today's society as a whole is a coin with two faces. On the one hand, sustainable development includes the necessity for humanity to grow food through agriculture as well as to maintain natural environments for ecological services other than food. The search for a balance between production and protection in land use is therefore a major challenge to future society at both local and global level. On the other hand, demographic growth and concentration of human population on flat areas of costs and internal valleys bring about a sealing process of former, fertile agricultural soils that creates unbalances of land use at both local and international levels, thus undermining the sustainability of society as a whole. If alternative land use encroaches on soil for food and feed nutrition, other agricultural areas in the world must provide food for local people. However, the enlarging ecological footprint for food nutrition of an increasing human population has its own limit in the physical boundaries of the planet (p. 22).

Akkari et al. (2019) lament that in Quebec (Canada) "in many instances, land use planning has contributed to the takeover of good quality agricultural land because many planners and local and regional politicians have placed their priorities on population growth in their communities, and related subdivision development as well as

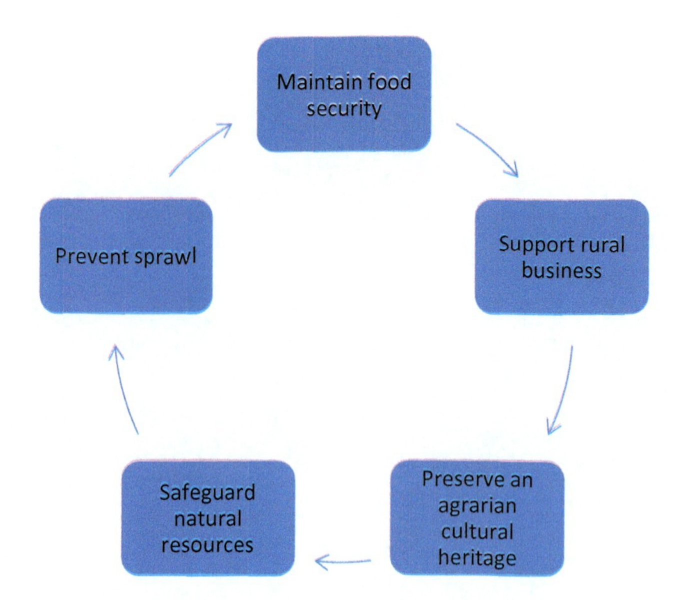

Fig. 6.19 Reasons for protecting farmland by law in USA. (Adapted from Hellerstein et al. 2002)

the development of industrial parks". They bitterly note how "money seems to be all that matters to many officials and professionals!"

In the United States, despite the relatively small fraction of the land in urban uses (3%), there is growing concern about the disappearance of farmland in some parts of the country. All 50 States have enacted farmland protection programs to help slow the conversion of farmland to developed uses (Hellerstein et al. 2002). This concern justifies the adoption of an expanding array of farmland protection programs by non-profit organisations and by county, State, and Federal governments. Interest in protecting farmland arises from the reasons reported in Fig. 6.19.

Another threat is the menace of losing soil, either quantitatively or qualitatively, because of either erosion or pollution caused by agricultural misuses. Most agricultural land in the world is losing soil at rates ranging from 13 to 40 tons ha^{-1} $year^{-1}$. Because soil forms very slowly, this means that soil is being lost 13–40 times faster than the rate of renewal (Pimentel and Kounang 1998). Erosion occurs when the soil lacks protective vegetative cover. Soil erosion reduces the productivity of the land by loss of water, soil organic matter, nutrients, biota, and depth of soil. In addition to degrading soil quality and reducing agronomic/biomass productivity on-site through a decrease in use-efficiency of inputs, off-site impacts of accelerated erosion include eutrophication and contamination, sedimentation of reservoirs and waterways, and emissions of greenhouse gases (e.g., CO_2, CH_4 and N_2O) (Lal 2014). The acceptance of the recommendations provided early by FAO (1976), as reported in Box 6.4, is a preliminary step to frame cooperation in accordance to the indispensable goal of protecting the soil as the base of a sustainable agriculture.

Hydrological soil erosion is a combined process yielded by four main factors represented in Fig. 6.20. Climate, soil, land, and man contribute to increase or

Box 6.4 Recommendations by FAO (1976, p. 3, Modified) for Soil Protection Involvement

SOIL VALUE

Soil in a basic resource for the present and the future

The value of its conservation extends beyond that which can be expressed in monetary terms.

The damage caused by severe soil erosion is frequently irreversible

It is consequently desirable to take conservation measures to prevent onset of erosion rather than acting after it has commenced.

AGENTS OF COOPERATION

Soil conservation is an interdisciplinary subject

Main disciplines involved are agronomy, soil science, range management, forestry, ecology, hydrology, engineering, geography, economics, and sociology.

Soil conservation is inclusive

It concerns all aspects of land use planning, development and management which contribute to the maintenance and improvement of soil resources.

reduction of the transfer of soil under the splash effect of rainfall that hits the soil surface and runs away with a speed that depends on the land slope and its state of cover. On arable land, the most effective control of erosion can be achieved with an appropriate crop management.

Close-growing crops, crop-residue management, and the use of sod-forming crops in rotations are the main agronomic practices for enhancing the crop potential for protecting the soil from erosion and improving its fertility. These are fundamental measures of soil protection dealt with later in the section of ecological intensification practices. According to Lal (2014), crop management practices are important to strengthen numerous ecosystem services, such as (a) improving water quality and renewability; (b) increasing below and above-ground biodiversity; (c) enhancing soil resilience to climate change and extreme events, and (d) mitigating climate change by sequestering C in soil and reducing the emission of CO_2, CH_4 and N_2O. An effective control of accelerated erosion through good practices is essential to sustainable development.

Cropland is highly susceptible to wind and water erosion because it is tilled repeatedly. Major practices that need tillage are:

(a) seed-bed preparation (ploughing, chiselling, harrowing);
(b) mechanical weed control;
(c) soil-fertility management (applying organic and mineral fertilisers, green manuring, etc.).

In addition, cropland often is bare between plantings for several months of the year. Tillage, in addition to mixing and stirring of soil, breaks up aggregates and exposes organo-mineral surfaces otherwise inaccessible to soil decomposers. Losses

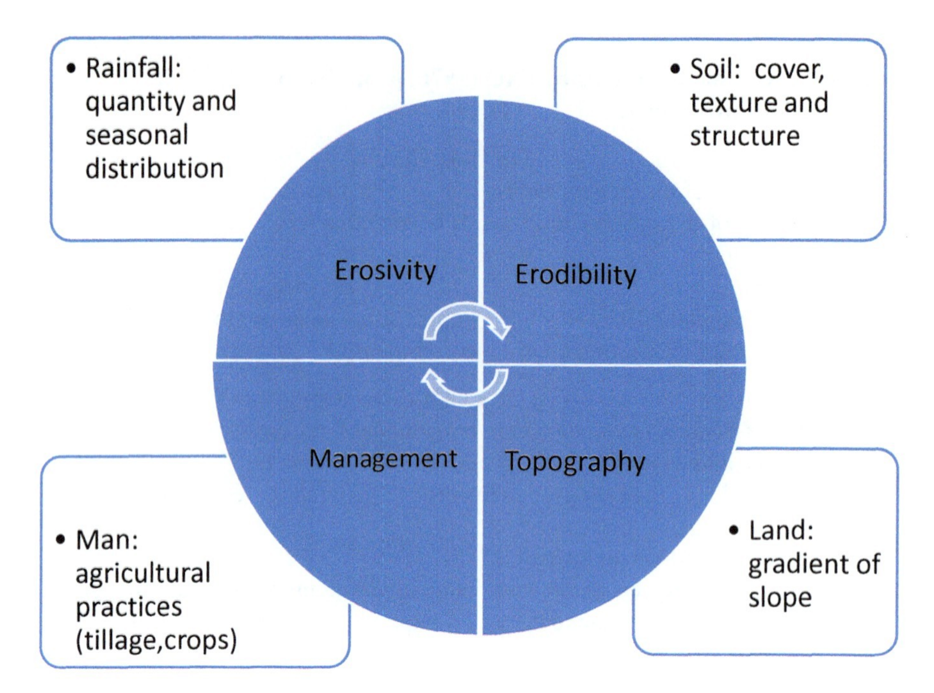

Fig. 6.20 Main factors involved in hydrogeological soil erosion. (After Hudson 1976, p. 169, modified)

of SOC (soil organic carbon) of as much as 50% in surface soils (20 cm) have been observed after cultivation for 30–50 years. Reductions averaged around 30% of the original amount in the top 100-cm (Post and Kwon 2000). Identification of the optimal tillage system requires a global consideration of soil management, rather than an analysis focusing on tillage alone, taking into account soil ecology. Organic residue management, the prevention of compaction, crop rotation and the timing of cultivation must all be considered together, taking also into account their impact on pest populations and on the natural enemies of pests and ecosystem engineers (Roger-Estrade et al. 2010).

6.4 Ecological Intensification Principles and Practices

The concept of *ecological intensification* is now emerging as an appropriate 'frame' to guide the agroecological transition toward a more sustainable agriculture. Its technological rationale is clear in the following definition by Tittonell and Giller (2013):

> Ecological intensification is now understood as a means of increasing agriculture outputs (food, fibre, agro-fuels and environmental services) while reducing the use and the need for external inputs (agrochemicals, fuel, and plastic), capitalising on ecological processes that support and regulate primary productivity in agroecosystem.

Ecological intensification is a knowledge intensive process that proceeds from two essential sources of information (a) in-depth knowledge of biological regulations in ecosystems, and (b) integration of traditional agricultural knowledge held by local farmers (Malèzieux 2012; Tscharntke et al. 2012). The role of science ensures innovation and the role of traditional farming ensures people empowerment, and therefore democracy, in a way that appropriate policy measures follow for cooperative outcomes (Fig. 6.21).

This state of active cooperation is a question of 'recognition' not yet performed by important stakeholders, such as scientists, as Tilman (2000) warns about the ethics of science:

> Science has much to contribute to dialogues on policy and ethics. Although academic institutions seem to value such contributions less than contributions to peer-reviewed journals, this is short sighted. Ultimately, society invests in science because advances in scientific knowledge benefit society. The ethics of science cannot eschew involvement in public discourse. Science must contribute, in an open, unbiased manner, to relevant issues.

More recently, as a consequence of other international assessments, specifically the Millennium Ecosystem Assessment and the Intergovernmental Panel on Climate Change (IPCC), the Intergovernmental Platform on Biodiversity and Ecosystem Services (IPBES) was established in 2012 as an independent intergovernmental body open to all member countries of the United Nations (Diaz et al. 2015). The goal of this new institution is that of "strengthening the science-policy interface for biodiversity and ecosystem services for the conservation and sustainable use of biodiversity, long-term human well-being and sustainable development" (http://www.ipbses.net). The conceptual framework, in the context of IPBES, concerns enhancement of the relationships between people and nature through the four specific functions reported in Fig. 6.22. Integrative conceptual frameworks have the ability to provide (a) a shared language and a common set of relationships and definitions among stakeholders, (b) useful tools in activities requiring interdisciplinary collaboration for making sense of

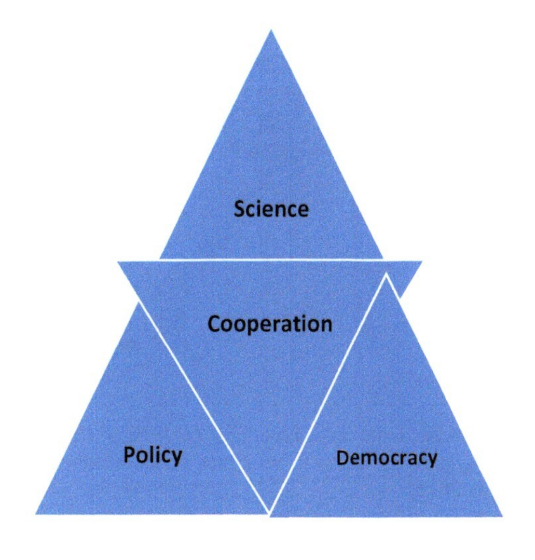

Fig. 6.21 Cooperation-based interaction in an institutional frame for ecological intensification in agriculture

complexity and supporting communication across disciplines and knowledge systems, and between knowledge and policy (Diaz et al. 2015).

IPBES has a fundamental element of ethics of knowledge sharing and construction, as described by Diaz et al. (2015):

> it explicitly embraces different scientific disciplines (natural, social, engineering sciences), as well as diverse stakeholders (the scientific community, governments, international organizations, and civil society at different levels), and their different knowledge systems (western science, indigenous, local and practitioners' knowledge).

Mutual recognition and enrichment among different disciplines and knowledge systems is an essential goal of IPBES in that all these knowledge systems can work in complementary and mutually enriching ways. IPBES emphasizes the need for co-production through the engagement and cooperation of different stakeholders, such as scientists from different disciplines, practitioners and disseminators according to a Multiple Evidence Based (MEB) approach. MEB highlights the complementarity, synergy and cross-fertilization of knowledge systems, rather than the integration of one system into another. Valuation frameworks applicable to diverse socio-cultural contexts would be a major contribution by IPBES to the knowledge-policy interface (Diaz et al. 2015).

"Capitalising on ecological processes that support and regulate primary productivity in agroecosystem"(Tittonell and Giller 2013) is the general principle underlying the transition to a more sustainable agriculture within the frame of ecological intensification. This framing is a kind of 'faith' in the scientific principles governing biological development and potentially applicable to agricultural design and

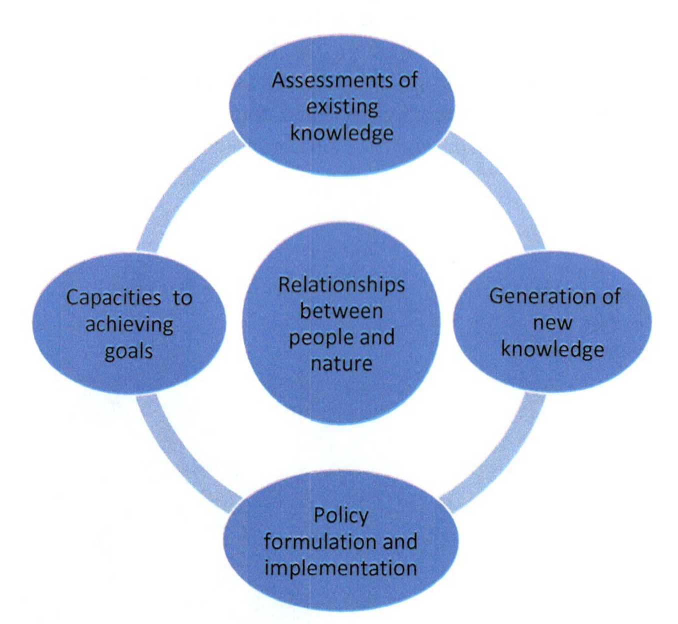

Fig. 6.22 Goal and functions of IPBES (Intergovernmental Platform on Biodiversity and Ecosystem Services) in an ethics of knowledge sharing and construction. (Adapted from Diaz et al. 2015)

management. Ecological intensification unveils a human attitude to trust 'mother nature' as both the 'written book' and the book reader (man) – a unity that needs recognition in its systemic ontology. Scientific knowledge is the result of the application of human mind to nature organisation patterns to discover regularities, while technological tools are the means that feedback on nature in order to adjust those patterns to human necessities. The co-evolutionary process between man and nature needs a governance that today requires human understanding, creativity and symbiotic behaviour. According to Harper (1982), human understanding of nature has progressed just with the development of agricultural science. Through the experiential learning provided by the conditions of cultivation (simplified and controlled environment), the plant ecologist has derived basic information as listed in Box 6.5.

Box 6.5 Basic Information for Ecological Science Derived from Agricultural Science (Harper 1982, Modified)

ON THE MEANINGS OF ADAPTATION

1. Adaptation is the change in a phenotype as a result of some environmental experience [e.g. cultivation].
2. The change is assumed to improve the ability to grow and leave descendants.
3. An adaptation is any aspect of form or behaviour that *at a reasonable guess* is the result of natural selection.
4. An adaptation is any feature of form or behaviour that can account for the ability of an organism to live and do.

ON THE ECOLOGICAL INFORMATION DRAWN FROM CULTIVATION

1. The effects of environmental variables (temperature, radiation, water supply, nutrients) on plant growth;
2. the effect of density on plant and crop performance (intraspecific competition);
3. the effect of two species interactions at a single trophic level (weed-crop interactions, legume-grass interactions, the behaviour of crop mixtures);
4. the effects of host-parasite interactions (crop diseases);
5. the effects of predator-prey interactions (grazing animal and sward; pests and monoculture);
6. the interactions of three species systems, plant-animal-animal (grazing effects on grass-legume pastures and the biological control of weeds by introduced insects);
7. symbiosis (legume-*Rhizobium* associations);
8. community assimilation, radiation capture and productivity;
9. genetic variation within and between populations.

According to Harper (1982), "in agricultural science the workings of simple systems begin to be understood at a level that makes predictive ecology possible". This assumption legitimates the good choice to take successful traditional farming systems as exemplars of sustainable agricultural systems to imitate and reproduce in their structural and functional organisation.

Ecological intensification is an agroecological strategy that relies on the intensification of both human and natural processes for favouring a multifunctional agriculture. Agricultural functions (or *services*, if considered as benefits for man and the biological community) concern:

(a) *social functions* (production of goods, such as food, fibres, etc.; employment and revenue; culture, such as ethical and aesthetical values);
(b) *environmental functions* (preservation of natural resources, such as soil, water, air and biodiversity; efficient use of imported resources, such as energy, materials, selected and manipulated organisms; preservation of rural landscape and its living community).

In relation to these multiple goals, intensification of natural processes means capitalising more on site-specificity, relying on gratuitous local resources such as solar radiation, atmospheric CO_2, nitrogen and rainfall, soil and biodiversity. A fortunate case with agriculture is that solar radiation, atmospheric CO_2 an N_2 are abundant and available, while rainfall is erratic, and soil and biodiversity conditions depends in large part by human planning and management. The challenge is having all necessary institutional support for sustaining the process of farming organisation in view of these goals. The core facts behind a multifunctional agriculture concern *production* and *protection*, thereby a process of ecological intensification aims at balancing the agroecosystem properties of *productivity* and *resilience*. Ecological intensification is at its roots a question of "differentiated land use" (DLU), a concept promoted by Wolfgang Haber (1989, 1990), a distinguished scholar of landscape ecology, who pointed out his theory in the following points:

1. the basic idea of DLU proceeds on the assumptions that each type of land-use (or *ecotope*) unavoidably causes environmental impacts and that their mitigation has inherent limits;
2. the prevailing land-use must be diversified within itself; homogeneity of large tracts of land must be avoided;
3. fine-scale diversity of intensive land-use provides diversification, and thus mitigation, of environmental impacts;
4. spatial and temporal partitioning of land-use will partition impacts at the same time, thus further mitigating impacts;
5. differentiated land-use fosters biological diversity by maintaining spatial heterogeneity and contributes to an important goal of nature conservation.

In practise, the concept of DLU contains the principles for carrying out ecological intensification of agroecosystems at every hierarchical level of organisation, field, cropping system, farming system, and local rural system. A simple model of ecological intensification of a farm with progression from monoculture (farmscape 1) to poly-culture land-use (farmscape 2 and 3) is reported in Fig. 6.23.

According to Haber (1989), the basic and critical spatial unit for both landscape and ecosystem ecology is the site:

> a small section of the earth's surface, determined to a great extent by the local geological situation (lithosphere) within the regional climate [...] The site can be regarded as the spatial representation of its ecosystem, a concept called *ecotope* [...] Each *ecotope* can be considered unique in its assemblage of living and non living components. However, similar *ecotopes* have recurring properties, allowing recognition of *ecotope* types (ecosystem types)-often represented by vegetation units (p. 217).

In this sense, each field planted to a crop can be regarded as an *ecotope*, a site belonging to a farm adapted to a specific vegetation unit constituted by the planted crop. "Like species diversity in an ecosystem, there is *ecotope* (-type) diversity in a landscape; common, frequent, or rare species have their analogy in common, frequent, or rare *ecotope* types" (Haber 1989, p. 218). Similarly, in a farmscape (Huyck and Francis 1995), there is a gradient of crop diversity, where a crop as an *ecotope* can be dominant or rare. DLU in agroecosystems happens in both space and time according to the classical agronomic classification of sole cropping, polyculture and its variants, and rotation. The condition of *sole cropping* happens when only one crop species is planted in a field at a time; the terms *intercropping* and *mixed cropping* apply when two or more crop species are grown together in a field. *Monoculture* is repetitive growing of the same sole crop on the same land and *rotation* is the orderly sequence of crops in the same field (Fig. 6.24).

The choice of the cropping system determines the gradient of agroecosystem biodiversity, both planned and associated, constraints management and affects both agroecosystem productivity and resilience. Biodiversity increases in space with the transition from sole cropping to intercropping systems, and in time with the transition from monoculture to crop rotation systems. Modern industrialised cropping systems are highly mechanised with big and powerful machines that require large, uniform fields for their operations. Therefore, simplified cropping systems like sole

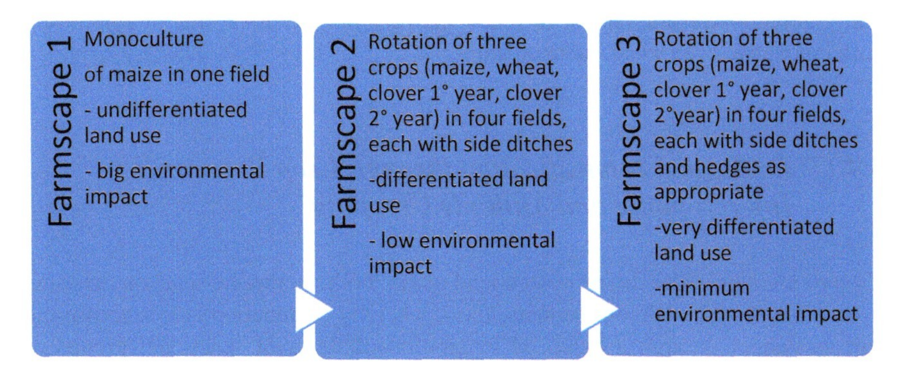

Fig. 6.23 Ecological intensification of a simplified farm model from monoculture to poly-culture land-use

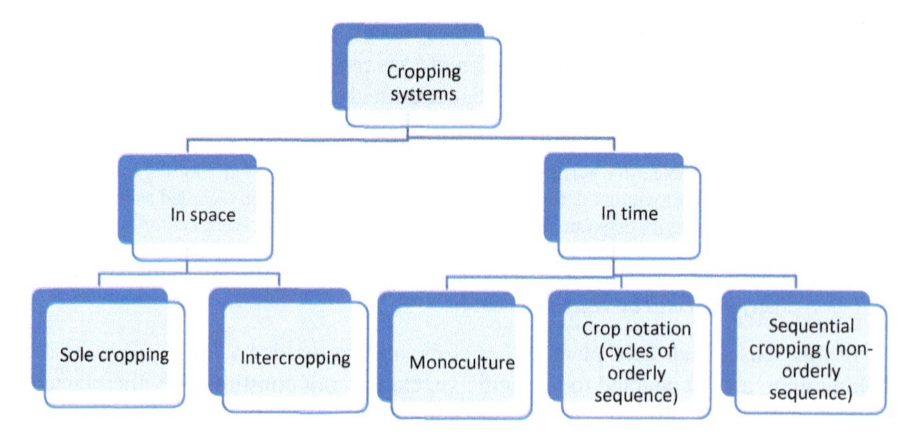

Fig. 6.24 Types of cropping systems in a spatial-temporal dimension. (After Francis 1989, modified)

crops and monocultures prevail. These systems are capital-intensive driven, and therefore, they expand in both poor and rich countries, wherever the capital prospers. The dramatic consequence is that biodiversity shrinks at both local and global levels, affecting not only species but also entire natural ecosystems and their package of support, regulation and cultural services.

Looking at the simple sketch of the cropping system history (Fig. 6.25), it is curious to observe how the entire historical process has been characterised by a progressive 'ecological intensification'. The site-specificity of agricultural settlement has been progressively strengthened with the intensification of crop establishment, first erratic (shifting agriculture), than permanent but extensive with bare fallow, and finally permanent with an intensive biological structure and functioning, involving many different crops and livestock. This process has recently undergone an inversion, the industrial revolution, which has replaced in agriculture the ecological services provided by biological components with mechanical, physical, and chemical means supported by a limited stock of fossil fuels energy. There is now the opportunity to reconsider the benefit of renewing the multifunctional role of agriculture with innovative ecological intensification.

6.4.1 *Ecological Intensification through Crop Rotation, Intercropping and Multiple Cropping*

Crop rotation provides differentiated land use on both a temporal scale on the same field and a spatial scale on the same farm, whereby its function is to stabilise the agroecosystem through a balanced integration of different functional groups of crops. According to the standard frame, rotations have the potential to perform all ecosystem services shown in Fig. 6.26. Innovative solutions of ecological

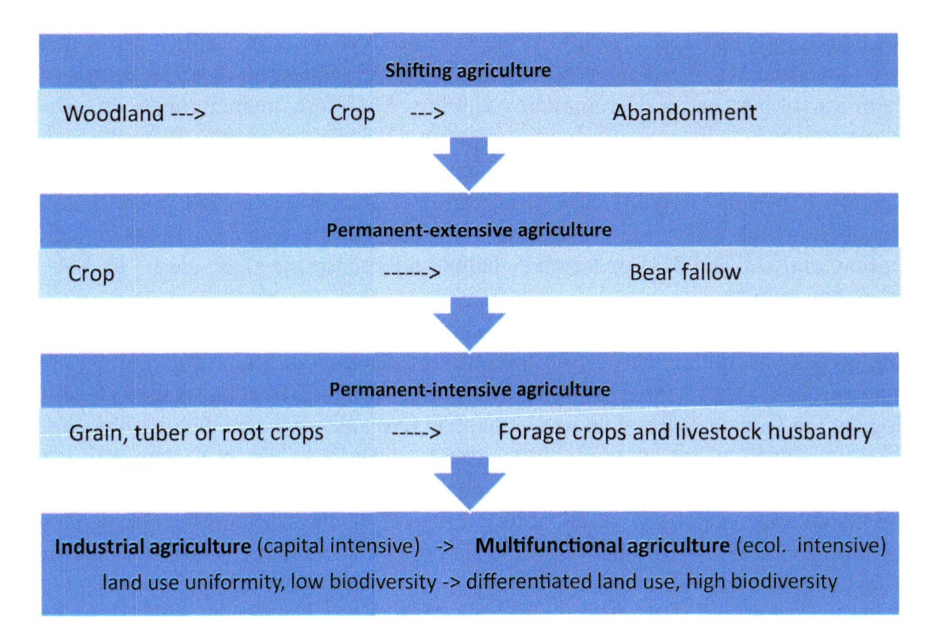

Fig. 6.25 Sketch of cropping system evolution

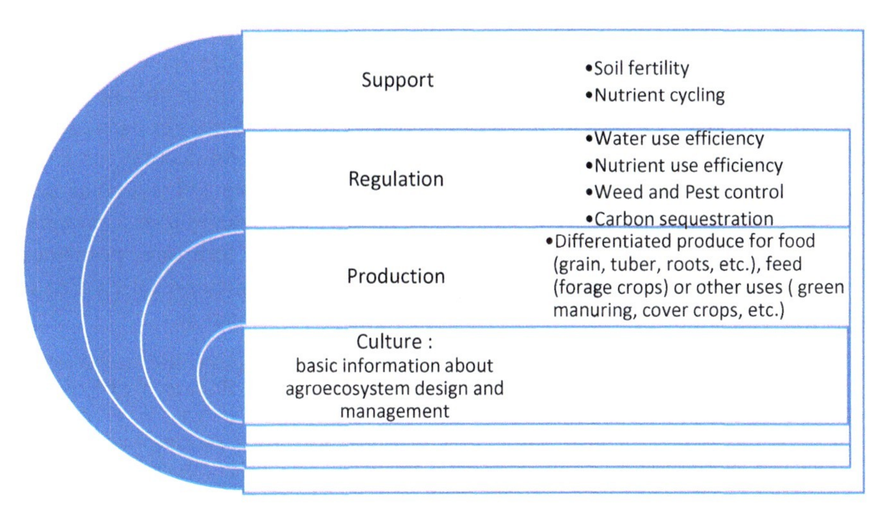

Fig. 6.26 Potential of crop rotations for delivering ecosystem services

intensification based on rotation are available for many rural sites in the world (MacRae et al. 1990; Karlen et al. 1994; Robson et al. 2002), but their implementation depends on an enabling socioeconomic context.

A crop rotation is part of a plan of farm organisation that transcends the farm limits themselves and clashes against economic, cultural and political barriers that constraint farmer's choices. No kind of sustainable agriculture will be operational without an assemblage of crops on the land that sustains soil fertility, water and nutrient efficiency, weed and pest regulation, carbon sequestration, and a culture of both care of land and rural community viability. Therefore, maintaining good environmental and socioeconomic conditions in rural lands is not *an ethical question* but *the ethical question* that underpins sustainable development of the entire human society. In a logic of ecological intensification (Bommarco et al. 2013), we should be able to point out a kind of benchmark for the best crop rotation patterns to help agricultural stakeholders understand, judge and choose the right means to make them operational.

Each cycle of a crop rotation is a planned, orderly assemblage of crops for building up and consuming soil fertility in a way to maintain or increasing it along the cycle sequence. In its most complete operational form, a crop rotation uses all ecological functional groups of an agroecosystem (crop as *producers*, livestock as *consumers* and soil microbial, meso- and macro-faunal populations as *decomposers*) in a complementary, synergistic pattern of interdependence between grazing and detritus chains that ensures self-sustainability. Table 6.5 displays a synopsis of principles of crop rotation and mechanisms of functioning for component crop in view of ecosystem services release.

In a recent paper, Weisshuhn et al. (2017) analyse the potential role and long-term effects of field perennial polycultures (mixtures) in agroecosystems, with the aim of reducing the trade-offs between provisioning and regulating ecosystem services. When poly-annual forage crops are included, crop rotations are a suitable tool for the assessment of the long-term effects, which are not visible at the single-crop level. Perennial polycultures enhance soil fertility, soil protection, climate regulation, pollination, pests and weed control, and landscape aesthetics compared with maize monoculture. However, they score lower for biomass production compared with maize, which confirms the trade-off between provisioning and regulating ecosystem services. Unfortunately, the additional positive ecosystem services provided by perennial polycultures, such as reduced costs for mineral fertilisers, pesticides, and soil tillage, and a significant preceding crop effect that increases the yields of subsequent crops, are not part of a praxis for a complete evaluation:

> The ES [ecosystem services] concept proved applicable for the comprehensive assessment of agroecosystems, and for highlighting trade-offs between services. However, the ES concept needs to be further developed in order to integrate agricultural inputs and the resulting negative externalities of farming practices. These include the mitigation or compensation of soil losses, greenhouse gas emissions, energy consumption, pesticide resistance, biodiversity losses, and the contamination of soil or water bodies, among other practices. Based on the ES evaluation, perennial polycultures as an agroecological strategy in cropping system design have the potential to contribute to the sustainable intensification of agricultural systems. The integration of perennial polycultures as a main component of crop rotation effectively provides diverse ES and opportunities for nature conservation. This calls for an adoption strategy to increase the cultivation of perennial polycultures and transition to more sustainable agricultural systems. To overcome the economic constraints for growing such crops, new incentives must be developed (Weisshuhn et al. 2017).

If each crop is meant as an *ecotope* – the smallest spatial object that has homogeneous properties (Golley 1993, p. 190) – cropping systems are mosaics or

Table 6.5 Crop rotation principles and mechanisms of functioning in a frame of ecosystem services (after Robson et al. 2002, modified)

Principles of crop rotation	Mechanisms of functioning
PRODUCTION	
Suitability to climate and soil	Acclimation
Appropriate sowing time (e.g. autumn or spring)	Seasonality of harvest
Balance of forage and cash crops	Availability of internal resources for feeding and market
SUPPORT	
Rotation of deep and shallow rooting crops	Improvement of soil structure, aeration, water-holding capacity and permeability
Alternating use of crop producing large and small root biomass	More root biomass left in the soil favours more mineralization and nutrient uptake by succeeding crops
Rotation of poly-annual, nitrogen fixing crops with nitrogen demanding crops	Availability of internal sources for enhancing soil organic matter and nitrogen for crop uptake
REGULATION	
Rotation of poly-annual forage crops with annual grain crops	Breaking of weeds, pest and diseases lifecycles; carbon sequestration
Alternating use of autumn and spring sown crops	" " " "
Plantation of cover crops	Maintaining living and dead mulches to compete against weeds and protect for erosion and nutrient leaching
CULTURE	
Integration of biological, physical and chemical knowledge in a socioeconomic context	Experiential learning, social communication, market transactions, good timing of management operations
Animal welfare	Biophilia
Landscape functionality and aesthetics	Personal and social appreciation

assemblages of *ecotopes* that constitute the basic elements of both the farmscape and the landscape, determining their functionality and release of ecosystem services. This kind of landscape-ecology approach has been used recently for characterising rural environment development in a framework of ecosystems services release (Smith et al. 2011; Landis 2017; Rosenzweig and Schipanski 2019). In general, there is consensus on the relevant fact that the current model of conventional agricultural produces high yields but also a loss of biodiversity, ecological functions, and critical ecosystem services in agricultural landscapes (Foley et al. 2005). A key consequence of agricultural industrialisation is landscape simplification, where once heterogeneous landscapes contains increasingly fewer crops and non-crop habitats. Landscape simplification exacerbates biodiversity loss, which leads to reduction in ecosystem services on which agriculture depends (Landis 2017). There is the necessity to provide changes toward ecological intensification alternatives. A very significant case of ecological intensification of rural land severely constrained by climatic conditions is that reported by Rosenzweig and Schipanski (2019), who used high-resolution satellite data to quantify dryland cropping patterns from 2008 to 2016 in the US High Plains. Their findings show how the High Plains witnessed a profound shift in cropping system patterns, as the historically dominant wheat fallow system was replaced by more intensified rotations (continuous cropping). Four crops (corn, sorghum, millet, and field peas) increased by nearly 0.4 million hectares, from 17% of dry cropland in 2008 to 26% by 2016. The authors estimate that, from 2008 to 2016, these patterns resulted in a 0.53 Tg increase (10%) in annual grain, an increase in annual net farm operating income, substantial reductions in herbicide use, and an increase in C sequestration that corresponds to greenhouse gas (reduction of 0.32 million metric tons of CO_2 equivalents per year). They conclude pointing out that cropping system intensification represents a rare win-win strategy for environment and economy. Their results suggest that at a time when the majority of counties in the High Plains are experiencing depopulation, cropping intensification could potentially reverse these trends by enabling profitability with smaller land areas. These considerations are valid for other regions undergoing similar transformations, and for policy that can support these shifts toward more sustainable cropping systems.

The strategy of ecological intensification of cropping systems by shifting from a crop-fallow to a continuous cropping scheme is potentially applicable, for instance, to all regions in the world with a Mediterranean-like climate, usually affected by long periods of summer draught. This change is technically feasible and does not depend exclusively on overcoming climatic restraints, but rather on the rising of a favourable 'political' climate, the only one capable of investing more in smaller, more labour-intensive, land property than in larger, more capital-intensive one. Italy experienced a historical, similar change rather recently, soon after the second world-war, when a first national law in 1948 permitted the development of a smallholder property from a landlord property, which was purchased by the State and passed to agricultural workers or family farmers with low interest mortgages. From 1948 to 1968, about 2 millions of ha, most part in Central and Southern Italy, undergone this procedure of compulsory expropriation (especially from size larger than 2500 ha, or between 1000 and 2500 ha). Redistribution of land, most of which of scanty fertility, was in small farm size (10–20 ha) for each family farmer, according to family workforce and available arable land.

Rotations that aim at sustainability need among their crop components some leguminous species that improve soil fertility through their symbiosis with *Rhizobium* spp., a bacterium able to fix atmospheric nitrogen. After a legume cultivation, its residues on and within the soil release through microbial mineralisation nitrogen compounds and other minerals that can benefit the nutrient uptake of the succeeding crops, usually nitrogen demanding, such as cereal crops. Legume species are very common in Mediterranean areas and widespread in any kind of soil, both as annual and poly-annual forms. Considering the adaptation to Mediterranean environment, very common are the so called annual self-reseeding legumes, such as many annual species of *Trifolium* and *Medicago*, that are used in short rotations with cereal crops in farming system with grazing sheep or cattle (*ley farming*[4]) (Webber et al. 1976; Caporali and Campiglia 2001). Ley farming is well-known in South Australia for having improved farming systems with the full integration of animal husbandry within cropping systems based on the pasture phase of the rotation. Ley farming modulates on the input of annual rainfall of the site, which decreases in South Australia with the distance from the cost (Box 6.6). Annual *Medicago* spp. are more adapted than *Trifolium subterraneum* to drier environment, in that they possess more *hard* seeds, i.e. seeds with coats resistant to the entry of water so preventing germination. Hard seeds remain non-germinated in the soil for the first year, when a cereal crop can grow; after the cereal crop, hard seeds germinate spontaneously, regenerating a phase of medic pasture of 1 or more years, according to the edaphic conditions. Unlike medics, subterranean clover sets fewer hard seeds and regenerates best in the year following seed set. Therefore, it is more convenient to prolong the pasture phase of rotation to 2–3 years, according to edaphic conditions and the necessity to accumulate more nitrogen and organic matter into the soil. Considering its vast range of ecosystem services, spanning from support to culture, ley farming is a brilliant solution of continuous cropping, or ecological intensification, that has largely improved the discontinuous cereal-fallow cropping system relying merely on improved knowledge application of natural resources (annual self-reseeding legumes) imported from the original Euro-African Mediterranean area. Australian researches have largely contributed to improvement of seed production of annual legumes by breeding new cultivars adapted for use in all Mediterranean areas of the world. This is the end of an instructing history starting with the first settlers arrived in South Australia in 1836 to initiate their cultivation:

> Cropping damaged soil structure and soon wind and water erosion led to violent dust storms in summer and deep gutters gouged into hillsides in winter. We now know that the first farmers exploited their resources: they did not realize that the environment had to be carefully maintained [...] Farmers found that their crops needed phosphorous: they applied superphosphate and yield increased [...]They found that pasture legumes (clovers and medics) improved soil structure and increased the amount of nitrogen in the soil. The combination superphosphate and pasture legumes raised cereal yields and provided feed for increasing numbers of livestock (Webber et al. 1976, p. 12).

Farmers' attitudes and social networking are crucial for the innovation and conservation of biodiversity in agricultural landscapes (Jackson et al. 2010). Multiple drivers affect the behaviour and cooperation of social actors (stakeholders and decision-makers). Research at the household and community-level has clearly demonstrated the complexity and value of participatory frameworks as, for example, the ley farming seed system in South Australia.

In Mediterranean environments where annual precipitations are more than 500 mm and less than 900 mm, like in most part of the Southern costs of Europe, other kind of legumes, both grain and forage legumes, can contribute to enhancement of rotations flexibility and performances according to different edaphic

---

[4] Ley farming is an integrated system of cereal and livestock production (Webber et al. 1976).

Box 6.6 Rotations in Ley Farming Systems of South Australia and Their Associated Ecosystem Services (Adapted from Webber et al. 1976)

RAIN FALL: 250–500 mm

Rotations in annual medic area (*Medicago* spp cvs)

R A I N F A L L: 250-500 mm

Rotations in annual medic area (*Medicago* spp cvs)

more fertile soil	less fertile soil
1 year of cereal (wheat or barley) 1 year of medic pasture	1 year of cereal (wheat or barley) 2-3 years of medic pasture

R A I N F A L L : > 500 mm

Rotations in clover area *(Trifolium subterraneum* cvs)

more fertile soil	less fertile soil
2 years of grain crops (wheat and barley) 2 years of clover pasture	3 years of grain crops (wheat, lupin, barley) 3 years of cloverpasture

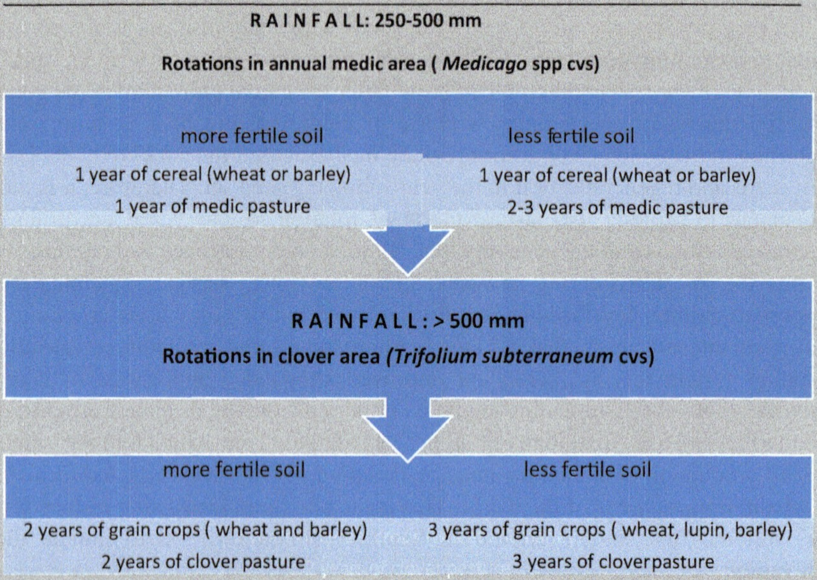

Ecosystem services provided

Support: increased soil fertility (more organic matter and more N-fixation), and improved soil structure.

Provision: increased herbage growth, better quality livestock production and more stable farm income; increased cereal crop yields and greater cropping flexibility.

Regulation: better control of soil erosion, particularly when combined with contour banking; tillage and N-fertilizers reduction, and more C sequestration.

Culture: enhancement of personal and institutional knowledge, skill and competence.

conditions. For instance, in sandy and acid soils of volcanic origin, like those of Central Italy, a kind of ley farming rotation between 1 year of wheat and 1 year of crimson clover (*Trifolium incarnatum*) as pasture for sheep is usually established

Fig. 6.27 Sheep grazing on a winter pasture of crimson clover with background of cork trees. Tuscania (VT), Central Italy

(Fig. 6.27). If soil changes to clayey or basic pH, Sweetvetch (*Hedysarum coronarium*) and Sainfoin (*Onobrychis viciaefolia*), respectively, replace crimson clover in ley farming rotations. In the presence of deep soils, Lucerne (*Medicago sativa*) is a common poly-annual legume for longer rotations, since it is the most productive fodder crop to be stored as hay. In this kind of rotation, 3–4 years of annual crops alternate with a 3 years Lucerne cultivation, for example with this sequence of crops: sunflower, wheat, Lucerne (3 years), wheat, oats. In the 3 years of Lucerne cultivation, the soil is at rest (unploughed) and therefore soil microorganisms provide for humification rather than for mineralisation thus rebuilding higher levels of organic matter and consequently of soil fertility for the benefit of the annual crops that follow. What is really astonishing with Lucerne cultivation in a Mediterranean environment (Fig. 6.28) is its capacity to persist vital and 'greening' during summer time, when arable land appears 'yellowish' after cereal harvest with only stubble cover on the soil. The great capability of Lucerne to penetrate a few meters of soil profile with its deep taproot in search of water accounts for its success in providing biomass for feeding and balancing the summer deficit of forage production due to prolonged draught period.

These farming systems are example of efficient ecological intensification of integrated cereal and livestock production based on local, well-adapted legume resources that contribute to an expansion of the range of ecosystems services in any kind of rural area. In Italy, given the fortunate situation of historical remains near

Fig. 6.28 Lucerne stand with wheat harvesting on background. Tuscania (VT), Central Italy

any town in the countryside, smallholder farms can provide some more direct cultural services to tourists as facilities for direct sale of their own agricultural products, room and board, restoration, and teaching farms.

Crop rotation in Mediterranean environment appears historically well documented in Latin literature by Cato, Varro, Virgil, etc. since the second century B.C., which is highly prized from a literary, scientific and cultural point of view. Later, the first work of great scientific and technical value on innovative crop rotations was about the introduction of a biennial clover fodder crop (*Trifolium pratense*) in a sequence of grain cereals, which dates back to the Italian Renaissance. It was Camillo Tarello's work "Ricordo di Agricoltura" (Memory of Agriculture), in which the Italian agronomist presented his innovative cropping system to the Senate of the Republic of Venice in 1565. He proposed to "teach the method of double income and be left over with a surplus of two thirds of grain fodder through the simple introduction of Red Clover (*Trifolium pratense*) rotated with cereal crops". His discussion made continual reference to Latin and medieval authors and represents therefore, a cultural and scientific outline of agriculture evolution based on ancient tradition, which according to Socrates is mother of all the arts. Being aware of the importance of his project for a new applied biotechnology, Tarello negotiated with the Senate of the Republic of Venice for a long time to receive compensation (a type of patent royalty) from farmers who were interested in applying this practice on their farms.

Intercropping is also one of the pillars of traditional agriculture in all continents and it is still widely practiced in Latin America, Asia and Africa as a means of increasing crop production per unit land area with limited capital investment and minimal risk of total crop failure (Francis 1989; Vandermeer 1989; Liebman and Dyck 1993).

> An enormous variety of intercropping systems exists, reflecting the range of crops and management practices farmers throughout the world use to meet their requirements for food, fiber, medicine, fuel, building materials, forage, and cash. Intercropping systems may involve mixtures of annual crops with other annuals, annuals with perennials, or perennials with perennials (Liebman and Dyck 1993).

Intercropping involves spatial diversification of cropping systems within the field of cultivation, and therefore, a high degree of interaction between the plants grown occurs for both planned and associated biodiversity. Intercropping has been dismissed in industrialised agriculture as being incompatible with 'modern' mechanisation; however, this could be only a bias masking a lack of attention from agricultural engineers, not an inherent incompatibility of intercropping with machinery (Liebman and Dyck 1993).

In his finest piece of work – 'Elementi D' Agricoltura' (Elements of Agriculture) – the distinguished Italian agronomist Filippo Re (1806) had already shown his appreciation for intercropping as follows:

> The tradition of dividing the country-side into many rectangular pieces (fields) and the rows of trees in the middle from which the vine hangs and joins like festoons in undefined symmetry shapes the gardens in our country-side [...] It seems to me to be a privilege granted to most of the Italian soil, considering its depth which is suitable for accommodating long roots and feeding them, and feeding both trees and grass at the same time (p. 158).

Intercropping is in fact the practice which consists of cultivating more than one crop in the same area at the same time as inferred by Re's last statement. This integrated cropping system between herbaceous and tree crops, which was the norm of cultivation in the 1800s and the first half of the 1900s, was swept away by the industrialisation process. Fields were enlarged; hedges destroyed; crops separated in fields for fruit trees, grain crops, and vegetables crops; livestock previously distributed in small stalls on a large territory was progressively confined in larger tying stalls with hundreds of heads. Landscape abruptly changed and lost most of his inspiring quality. Not only crops, but also an entire kind of civilisation changed and its values almost entirely forgotten. The greater diversity of structural components (fields, crops, livestock species, hedges and woodland among fields) corresponded to greater functional integration, major autonomy of the system, lesser need for external inputs of energy-matter, less release of outputs affecting the environment. The processes for the operation of the system were more cyclic and less linear, the internal system being intended to guarantee a major stability of production over time. The input of information was enough to ensure to agroecosystems a cybernetic state, or state of self-control and sustainable development. Diversity as a whole produces an output of information that an external observer can appreciate. A context of farms or group of farms analysed from a good point of observation provides a global framework of apprehension that we define as *landscape*. This framework is able to arouse both elements of cognitive affection for aesthetic appraisal and interpretative cues. The landscape is considered a cultural good (Antrop 2000) since it contains signs on the territory referring to the organising mind as well as to a number of functions occurring in it. The landscape evolves continuously due to internal and external factors acting concomitantly because of the direct action of farmers and, in general, because of all those individuals and institutions that develop ideas for the change of the rural infrastructures at the local level. The external factors are above all indirect. The trends of international, national and regional policy may sensitively affect landscape features in the long run. As the industrialization paradigm caught on in Western countries, the agricultural landscape has progressively changed and many structural and functional features typical of the traditional agro-ecosystem, or mixed agriculture, have been erased from the territory. As industrialized agriculture caught on, farms became bigger, more specialised with a shift in favour of rotations devoted to economically more profitable annual cash crops and subsidies envisaged by the agricultural policy. As a result, the agricultural landscape moved from heterogeneity (diversified crops) to uniformity (sole cropping and monoculture). This structural simplification led to a sensitive aesthetic levelling out and a remarkable loss of sustainability due to the increased opening of bio-geochemical cycles within the agro-ecosystem, the increase in negative environmental impact resulting from eutrophication, water pollution and the subsequent effects on food chains and, consequently, on the health of both the agro-ecosystem and man. Figure 6.29 shows an example of landscape evolution from mixed to industrialized farming patterns in one of the most typical agricultural areas of Italy, the Chianti area, which is internationally recognised for its valuable quality in terms of landscape, products (wine, oil, cheese) and production processes.

After some decades of experiential learning in industrialised agriculture, we have learnt some lessons. Sole crops result from the need of specialization for industrializing agriculture, for homogenizing and simplifying the cropping systems according to work-time and characteristics of big machinery, and technical itineraries, such as sowing or planting time, harvesting, manuring, weeding and control against plant eating insects and plant pathogens. Due to their planned constitution (one species in the field), sole crops are an easy prey to infestations. The availability of open ecological niches inherent in their field arrangement (empty spaces between rows available to weed colonisation, availability of homogeneous biomass for pests and predators) paradoxically facilitates infestations, rendering the interventions of pest control by means of pesticide-based treatments indispensable.

When compared to sole crops, well-balanced intercropping systems are able to more effectively use native resources for productivity purposes and ensure increased protection against weeds, plant pathogens and pests (Table 6.6).

Each specific combination of intercropping system can provide many or some of the ecosystem services listed in Table 6.6 and each site of application requires intercropping systems tailored on its environmental quality. Theoretically, in absence of socioeconomic constraints, the best technical solutions for each site may be projected, investigated and applied. Unfortunately, the current societal organisation

Fig. 6.29 Evolution from traditional to industrialised farming systems in the Chianti landscape (Tuscany, Italy). Larger, up and down tilled vine fields replace old terraces with intercropped olive and vine that remain isolated around old villages and villas

does not operate in favour of this rationale, and constitutes a barrier from an ethical perspective that cares for the common good. Among the most common intercropping systems, it is possible to distinguish

(a) *mixed intercropping* providing for the simultaneous cultivation of two or more species on the same field with a random adjustment of the individuals which is usually obtained through the broadcast sowing of a seed mixture. It is the typical case of fodder crops which have more species and which also accounts for one of the most common cases of intercropping observed in today's agricultural practice;
(b) *intercropping in rows* which provides for the cultivation of two or more crops simultaneously on the same field where one or more crops are planted in rows. This scheme can be adopted both for tree crops, as in the case of one or more rows of vine alternated with one or more rows of olive, and for tree crops combined with herbaceous crops. The last case typically concerns mixed crop agriculture where, for example, rows of vine mixed with elm-tree co-exist on the edges of the same field, while herbaceous crops rotate in the field;
(c) *strip intercropping* which provides for the cultivation of two or more crops on the same field in different strips the width of which is such to allow for both tillage and individual treatments but also for the agronomical interaction between the two crops which are usually herbaceous crops;

Table 6.6 Range of ecosystem services provided by intercropping systems

Type of ecosystem service	Mechanism of functioning
SUPPORT	
Physical support	Intercrops interact mechanically
Chemical support	Intercrops interact nutritionally
Biological support	Habitat and niche differentiation
PROVISION	
Increased yield	Enhanced use of resources
Increased yield stability	Broader tolerance to environmental changes
Reduced production costs	Fewer agrochemical treatments
REGULATION	
Reduced soil erosion	More continuous soil plant cover
Nutrient conservation	More diversified exploration of the soil profile
C-sequestration	Reduced tillage
Biological control of weeds, pests and diseases	Reduced susceptibility to infestations due to crop heterogeneity and physical barriers to diffusion
CULTURE	
Reliance on biophilia hypothesis	Complementarity in the use of native resources
Landscape improvement	Agroecosystem as a community of life

(d) *relay intercropping* providing for the cultivation of two or more crops simultaneously on the same field during part of the growth cycle of each of them. For this situation to be realized, a second crop should be sown within an existing crop at a given time, which may occur by re-sowing a herbaceous crop either within another herbaceous crop or within a tree crop with one of the modes previously envisaged in points (a), (b) and (c).

In short, the scientific reason which provides an explanation for concrete advantages in the systems of intercropping identifies itself with the complementary use of native resources which takes place through different structural and functional features of associated crops and meets the crop requirements (in particular in terms of water and nutrients) at different times. Two examples concerning the use of self-reseeding leguminous crops, the one related to intercropping with tree crops and the other to intercropping with herbaceous crops will be analysed in order to illustrate this general indication.

Self-reseeding annual legumes are actually Mediterranean biological resources, i.e. species that have perfectly adapted to the Mediterranean climatic environment. In such a climate, the favourable period for plant growth is from autumn to spring (humid and relatively mild period) whereas the unfavourable period is from spring to autumn (dry and hot period). In this condition, an annual plant germinates when the first autumn rain falls, grows until spring, flowers and bears fruit before summer. In the case of self-reseeding legumes like subterranean clover, the seed is actually re-sown by the same parent plant that buries *capitula*, each of which has four seeds. The seed spends the adverse period in the soil and does not germinate until the subsequent autumn rainfall. A more or less conspicuous part of the seed is not likely to germinate (hard seeds), above all in the case of cultivars which are not genetically improved to reduce dormancy, and therefore it can remain in the soil for decades in the state of latent life. This typical seasonal cycle of self-reseeding legumes allows for their versatile use within both tree and herbaceous crops.

Specialised tree crops in Italy, such as vine and fruit crops in general, develop their phases of flowering and fruitification during spring and summer time, while self-reseeding legumes close their cycle with setting seed in spring. This is a fortunate case for planning an intercropping system with tree crops and self-reseeding legumes, exploiting the complementarity in time of their development cycles. In practise, under-sowing subclover early in September provides a cover crop for a tree crop from autumn to spring, useful for carbon sequestration, nitrogen fixation, water-cycle regulation, weed control and soil protection against erosion. The subclover biomass serves potentially as pasture or cutting as appropriate. The intercropped system is *productive* all year-round, in that during winter is the herbaceous cover that operates photosynthesis, while during summer is the tree cover that does it. This is a kind off win-win strategy for both production and protection based on ecological intensification (Caporali 2010).

If the ecological intensification aim is to enhance the content of soil organic matter while providing nutrients to the tree crop, many vineyard managers grow a winter annual legume cover crop, such as bean (Fig. 6.30), which is sown mid-row in autumn, and in spring mulched off and rotary-hoed into the ground as a green manure. The effects of a mid-row cover crop, as reported by White (2009), are the following:

(a) protecting the soil from raindrop impact, which initiates structural break-down and consequent erosion;
(b) improving soil structure, infiltration, soil strength, and drainage;
(c) building up soil organic matter and enhancing biological activity,
(d) providing a habitat for beneficial insects and potential predators of grapevine pests;
(e) suppressing weeds.

Cover-crop mixtures composed of legume and non-legume species could provide both catch crop and green manure services simultaneously by combining advantages of both sole crop species. The green manure service of cover crop mixtures generally lies between those of non-legume sole crops and legume sole crops because mixtures have lower C:N ratios and acquire more N than non-legume sole crops (Couëdel et al. 2018).

As to the ecological intensification of conventional crop rotations in Central Italy, the most common cash crop rotation is sunflower-wheat, where a long period of bare fallow occurs between the wheat harvest (June) and the sunflower sowing (March), and a short period of tilled fallow between the harvest of sunflower (August) and the sowing od wheat (November). A self-reseeding legume as subclover can ecologically intensify it performing different important roles as shown in the following scheme:

Fig. 6.30 Vineyards with a faba bean cover crop before green manuring. Colle Val d'Elsa (Central Italy)

Conventional rotation

Main crop--Main crop-----------------------------------Main crop

(wheat) ---bare fallow(8 months)---------------(sunflower)---bare fallow(2 months)-------(wheat)

Ecologically intensified rotation

Main intercrop	-------------------------------------	(Main crop)---bare fallow (2 months)---	Main intercrop
(wheat +	subclover cover crop +	(sunflower)	(wheat +
subclover)	green manuring or		subclover)
	sod seeding		

Details on patterns of sowing, management and harvesting of this intercrop system are in Caporali and Campiglia (2001). The sowing of a wheat/subterranean clover intercropping can be effected through a modified grain drill where a double row of wheat (two rows at a distance of 15 cm) alternates with a strip of clover (30 cm wide). After harvesting wheat, the clover self-reseeds and re-generates after the first rain in autumn as a cover crop that can finally be used as green manuring for the subsequent sunflower crop. The subterranean clover has multiple functions in this crop system, as intercrop in wheat, cover crop after harvesting wheat and green manuring crop before sowing sunflower. In more stressed climatic conditions, the sunflower crop can be replaced by crops which are more resistant to drought (for example, sorghum) or, in well-irrigated conditions, by vegetables (for example, tomato).

In general, cropping systems in the form of crop rotations, intercropping systems, and multiple cropping systems (cover crops, living and dead mulches, catch crops and green manuring) may contribute enormously to the maintenance of planned agrobiodiversity in farming systems using modern varieties and cultivars. Globally, however, the highest levels of agrobiodiversity still occur in farm ecosystems and the surrounding landscapes of smallholder and indigenous populations, where they maintain a heritage of local varieties, landraces and ecotypes that are the current result of the domestication process (Zimmerer and de Haan 2017).

Genetic amelioration of both plants grown and animals raised is as old as agriculture, since it has been historically realized by farmers through mass selection, that is choosing the seeds which prove to be more suitable (bigger and well-shaped) from the cultivated fields in order to be re-sown, and choosing the best animals from livestock for reproduction. In this way farmers have selected a huge variety of local species (*landraces*) for each crop or animal species raised all over the world. Such a selection took thousands of years through hundreds of generations of farmers and led to a real explosion of biodiversity. Local species have the great advantage of being adapted to the local situations of cultivation and animal breeding as products of a co-evolutionary process involving the biophysical and the cultural environment at the same time, a genuine expression of man/nature co-evolution. Becoming adapted to local conditions means reaching the top production in relation to environmentally available resources and continuously producing genotypes that have become resistant to any local adversity. Furthermore, the availability of many genotypes implies a stable production and therefore a sustainable agricultural activity. Just to provide a numerical indication, before the advent of the "green revolution" there were 30,000 local species of rice in India. Only 40–50 will be left in the near future (FAO 1995). The drastic impoverishment of this genetic and phenotypic diversity that has been accumulated through local agricultural activity is a further expression of biodiversity erosion, which is added to the loss of entire ecosystems and of the species living in them. All that due to the causes of uncontrolled human activity. The main cause for the eradication of local species was the adoption of the industrialized agriculture paradigm in western countries, a model to export later to each part of the world. In this paradigm, the genetic amelioration carried out in the laboratories of seed-companies provides for the constitution of genotypes that give the maximum productivity for crop environment presenting the minimum limiting factors. Consequently, crop environment modification became the rule in such a way as to add what was missing through fertilisers, irrigation water, and through chemical control for the eradication or prevention of pests. This style of agriculture produced a high increase in unit yields per hectare of cultivated soil and per head of livestock reared. However, no comment arouse, until recently, both for the aspects concerning the amount of energy-matter inputs entered to obtain these high outputs in terms of yield and for the aspects of environmental impact concerning pollution, eutrophication and the loss of biological and social balances. In this regard, Gonzales (2000) refers to the *culture of the commercial seed*, which opposes and tries to eradicate the *culture of the native seed*. The commercial seed, most of which is under the control of trans-national (multinational) organizations, results from scientific manipulation in the laboratory and in experimental plots and only fosters the genetic features that are significant for production-related aspects. The native seed is a result of a wider and involved process of breeding on the spot that directly concerns the local population, its social organization, its crop and religious systems and its food preferences, where natural and cultural biodiversity are united. Currently, about 1.4 billion people in the world live in rural communities that are widely self-sufficient for the seed they need and therefore, still depend on this culture of the native seed. Today this circumstance is recognised as an opportunity by international organisations such as IPGRI (International Plant Genetic Resource Institute) which are devoted to the preservation of biodiversity *in situ*, that is within the same farms which make use of the seed instead of germoplasm banks (seed conservation *ex situ*, in warehouses and at low temperature). The need for the preservation of local species became an international priority and was ratified by more than 150 countries which signed the Convention on Biodiversity in Rio (1992) through which they committed themselves to carry out a policy targeted to recover all endangered local species. The possibility of obtaining, in the future, some genetic material useful to face any event related to environmental changes, of natural or anthropic origin, that is the possibility to guarantee long-term agriculture sustainability, also depends on the preservation of local species *in situ*. That allows for the accomplishment of the evolutionary process of the genetic material directly in the field (whereas this is not obviously possible in the case of *ex situ* conservation). A coordinated global partnership of researchers in eight countries and on five continents has measured the amount and distribution of genetic diversity present in farmers' fields of 27 crop species (Jarvis et al. 2008). The findings show that considerable crop genetic diversity continues to be maintained on farm, in the form of traditional crop varieties. This suggests that diversity may persist locally as an insurance to meet future environmental changes or social and economic needs.

Unfortunately, the highly productive varieties of the artificially enriched and chemically protected crop environment of conventional farming are not suitable for low-input environments, where it is necessary to only rely on the native resources of the system that also follow the natural rhythms of the release of nutrients (Janssens et al. 1990). For instance, the grain cereal ideotype for high-input regime has a tendency to be smaller-sized, with a lower leaf system and a reduced and less branched root system. Its fruit stalk bears only one ear, which accounts for more than 50% of dry weight of the whole resulting biomass and therefore shows a high *harvest index*, i.e. a high ratio of biomass which is removed through harvest as compared to the biomass remaining in the field with crop residues. Such an ideotype proves to be useful in the presence of sole crops made up of very close individuals which are not so competitive, and which need water and nutrients available in their limited root expansion profile to grow and which are not able to compete with weeds that need chemical control.

On the contrary, the low-input ideotype should be perhaps much closer to ancestral features that we can still observe in local species, such as (a) high size and high leaf area index; (b) high tillering capacity with many fertile stalks; (c) extended and branched root system able to explore deeper soil horizons, get water and nutrients more autonomously. Consequently, this ideotype should be more resistant to adverse conditions like drought and more competitive towards weeds. These vigorous individuals should need a lower density of investment (fewer plants per unit area) and therefore a lower sowing density, which also means greater economical saving. Moreover, a research gap to bridge is breeding for more efficient crop polycultures (Milla et al. 2017). Because breeding is mostly carried out in presence of high inputs, it has systematically missed the opportunity to exploit genetic differences at low levels of inputs, and very few breeders select in sub-optimal or stress conditions. It is suggested that the best avenue to a sustainable increase of agricultural production in low-input agricultural systems is through locally based breeding programs (Ceccarelli 1996).

This comparison between contrasting crop ideotypes indicates that the paradigm of industrialized agriculture has been useful especially to those who have taken economical advantage by selling seeds, machinery, fertilisers and pesticides. Farmers can render agriculture productive and more environmentally friendly through a strategy that is different from the strategy adopted by conventional agriculture. Local species represent a great opportunity to be re-examined for this purpose and therefore the regional, national, international action supporting *in situ* preservation and the enhancement of their genotypes must be promoted through research which is to be radically redefined for its final multiple objectives, such as resistance and yield stability rather than maximum yield.

6.4.2 Ecological Intensification: The Role of Hedgerows

"One of the most prominent features in the moist temperate landscape is a hedgerow bordering a pasture or cultivated field. Agriculture and hedgerows apparently developed together, and today coexist over about 10% of our planet's land surface" (Forman and Baudry 1984). Hedgerows constitute significant components of agroecosystems and their existence depends exclusively on man willingness to maintain or regenerate them. Their origin dates back to human need of ordering landscape with physical barriers delimiting individual properties from communal land uses. According to Caborn (1971), their introduction "marks the transformation from a collective, subsistence style of farming to commercial husbandry yielding a marketable surplus". Forman and Godron (1986) recognise three predominant types of hedgerow origins: *planted*, *spontaneous*, and *remnant*. Usually, shrubs or trees are planted in a single row and tend to have equal-aged dominants, relative homogeneity in vertical and horizontal structure, and rather low species diversity. In spontaneous hedgerows, trees and shrubs grow along fences, stone walls, or ditches, from seeds dispersed by animals and the wind. Forest remnant hedgerows result from the process of forest-clearing, such as a row of trees and shrubs commonly left along a property line (Forman and Baudry 1984). Their persistence in a farm, region or landscape is dependent on the recognition of their potential for ecosystem services release as presented in Fig. 6.31.

The consolidation of the sustainable agriculture paradigm depends on the acceptance of making a balance of trade-offs among competing economic and environmental goals. Hedgerows occupy a share of potentially productive land from an agricultural perspective that is set-aside for not marketable protection purposes. A strict economic evaluation that considers only provision services while neglecting support, regulation and cultural services will result in policy of hedgerow dismantle. For instance, this happened rather recently in Great Britain, when hedges have been removed at a rate of 4500 miles per year during the period 1946–70, with an annual loss of 0.73% of the total (Caborn 1971). This happened disregarding the fact that (a) hedgerows were a heritage of Enclosure Acts involving over seven three million acres and (b) between 1750 and 1850 over million km of new hedges were planted (Thomas 2003). The great cultural value of hedgerows as exemplars of nature/nurture co-productions has been object of specific interview investigation through the collection and exploration of different stakeholder perspectives that have implications for managing future hedged landscapes:

> Hedgerows were viewed by all the participants as part of the English cultural landscape. They were not just a means for conserving biodiversity in the countryside but part of England's history and national identity. They were appreciated for providing signs of the changing seasons; for the way they break up the landscape; for their sense of mystery and intimacy; for their connections with the past and childhood memories; and for their contribution to a sense of place (Oreszczyn and Lane 2000).

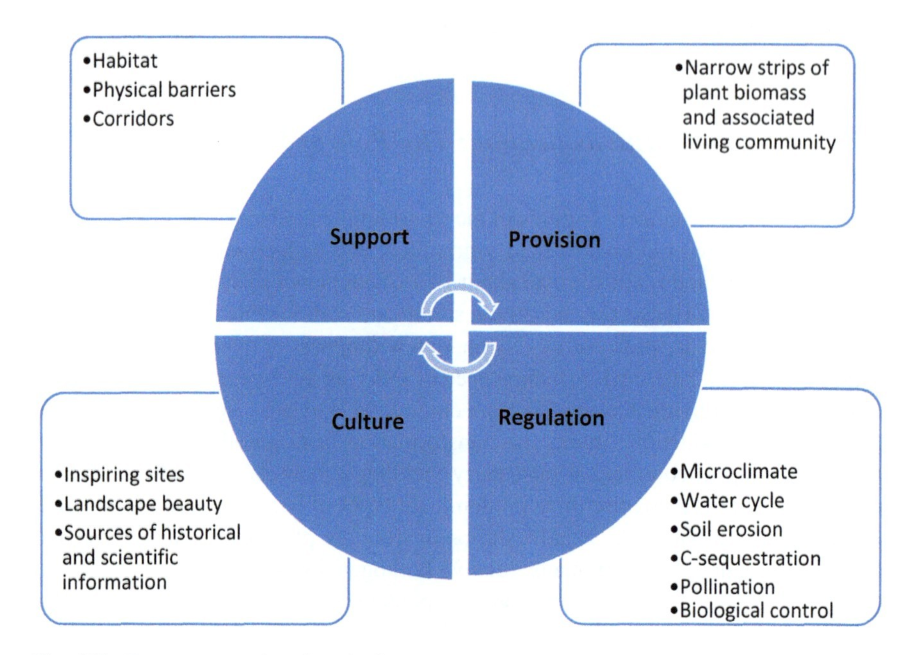

Fig. 6.31 Ecosystem services from hedgerows

Moreover, that analysis revealed a wide divide between expert and lay persons' appreciation of landscape. Participation of all stakeholder perspectives, on an equal basis, could lead to more holistic landscape research, policy and decision-making processes that take account of the less measurable aspects of landscape.

Cultural appreciation of the intangible values that even a single hedge can inspire, as a 'corridor' of meaning transfer, comes from the under reported poem "The infinite" (L'Infinito) of the Italian romantic poet Giacomo Leopardi (1798–1837) (Fig. 6.32).

The multiple functions of hedgerows and the relative ecosystem services that they deliver are well-known in scientific literature concerning landscape ecology, agronomic performances of crops in relation to productive and protective aspects, and environmental impacts (Forman and Godron 1986; Borin et al. 2010; Walton et al. 2014; Dainese et al. 2016). According to Forman and Godron (1986), hedgerows are line corridors bordering fields, streams, and roads that act as physical barriers and filters intercepting air, water, and animal movements. Therefore, they affect climate, biogeochemical cycles, transfer of soil, agrochemicals and biological agents, playing a role in carbon sequestration and pollution mitigation. As a living community that prospers in a habitat moulded by man with recurrent pruning when well kept, a hedge manifests the typical character of an *ecotone*, where the biodiversity gradient is very high, with species coming from the adjacent environment, be it field, woodland or forest. Even in a few meters of width, a rather composite assemblage of vascular plants (trees, shrubs and herbs) can occur, creating habitats and ecological niches available for any kind of micro, meso-and macro-fauna. For

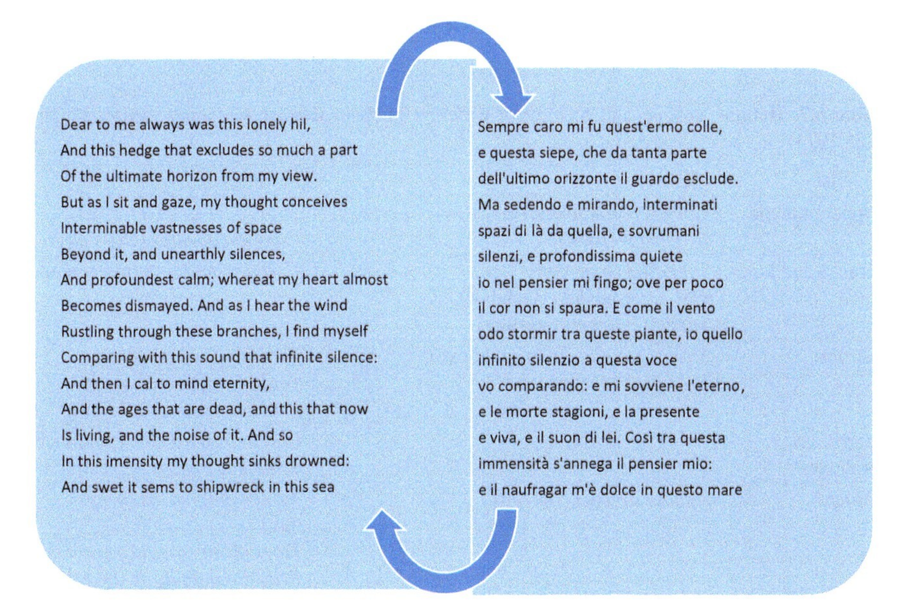

Fig. 6.32 Giacomo Leopardi's "L'Infinito" with translation by R.C. Trevelyan (1941)

instance, Pollard et al. (1974) points out that the following birds use different portions of English hedgerows for different functions (Table 6.7):

The interaction between birds and hedges accounts for the dispersal of the hedge plants into the rural environment through the diffusion of seeds after bird ingestion. A relationship of reciprocal sustenance exists, with hedge plants giving shelter to birds and birds facilitating plant and hedge regeneration.

As to the hedge roles of physical barrier and biological filter, the first is about climatic conditioning and occurs aboveground, and the second concerns water, nutrient and agrochemical uptake that occurs belowground. Due to their nature of porous barrier aboveground, hedges intercept air movements reducing wind speed and its effects, and consequently, reducing evaporation downwind and conservation of water into the soil for the benefit of crops. The protective effects of windbreak or shelterbelt on soil erosion are well-known and described (Loomis and Connor 1992, p. 345) as follows:

> Windbreaks ('shelterbelts') of trees and shrubs planted in rows at right angles to prevailing winds are an ancient solution to wind erosion that is still employed in many regions. Hedgerows, with about 50% porosity to the wind, control erosion by creating a layer of slow-moving air near the ground. Tallness is important; as a general rule, a permeable windbreak will reduce the relative wind velocity near the ground by 50% for a distance downwind of 10 times its height. A 10 m tall break thus protect 100 m of cropland but the break itself will occupy perhaps a 10 m wide strip.

If hedgerows follow contours down the slope, they intercept both runoff and soil sediment, conserving both water, nutrients and soil, and reducing even slope gradient in the long term. Results of a research carried out on the catchment of the Yangtze River in China with contour hedgerows to control soil erosion (Bu et al. 2008) showed the following effects after 4 years of cultivation and crop planting:

Table 6.7 Habitat, nesting and feeding of different birds in English hedgerows (after Pollard et al. 1974)

Habitat	Nesting	Feeding
Upper branches	*Corvus coronae* (carrion crow), *Corvus frugilegus* (rook)	*Parus caeruleus* (blue tit) *Fringilla coelebs* (chaffink)
Trunks and holes	*Tyto alba* (barn owl) *Corvus monedula* (jackdaw) *Sturnus vulgaris* (starling)	*Certhia familiaris* (treecreeper)
Shrubs	*Streptopelia turtur* (turtle dove) *Pica pica* (magpie)	*Turdus pilaris* (fieldfare) *Turdus iliacus* (redwing) *Erithacus rubecola* (robin)
Herbs, law brambles	–	*Carduelis carduelis* (goldfinch) *Carduelis chloris* (greenfinch)
Ground	*Alauda arvensis* (skylark)	*Turdus philomelos* (song thrush) *Troglodytes troglodytes* (wren) *Prunella modularis* (hedge sparrow)

(a) soil fertility increased dramatically in the hedgerow plots;
(b) soil organic matter, total nitrogen, and total phosphorus contents in the hedgerow plots were five to nine times higher than that in the control plot;
(c) all hedgerow plots showed a major effect on reducing soil loss and surface runoff;
(d) overland flow velocity along the upper portion of the hedgerow plots was greatly reduced due to hedgerow resistance.

A recent report by DEFRA (Department for Environment, Food & Rural Affairs, Gov. UK) (Walton et al. 2014) explicitly recognises that hedgerow functions are to be regarded as ecosystem services of regulation, specifically important for both ecosystem resilience, and crop production and protection (Table 6.8).

The role of hedgerows as important drivers of ecological intensification of agroecosystem at both farm and landscape levels needs to be more understood and implemented.

Campi et al. (2009) report on the effects of a tree windbreak (*Cupressus arizonica*, 3 m in height, bordered at North the experimental field) on microclimate and durum wheat production in a Mediterranean environment. Their findings show that, when wind blew from the North direction, the windbreak presence influenced the wind speed until the distance 12.7*H* (*H* is the windbreak height), and temperature increased in a distance of 4.7*H* from the barrier. The windbreak mitigated ET (evapotraspiration) for a distance of 12.7 times its height. Out of this area, the ET was 16% higher than the ET measured near the windbreak belt (<4.7*H*). Yield performances changed accordingly the distance from the windbreak. Within the distance of 18 times the windbreak height, wheat productions were higher than those obtained in the zone not influenced by the windbreaks. The authors conclude that, since windbreaks reduce ET, farms of the Mediterranean environments should be redesigned in order to consider the windbreaks as possible issue of sustainability.

The biological filter effect, or *buffering effect,* of a hedge is mainly explained by the action of its plant root mass that extends below the soil surface and effectively intercepts water that flows in the sub-soil towards drainage ditches. This process takes place in any kind of climate, from temperate (Lowrance 1998) to desert climates (Schade et al. 2002). There are two types of mechanisms involved in the process. First, the hedge plant roots absorb part of the nutrients useful for the

Table 6.8 Multiple hedgerow functions for ecosystem resilience, crop advantages and environmental safeguard (adapted from Walton et al. 2014)

Ecosystem resilience	Crop and environmental advantages
Climate mitigation	Crop stress reduction Crop yield increase
Soil loss reduction	Higher soil profile
Flood risk reduction	Land resilience enhancement
Water quality improvements	Reduction of eutrophication and pollution
Increased biodiversity	Crop pollination improvement Crop pest reduction

development of plant structures, and considerably reduce the load of nutrients flowing into ditches, in particular in terms of N and P. Second, the slowing down of the water flowing into the sub-soil near the hedge facilitates denitrification, hence part of water nitrates is converted into nitrogen that returns to the atmosphere. As a whole, the hedge serves as a filtering and purifying structure for water and it is suitable for playing a protective role against eutrophication events. This function can become more capillary in the territory as the network of hedges continues to extend. As to the multiple functions of buffer strips (BS) in farming areas, Borin et al. (2010) report that young BS reduced total runoff by 33%, losses of N by 44% and P by 50% compared to no-BS in an experimentation carried out in Veneto (North-East Italy). A mature BS was able to abate both NO_3–N and dissolved phosphorus concentrations by almost 100%, in most cases having exiting water that satisfied the limit for avoiding eutrophication. The BS also proved to be a useful barrier for herbicides, with concentrations abated by 60% and 90%, depending on the chemical and the time elapsed since application. Considering the CO_2 immobilized in the wood and soil together, the different BS monitored stored up to 80 t ha^{-1} $year^{-1}$. As to the effects on crop, the BS caused negligible disturbance to maize, soybean and sugar-beet yields. The hedgerows, particularly if composed of trees taller than 6 m, positively influenced the aesthetic value of the territory. Lastly, through a multi-objective analysis, opportunity costs were estimated to support the public decision-maker in determining the subsides to be paid to encourage farmers to plant BS.

As to the importance of hedgerows for the reduction of flood risk in Great Britain, Walton et al. (2014) summarise their conclusions as follows:

> Strong evidence exists to show that individual hedges (and other forms of buffer strip) along contours or fringing water courses have the potential to reduce the volume of water reaching streams and rivers, and the speed with which it does so, following storms. Lines of shrubs or trees of species commonly found in British hedges can greatly increase infiltration of water into the soil even when only a metre or two wide – by a factor of 60–70 times compared to compacted upland sheep pasture. Hedges also reduce soil water levels in and beyond the hedge root zone during the summer, so it takes longer to become saturated during the autumn, providing a buffer against flooding events (p. 5).

All the functions of hedgerows can be further strengthened if their vegetation is associated with a 2–3 m wide herbaceous strip located at the margin of the cultivated field that is set up and managed in various ways (Marshall and Moonen 2002). These strips can be useful for agronomic, environmental and recreational effects since they allow for better access, use and enjoyment of the rural environment.

Ecological intensification advocates the harnessing of regulating and supporting ecosystem services to promote more sustainable food production, and this relies on effective management of non-cropped habitats. Hedgerows are an important component of the landscape in many farming systems across the world, management of which provides a potential mechanism to enhance ecological intensification (Garratt et al. 2017). Hedgerows are both a source habitat for natural enemies that spill over into neighbouring fields and a valuable forage resource and corridor for movement of pollinators (Fig. 6.33).

Differentiated seasonal flowering of hedgerows is the basic source of regulating ecosystem services of both pollination and biological control through predators or parasitoids. This is why Garratt et al. (2017) value highly hedgerow quality in that continuous unbroken hedgerows, with a high diversity of woody species, are more valuable for the some pollinators and natural enemies.

A specific research on these topics was conducted in eight organic cider-apple orchards in Asturias (North-West Spain) aiming to (i) identify the native flowering plants in the surrounding hedgerows and (ii) assess the attractiveness of those flowers for beneficial insects, such as pollinators and natural enemies of pests (Miñarro and Prida 2013). Sixty-three flowering plant species were recorded in the hedgerows from May to September 2012. Flower abundance and species richness decreased as the season progressed. *Hymenoptera* pollinators (honey bees, bumblebees and wild bees) accounted for 37.8% of the total insects recorded visiting flowers, whereas predatory hoverflies (14.9%) were the dominant natural enemies. The attractiveness for insects was assessed for 21 of the flowering plant species identified in the hedgerows. The authors concluded that the knowledge of the hedgerow floral resources could provide potential benefits for farmers by improving ecosystem services of pollination and biological control of pests.

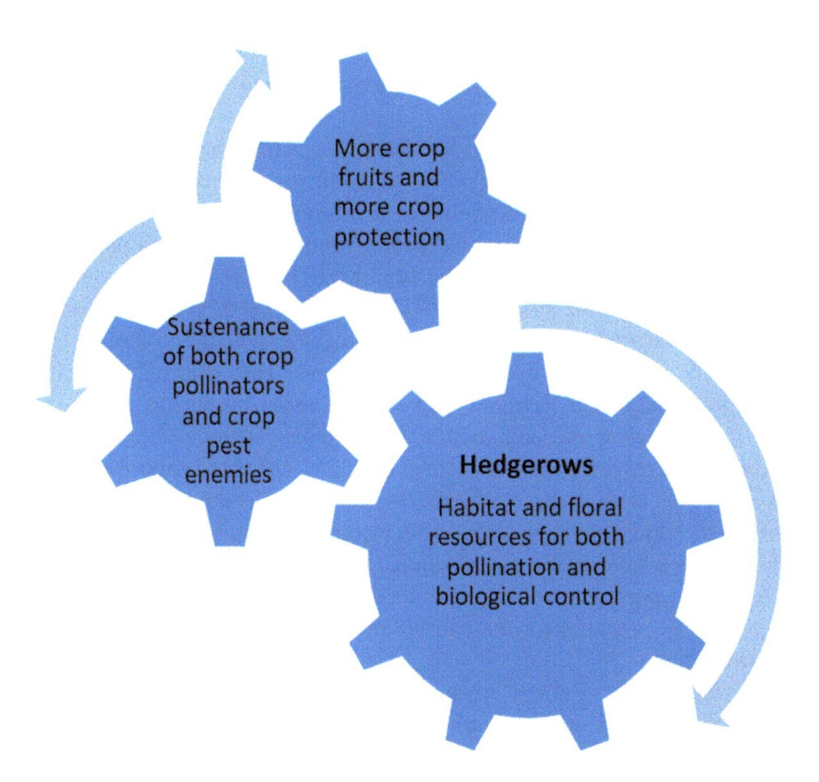

Fig. 6.33 Hedgerows sustain both crop pollination and biological control of crop pests. (Adapted from Garratt et al. 2017)

Morandin and Kremen (2013) investigated whether field edges restored with native perennial plants in California's Central Valley agricultural region increased flora abundance and potential bee nesting sites, and native bee and syrphid fly abundance and diversity, in comparison to relatively unmanaged edges. Native bees and syrphid flies collected from flowers were more abundant, species-rich, and diverse at hedgerow sites than in weedy, unmanaged edges. Hedgerows were especially important for supporting less-common species of native bees in intensive agricultural landscape and they acted as net exporters of native bees into adjacent fields. According to the authors, within-farm habitat restoration such as hedgerow creation may be essential for enhancing native pollinator abundance and diversity, and for pollination services to adjacent crops.

In an experiment concerning field margins in North-Eastern Italy, Dainese et al. (2016) assessed the effect of increased field-margin complexity at the local scale and increasing cover of hedgerows in the landscape on the provision of pest control and potential pollination. They found that high cover of hedgerows in the landscape enhanced aphid parasitism (from 12% to 18%) in cereal fields (maize and wheat) and potential pollination (visitation rate and seed set increased up to 70%) on *Raphanus sativus* (radish) as a phytometer plant. They concluded that hedgerows serve to develop a network of ecological corridors that can facilitate the movement of beneficial organisms, such as pollinators and natural enemies in the agricultural matrix. This suggests a need for new policies that pay particular attention to the conservation of hedgerows at large scales for promoting multiple ecosystem services in agroecosystems.

Gurr et al. (2016) report on multi-site field studies replicated in Thailand, China and Vietnam over a period of 4 years, in which they grew nectar producing plants around rice fields, and monitored levels of pest infestation, insecticide use and yields. They found that this inexpensive intervention significantly reduced populations of two key pests, reduced insecticide applications by 70%, increased grain yields by 5% and delivered an economic advantage of 7.5%. Additional field studies showed that predators and parasitoids of the main rice pests, together with detritivores, were more abundant in the presence of nectar-producing plants. They conclude that a simple diversification approach, in this case the growth of nectar-producing plants, can contribute to the ecological intensification of agricultural systems.

According to a review paper (Bianchi et al. 2006), enhanced natural enemy activity was associated with herbaceous habitats in 80% of the cases (e.g. fallows, field margins), and somewhat less often with wooded habitats (71%) and landscape patchiness (70%). The similar contributions of these landscape factors suggest that all are equally important in enhancing natural enemy populations. Diversified landscapes hold most potential for the conservation of biodiversity and sustaining the pest control function.

In summary, agroecosystem enhancement through hedgerows and other non-cropped habitat support greater numbers of beneficial arthropods than do sole crop fields and simple landscapes, generating a cascade of after-effects that interest the whole hierarchy of agroecosystem organisation (Fig. 6.34). According to Wratten et al. (2012), the strategy of ecological intensification through habitat enhancement is attractive for both farmers and society:

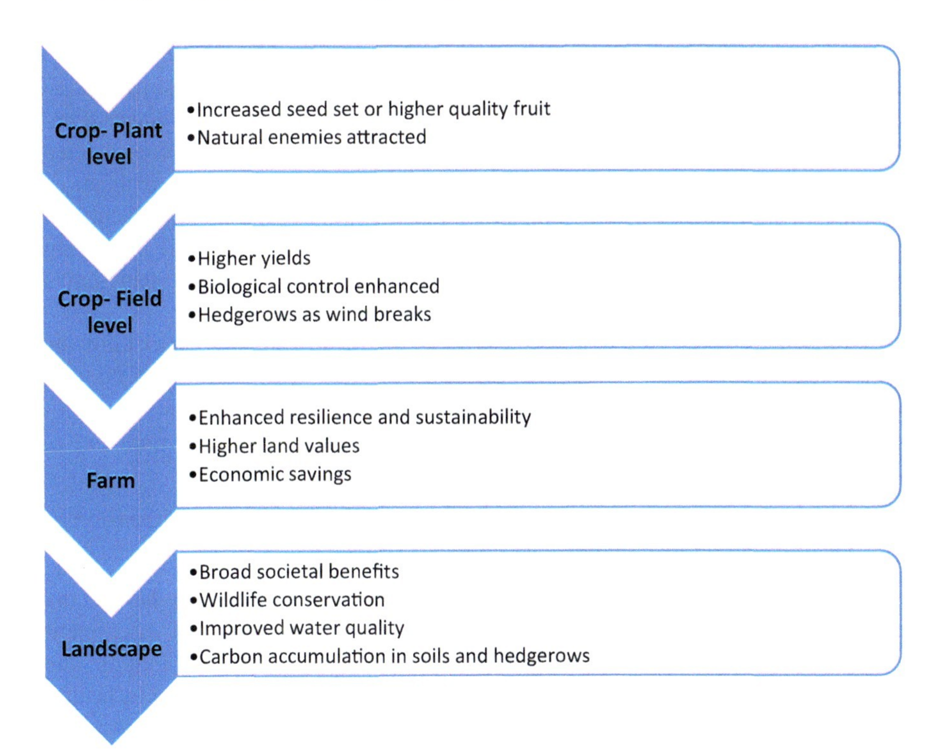

Fig. 6.34 Cascade of after-effects in the agroecosystem hierarchy through hedgerows and other pollinator habitat enhancement. (Adapted from Wratten et al. 2012)

> Thus, making small adjustments in the management of non-cropped habitats, such as adding nectar-producing plants, can combine the benefits of biodiversity and pest-control, thereby optimizing overall ecological services.

6.4.3 Valuing the Performances of Agroecosystem Ecological Intensification

The hierarchical organisation of agroecosystems (from the crop-plant to the landscape levels) requires a new methodology of investigation that takes in account:

(a) the relationships among the components within each level of organisation;
(b) the relationships among the levels of organisation;
(c) the general representation of agroecosystem as a process, i.e. an ongoing event fluctuating in a spatial and temporal scale;
(d) the constant presence of human agency as the prevalent constraint and responsible driver of the agroecosystem performances.

These four epistemological pillars of agroecology should guide in observing, measuring and valuing the performances of ecological intensification practices. Agroecosystems being socio-ecological systems, their performances require comparisons in a systemic way, considering the agroecosystem services that are expected and actually provided at each level of organisation, and for each site of application. Ecological intensification holds not only for natural processes but also for cultural processes, such as knowledge capacity building for appreciating and improving ecological services in favour of both man and the living community to which man belongs.

Within an agroecological frame, Caron et al. (2014) suggest "revisiting the notion of performance in agriculture" through a more complex analysis of interdependencies that concern:

(a) understanding bio-ecological mechanisms at work in agroecosystems;
(b) using biodiversity as leverage for ecological intensification;
(c) exploring the multiple and complex links between the technical dimension of practises and socio-economic and political changes.

Addressing current and future challenges and expectations requires, beyond the classical assessments of land and labour productivity, multi-criteria assessment methods that account for the inseparable relationships between agriculture, the environment, and the other human activity systems concerning settlements, health, nutrition, energy, employment, economy, policy and culture (Fig. 6.35).

The big frame for evaluating agricultural performances is absolutely changed and now well established at the international level, which is a guaranty for local recognition. The concepts of multi-functionality and ecosystem services, by now firmly assigned to agriculture as a human activity system, are the new key words that serve for observing, measuring and evaluating agricultural performances. In this frame, ecological intensification is a means to empower agriculture with the capability to meet most of the functions and ecosystem services that a knowledge society requires for a sustainable development. With ecological intensification, the role of agriculture in society changes radically, in that it becomes proactive instead of subservient to private interests, a genuine new driver for sustainable development based on a cooperative living community, including man and the other living beings.

Multi-functionality in agriculture means that ecosystem services are in strict connection, bring about a cascade of effects and need an integrative evaluation. As fairly noted by Bommarco et al. (2013), the importance of supporting (e.g., soil fertility) and regulating (e.g., pest control and crop pollination) services remains often undervalued, even if crop yield (the valued provisioning service) depends largely on these services:

> The challenge for ecological intensification would be to replace the reliance on external inputs by re-establishment of ecosystems services generated in the soil and the landscape surrounding the cultivated field, while maintaining high, stable productivity levels.

Unfortunately, current understanding and measurements of both ecological interdependencies between land use, biodiversity, and ecosystem services, and the

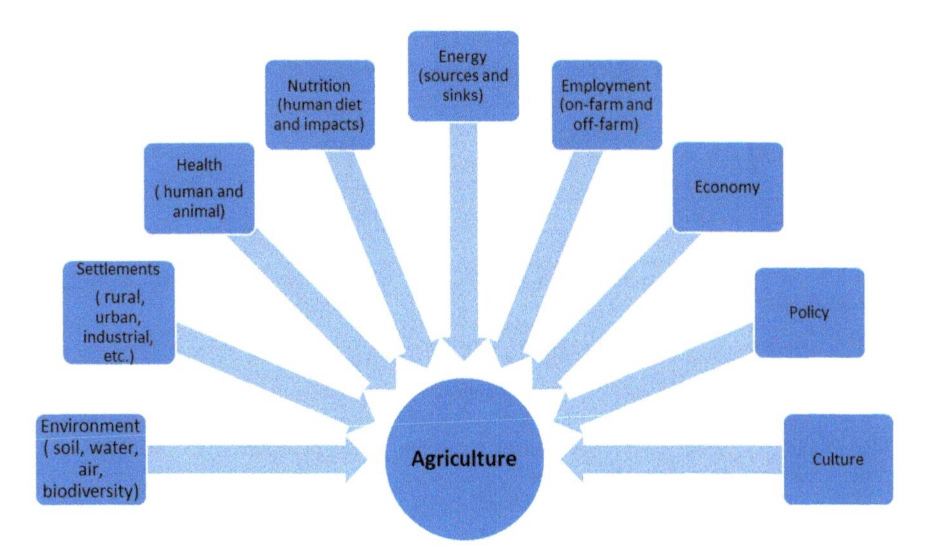

Fig. 6.35 The vast horizon of agriculture relationships, interdependencies and performances

contribution to yield formation from single and combined services, are lacking (Bommarco et al. 2013). To bridge this knowledge gap, the challenge for research on ecological intensification is reconnecting specialised and fragmented competences within multi-disciplinary, site-specific, goal-oriented and participatory programmes. Beside an ecological intensification of the rural environment, a knowledge intensification of the cultural environment is indispensable for achieving tangible progresses toward agriculture sustainability. "Science can play an important role by highlighting the drivers, trade-offs and conditions for a multi-scalar development process" (Caron et al. 2014). The first step is tentatively examining how the integration of different scale of intervention can happen and what kind of methods and contents is appropriate to focus on. A briefing of performances of ecological intensification practices and their interest for both farmers and the public in relation to the contribution to ecosystem services provision are reported in Table 6.9. This kind of outlook emerges from a necessary reframing of the question of "scaling up" as proposed by Caron et al. (2014):

> It is therefore necessary from a research perspective to look at transition toward ecological intensification as the result of multi-scalar processes, which all follow biological, ecological, managerial or political own rules and patterns, generate trade-offs and call for renewing the conception of the notion of performance. Understanding such interactions constitute a major gap that implies better integrating agronomic with human and political sciences in a way that goes beyond inter-disciplinarity.

From the structure of the Table 6.9 emerge both the topological and chorological dimension of the landscape of enquiry, which is the continuous environment where processes happen. The topological dimension refers to the vertical composition of each ecotope, linked to the organism-ecosystem functions and represented in

Table 6.9 Ecological intensification practices and services as a topic of private and public interest in a knowledge society (++ indicates strong interest, + indicate moderate interest, 0 indicates no interest) (☛) (SOM = Soil Organic Matter)

Ecological intensification practises	Field level	Farm level	Landscape level	Societal level
	Farmer's interest		Public interest	
On-field				
Soil conservation measures	Water retention ++	Water retention ++	Landscape resilience ++	Social security ++
Tillage	SOM maintenance ++	SOM maintenance ++	Bare soil reduction ++	Capitalising on solar energy ++
Manure and residue addition	SOM enhancement ++	Nutrient recycling ++	Water protection ++	Circular economy ++
Diversified crop rotation	Biodiversity enhancement ++	Biodiversity enhancement ++	Biodiversity enhancement ++	Biodiversity enhancement ++
Intercropping, cover cropping, green manuring	Complementary use of natural resources +(0)	Complementary use of natural resources +(0)	Complementary use of natural resources ++(+)	Complementary use of natural resources ++(+)
Off-field				
Hedgerows, field margins	Crop protection +(0)	Crop protection +(0)	Landscape beauty and functionality ++	Environmental and human health ++
Agroforestry	Crop and forestry integration +(0)	Farming and forestry integration +(0)	Cropland and forestry integration ++(+)	Circular economy, environmental and human health ++(+)

(☛) The gradient of interest for ecological intensification practices is meant for farmers and the public supporting the paradigm of sustainable agriculture, and could be rejected by farmers and stakeholders devoted to conventional agriculture

Table 6.9 by the column dimension, from soil to agroforestry structure. The chorological dimension refers instead to the larger spatial extension regarded as mosaics of ecotopes from field to landscape and societal levels, including societal components with their cultural requirements, and represented by the row dimension. At the intersection of each point of this double matrix, we find the ecosystem service provided and the estimated appreciation level for both the farmer and the public interest. This tabular matrix could in theory function as a plan for investigating, connecting, understanding and explaining how the ecological intensification strategy impinges on the organisation of cropping and farming systems, landscape, local and regional economy, and rural and urban community. At the same time, Table 6.9 accounts for systemic methodology application and contents for innovative social learning.

From a methodological point of view, it refers to a valuable tool for implementing change through a systemic and learning approach, such as "Systemic Action Research", that proved useful in research training aimed at improving the practical situations in agriculture and rural development (Packham and Srikandarajah 2005). For producing change, it matters:

1. understanding what constitutes the system of action research and its context;
2. defining the purposes of the system;
3. defining what would be considered to be a system improvement;
4. involving the affected stakeholders in the action research.

The need for a paradigm shift in agriculture entails, therefore, a shift in education for innovative social learning and participatory research (Mendez et al. 2013, 2020). In the case of ecological intensification, the improvement concerns not only an innovative technology for enhancing crop performances at field and farming levels, but also the establishment of an enabling larger environment involving both landscape and people, fields and institutions. A topological improvement at field level, such as water retention or SOM enhancement, affects and is affected by ecosystem processes at chorological level, whereby a successful change intervenes whether the same goal – the sustainability of agriculture – is coherently pursued and sustained at any level of territorial organisation. What is important is that the assumptions being held by all members of the community of concern are drawn out and understood, and that all those affected by proposed changes (improvements) are included, following a participatory approach (Packham and Srikandarajah 2005; Mendez et al. 2020).

A basic element for the success of the ecological intensification strategy is the synergistic action of the proposed practices. For instance, all ecological intensification practices listed in column 1 of Table 6.9, both for on-field and off-field ecosystem service enhancement, are highly complementary and reciprocally reinforcing. They constitute a very package of ecological intensification practices interesting any ecosystem component (air, water, soil, and biodiversity) which is applicable, if appropriately tailored, to any kind of environment. Kremen and Miles (2012) advance a similar conviction in their review paper on ecosystem services in biologically diversified versus conventional farming systems. They examined 12 ecosystem services: biodiversity; soil quality; nutrient management; water-holding capacity; control of weeds, diseases, and pests; pollination services; carbon sequestration; energy efficiency and reduction of warming potential; resistance and resilience to climate change; and crop productivity. They found that biologically diversified farming systems performed better when compared with conventional ones except for the crop yield. In their opinion, despite significantly less public funding applied to agroecological research, diversified farming system had only somewhat reduced mean crop productivity while reducing environmental and social harms. More recently, Tamburini et al. (2020) focus more on-meta-analyses rather than on original studies and ascertain how agricultural diversification promotes multiple ecosystem services without compromising yield. Accordingly, they maintain that "widespread adoption of diversification practices shows promise to contribute to biodiversity conservation and food security from local to global scale". Both farmers and the public in general receive benefits from ecological intensification, even if the gradient of appreciation can be different according to the dominant paradigm that govern agriculture development. Today, most farmers follow the conventional pattern of industrialised agriculture, where the indicators of performance concern yield per unity of land, labour and capital. Interest for economic performances largely overwhelms interest for environmental protection and therefore many ecological intensification practices are often lacking in conventional agriculture schemes. On the contrary, farmers that follow agroecological principles and organic paradigms (Lin et al. 2011; Garibaldi et al. 2017) advocate ecological intensification practises as an avenue of "Systemic Development" towards sustainable agriculture. According to Packham and Srikandarajah (2005) "Systemic Development is a set of ideas that promotes thinking and acting that will ensure the continued development of the organisation (system) through participatory learning". Whereby, the task of the (farming) system's managers is:

> continually learn about the nature of the system's environment and plan strategies, allocate resources and ensure operations are carried out that will enable the organisation and the farming system to survive and flourish. This is best achieved through *experiential learning*, coupled with the individual learners' own insights through what is termed *inspirational learning*.

The ecological intensification package of Table 6.9 has the quality to constitute an *action research project*, regarding it as a strategic action susceptible to implementation and improvement at local level. The project would proceed through a spiral of cycles of planning, acting, observing and reflecting, each of these steps being self-critically implemented in an integrated way among stakeholders, as proposed by Garibaldi et al. (2017) in the framework for an Evidence-Based Assessment of the Ecological, Social, and Economic Performance of Farming Systems (Fig. 6.36). In this framework, an effective tool for building the evidence base for farming systems in a participatory way should:

1. be based on a strong theoretical grounding considering farming systems as integrated socioecological systems;
2. incorporate ecological, social, and economic aspect of sustainability;
3. integrate quantitative and qualitative research methods;
4. foster notions of participation, learning and empowerment;
5. be a simple, self-assessment tool targeted at the individual or household level, but which considers multi-scalar interactions;
6. provide data and assessment that allow comparability between sites.

Moreover, the tool would connect farmers, researchers and policy-makers through a knowledge sharing platform where communication, discussion, asking questions and sharing answers are practised. Concerning the steps of implementation (comparison, measurement, outcomes, and actions), they would follow sticking to the research paradigm concerned with the improvement of situations through the taking of informed action and the development of relevant theory, which is then used to guide further action. According to Packham and Srikandarajah (2005):

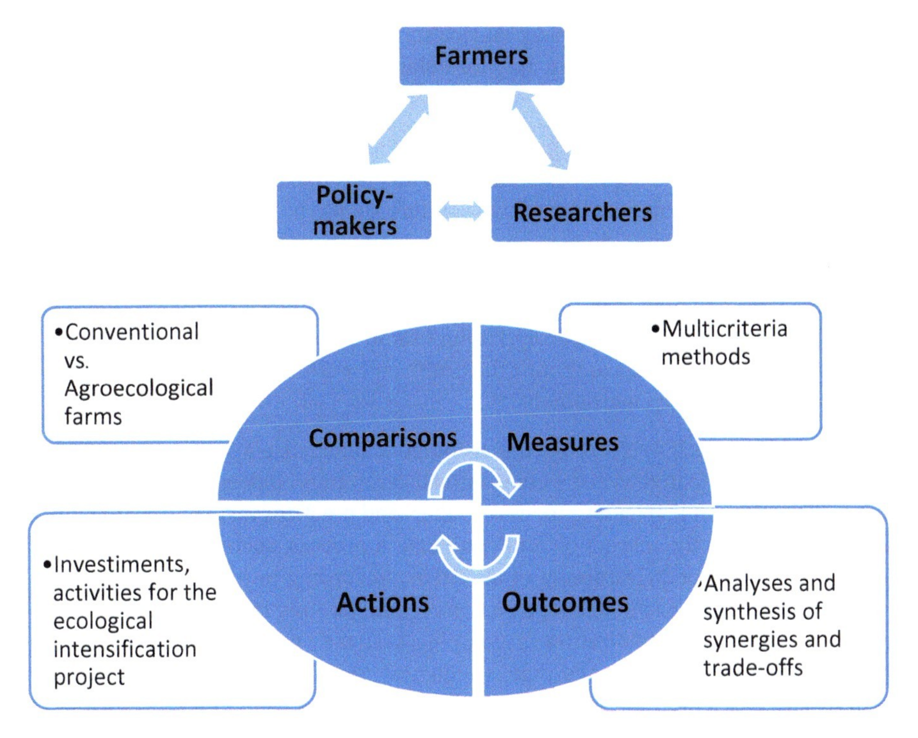

Fig. 6.36 Action research project (e.g. ecological intensification) in the framework of an Evidence-Based Assessment. (Adapted by Garibaldi et al. 2017)

> The intention of action research is to give people the power to act in order to bring about change (action) by generating knowledge through rational reflection on personal experience (research), so above all it is a learning process.

Performance comparison between farms (and farmers) that follow different farming paradigms is the first step of an informed action research project. Conventional farms and organic farms are often compared in that they represent the two extreme of a range of farming paradigms, with the former more interested in economic performances (mercantilist ethics) and the latter in agroecological performances (environmental ethics), even if a phenomenon of 'conventionalisation' of organic agriculture is emerging (Darnhofer et al. 2010). According to Reganold and Wachter (2016), "organic farming systems produce lower yields compared with conventional agriculture, however, they are more profitable and environmentally friendly, and deliver equally or more nutritious foods that contain less (or no) pesticide residues, compared with conventional farming. Moreover, initial evidence indicates that organic agricultural systems deliver greater ecosystem services and social benefits". Organic farming is considered one of the largest existing models of ecological intensification worldwide and may be seen as a laboratory for ecological innovations (Tittonell 2014). Organic farming has set out to be an alternative to conventional agriculture and food chains and through its representative body, the

International Federation of Organic Agriculture Movement (IFOAM), after a concerted and participatory process, has formulated four principles to inspire action (IFOAM 2005):

- the principle of health: "Organic agriculture should sustain and enhance the health of soil, plant, animal and human as one and indivisible";
- the principle of ecology: "Organic agriculture should be based on living ecological systems and cycles, work with them, emulate them and help sustain them";
- the principle of fairness: "Organic agriculture should build on relationships that ensure fairness with regard to the common environment and life opportunities";
- the principle of care: "Organic agriculture should be managed in a precautionary and responsible manner to protect the health and well-being of current and future generations and the environment".

These principles have a strong ethical component and display a much wider view of agriculture compared with the conventional agriculture. Indeed, the organic *standards* (normative rules) tend to focus on values and practices that are easy to codify and audit through the inspection and certification process, such as what inputs are permitted or excluded (Darnhofer et al. 2010). According to the principle of ecology, sustainable design and management in organic farming mean re-building structural and functional organisation of cropping systems and animal husbandry in a way that the natural processes that sustain soil fertility and crop protection are restored. This is a task not only for farmers, that are responsible for their own farm organisation, but also for all stakeholders that advocate the sustainable agriculture paradigm, and are responsible for creating an enabling environmental from both the biophysical and socioeconomic perspectives. The liver to drive the transition from conventional agriculture to sustainable agriculture resides in enhancing biodiversity at level of field, farm and landscape, in that biodiversity is the axis connecting all ecological intensification practices (both on-farm and off-farm) listed in Table 6.9. Ecological intensification through agroecology relies largely on the concept of spatiotemporal diversification (of species, processes and functions) and on the emergent patterns that result from that (Tscharntke et al. 2005; Ratnadass et al. 2012; Tittonell 2014; Struik and Kuyper 2017). Tittonell (2014) appropriately marks the difference between the concepts of sustainable intensification (Tilman et al. 2011; Wezel et al. 2015; Struik and Kuyper 2017) and ecological intensification as follows:

> The difference between sustainable and ecological intensification goes beyond pure semantics [...] Since the ecological processes that underpin support and regulation services operate beyond the boundaries of a single farm, the scales of analysis and design also differ. While sustainable intensification – and/or eco-efficient – solutions are still designed by reasoning at the scale of a single crop or agricultural field, ecological intensification needs to embrace the complexity of the landscape. As a consequence, actions to support ecological intensification may often require collective decision-making, which calls also for institutional innovation.

To clarify this point with a meaningful example, genetic modified (GM) crops are often reported as plausible strategic components in the process of sustainable intensification, as testified by the following citations:

> Geneticaly modified (GM) crops are already contributing to sustainable intensification trough higher yield and lower environmental impact and have potential to deliver significant improvements (Raybould and Poppy 2012);
>
> The role of genetic engineering in more sustainable crop production as well as natural resources conservation, including biodiversity, is plausible (Tsatsakis et al. 2017).

Instead, in a report of the European Environmental Agency, Quist et al. (2013) regard GM crops as components of an agricultural system alternative to agroecological innovations:

> We have chosen to contrast genetically modified (GM) crops and agroecological methods as two examples of innovation outputs and strategies that have very different outcomes in the way we produced food [...] The former is driven by production goals and short-term profit maximisation incentives, where the predominant types of GM crops developed thus far are economically profitable within a system of high-input industrialised monoculture that is largely unsustainable in its reliance on external, non-renewable inputs (p. 460).

Whether GM crops can contribute in the future to the process of ecological intensification depends on a change of framework that sees them as components of an agroecological system how tentatively suggested in Box 6.7.

According to Struik and Kuyper (2017), ecological intensification means agroecosystems be designed for benefitting from ecological processes and functions, and therefore requires an '*ecology-intensive* agronomy'. Caporali (2015a) has pointed out how ecology meets agronomy at the farm level by offering basic ecological principles (adaptation, integration, participation, evolution) for pursuing a sustainable management of both the whole farm systems and its subsystems, i.e. cropping and livestock systems. The benefits of constructing sustainable farming systems result at the higher levels of landscape and rural community as cultural evolution, as represented in Fig. 6.37.

Kleijn et al. (2019) agree that ecological intensification of mainstream farming can safeguard food production, with accompanying environmental benefits; however, they argue, this approach is rarely adopted by farmers in that there are mismatches in perceived benefits of ecological intensification between scientists and farmers, which hinders its uptake. Researchers are more concentrated on processes and farmers are more interested in outcomes (i.e. profits). In their view, "the studies

Box 6.7 A Tentative Framework for Ecological Intensification of GM Crops R & D

RATIONALE

Agroecology is fundamentally different from other approaches to sustainable development. It is based on bottom-up and territorial processes, helping to deliver contextualised solutions to local problems. Agroecological innovations are based on the co-creation of knowledge, combining science with the traditional, practical and local knowledge of producers. By enhancing their autonomy and adaptive capacity, agroecology empowers producers and communities as key agent of change (FAO 2018a).

DRIVERS

It seems plausible to assume that the GM story would have unfolded in a very different way if the first GM crop had been a GM food crop developed by the public sector (as opposed to *cash crops* developed by industry). One key element in the public debate about GM crops is therefore the danger arising from monopolistic or oligopolistic control of the technology. The concern is that large private seed producers will put themselves in a position where they are able to exploit small farmers to the disadvantages of the latter (Weal 2010).

SOURCES: INTRAGENIC OR TRANSGENIC?

There are three reasons for introducing different regulatory policies for intragenic and transgenic organisms. First, I argue that lay people's arguments against crossing species barriers should be regarded as important and relevant expressions of an alternative world-view. The only way to do this is by including this perspective in the procedures for assessing new GMOs, because the traditional arguments against it are based on a scientific world-view that is challenged in this lay perspective. Second, I follow Nielsen in arguing that the risks and scientific uncertainties involved in the production and release of intragenic organisms will probably be reduced, since the introduced trait is already present in the organism. Third, staying within species barriers expresses respect for the otherness of nature as something that we cannot and should not attempt to control completely (Myskja 2006; Nielsen 2003).

GOALS

The amount of arable land has not changed appreciably in more than half a century, increasing by only about 10%. And it is not likely to increase much in the future because we are losing it to urbanization, salinization and desertification as fast or faster than we are adding it [...] It would therefore be wise to increase investment in research on alternative forms of agriculture based on plants not now used in agriculture, but capable of growing at higher temperatures and using brackish or salt water for irrigation. Indeed, the array of molecular tools and knowledge available today might make it possible to design a wholly new kind of agriculture for a more arid, hotter world (Fedoroff 2010).

that have been carried out so far suggest that in the majority of crops and under the current economic paradigm it will be difficult for ecological intensification to achieve higher profits than under conventional intensification". The authors provide recommendations for overcoming these mismatches and highlight important additional factors, such as social acceptability of farming based on ecological intensification processes. They propose that "there are three complementary pathways towards wide-scale adoption of ecological intensification: through market-driven processes, regulatory instruments, and through reputational concerns". Figure 6.38 shows the main components of this social process that recognises the ecological intensification of agriculture as an ethical act. The authors' conclusion is that "future

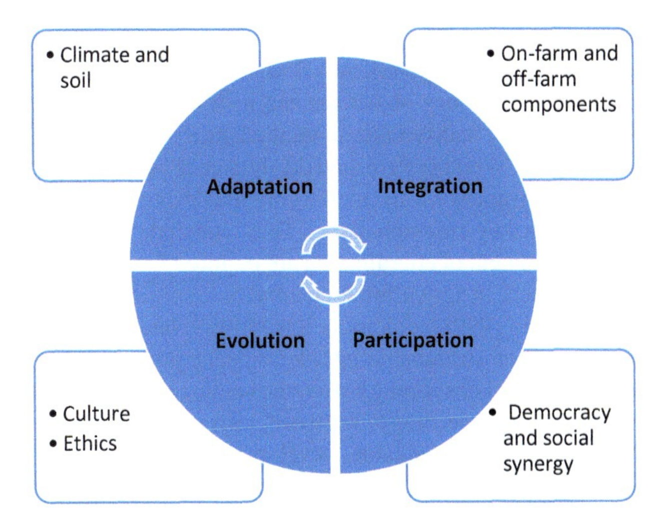

Fig. 6.37 Ecological principles for blending ecology and agronomy into agroecology

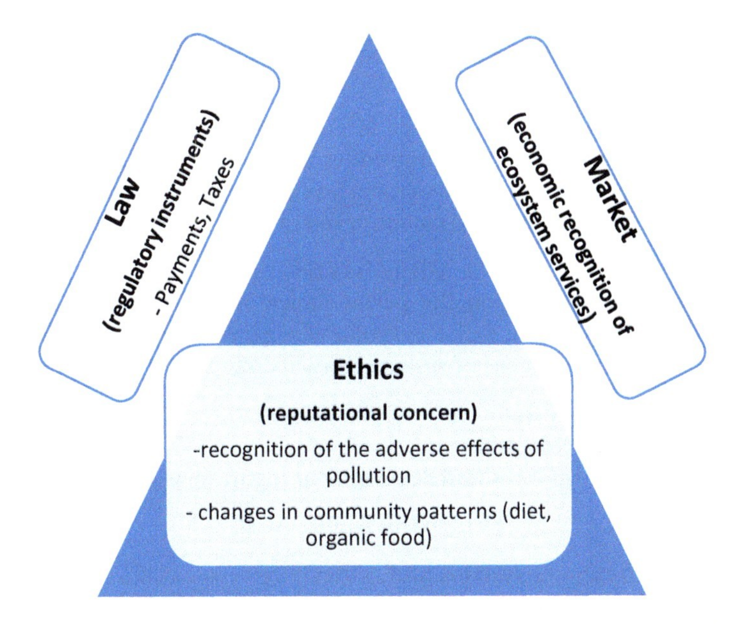

Fig. 6.38 Socio-ecological components of the recognition process of agriculture ecological intensification as an ethical act. (Adapted from Kleijn et al. 2019)

research should therefore not only address ecological, agronomic, and economic aspects of ecological intensification but also the sociological aspects".

Recent papers (Lescourret et al. 2015; Pasqual et al. 2017) stress the importance of valuation of nature's contributions to people in decision making, which is a discriminant factor in moulding the attitude toward two contrasting ethics, *utlitarian ethics* and *value ethics*. The first refers to an "individual self-interested behaviour, often associated with a belief in material economic growth as the basis for a good quality of life"; the second embraces "value pluralism by acknowledging the diversity of worldviews and values [and] may lead to a different iterative approach regarding identification of policy objectives and instrument" (Pasqual et al. 2017). By representing an explicit and symmetric relationships between the ecosystem (or system of nature) and the social system, and the dynamics between them, Lescourret et al. (2015) provide a framework for identifying the design of collective multiservice management as a key research issue. They conclude that new stakeholder organisations and instrument of coordination are required for the implementation of multiservice management in agricultural territory.

According to Thiele (1999), ecology is inherently related to sustainability and ethics in that "its focus on sustainable interdependence proves highly amenable to ethical formulations". FAO has recently played a leading role in "facilitating global and regional dialogue on agroecology" regarded "as a strategic approach to achieve Zero Hunger and the other Sustainable Development Goals (SDGs)"(FAO 2019, p. 2), which indeed are ethical goals. In order to implement this strategic approach, FAO has developed a global analytical framework for the multidimensional assessment of the performance of agroecology: the Tool for Agroecology Performance

Box 6.8 Main Characters and Procedure of TAPE (Adapted from FAO 2019)

TARGET AUDIENCE

The global and regional communities of practice on agroecology: scientists, advocates, producers, extension workers, policy makers, staff from NGOs and international organizations or funding institutions.

OBJECTIVES

To produce evidence on the performance of agroecological systems across the environmental, social and cultural, economic, health and nutrition, and governance dimensions of sustainability in order to:

(a) **build knowledge and empower producers**;
(b) **support agroecological transition**,
(c) **inform policy makers and development institutions**.

STEPWISE APPROACH

Step 0. **Description of systems and context**: agroecological zones, production systems, type of household, existing policies, climate change, enabling environment.

Step 1. **Characterisation of agroecological transition (CAET)**. On farm/household survey: current status, based on FAO's 10 elements of agroecology, can be self assessment by producer.

Step 2. **Criteria of performance**. On farm/household survey: measure progress and quantity impact, addressing 5 key dimensions (Environment & climate change, Health & nutrition, Society & Culture, Economy, Governance), 10 criteria (Land tenure, Productivity, Income, Added value,

Exposure to pesticides, Dietary diversity, Women's empowerment, Youth employment, Agricultural biodiversity, Soil health) for each dimension, and then appropriate indicators for each criterion, in order to generate evidence on the multidimensional performance of agroecology.

Step 3. **Analysis and participatory interpretation**. At territory/community scale: review CAET results, explain with context, enabling environment; review performance result and explain with CAET; analyse contribution to SDGs.

Evaluation (TAPE), as a standard methodology for the *characterization of agroecological transition* (CAET) (Box 6.8).

FAO (2019) reports on the state of TAPE as follows:

> In the second half of 2019, TAPE has started to be tested in a number of pilots or case studies, using the guidelines for application presented in this document. The purpose of the pilot studies is to validate or improve TAPE, with a particular emphasis on (i) the overall stepwise approach and (ii) the CAET and the Core Criteria of Performance. This is being done in a systematic way in each of FAO regions, starting with a workshop involving governments, scientists, producers' organizations and NGOs, in order to identify, strengthen and engage a community of practice on the process and utilization of the tool (p. 50).

6.5 Institutional Patterns of Ecological Intensification of Agriculture in Europe: The Role of Universities

With the Sorbonne Declaration[5] and the Magna Charta Universitatum (1988) a new course of Education Policy in Europe has been established with an unprecedented financial investment that still lasts today. At the same time, a radical change in the Agricultural Policy has occurred, which has recognised the necessity of a turning point towards a multifunctional agriculture, the start of an epochal process of change still lasting under the name of "agro-ecological transition". Similar signs of change were initiated in USA with the publication of the volume "Alternative Agriculture" by the National Research Council (1989) and the following declaration in the back cover:

> Alternative farming methods are practical and economical ways to maintain yields, conserve soil, maintain water quality, and lower operating costs through improved farm management and reduced use of fertilizers and pesticides […] These methods have not been widely adopted, however, for two reasons: 1) research on alternative agricultural systems is

[5] **The Sorbonne Declaration was signed in 1998**, by the ministers of four countries, namely France, Germany, UK and Italy. The aim of the Declaration was **to create a common frame of reference within the intended European Higher Education Area**, where **mobility should be promoted** both for students and graduates, as well as for the teaching staff. Also, it was meant **to ensure the promotion of qualifications**, with regard to the job market.

lacking, and 2) federal farm programs discourage their implementation – in some cases financially penalizing farmers for using alternative techniques such as crop rotation.

In consonance with the international framework of renewal emerged from the "Bruntland Report" of the World Commission on Environment and Development (1987), stands the critical role of European universities in society as defined in the preamble of the Magna Charta Universitatum:

1. at the approaching end of this millennium the future of mankind depends largely on cultural, scientific and technical development; this is build up in centres of culture, knowledge and research as represented by true universities;
2. the universities' task of spreading knowledge among the younger generations implies that, in today's world, they must also serve society as a whole; and that the cultural, social and economic future of society requires, in particular, a considerable investment in continuing education;
3. universities must give future generations education and training that will teach them, and through them others, to respect the great harmonies of their natural environment and of life itself.

The means to achieve these goals were really innovative for setting up the basis for a true European knowledge community of researchers, teachers and students firmly grounded on mobility of both students and teachers, and a general policy of equivalent status, titles, examinations, and award of scholarships. The character of *ecological intensification of education and training* is fully clear in the ethical command for present and future generations of "respect" for "the great harmonies of their natural environment and of life itself".

In the sector of agricultural policy, the increasing impact of agriculture on the environment had called for a new institutional concern for control, preservation and guide, that found its first application in the Regulation 85/797 (EEC 1985) on "improving the efficiency of agricultural systems". Article 19 authorized Member States to introduce national aid schemes in "environmental sensitive areas" to encourage farming practices compatible with the requirements of conserving natural resources and ensuring an adequate income for farmers. Payments to farmers took the form of an annual premium per hectare. Since then, the European Union's Common Agricultural Policy (CAP) has maintained its effort to invest for environmental protection and biodiversity conservation in agricultural land, as briefly summarized in Box 6.9.

The new agreement on CAP reform reached in 2013 maintains the two traditional pillars, *Direct Payments* to farmers and *Rural Development Policy,* but increases the links between them, thus offering a more holistic and integrated approach to policy support (EU 2013; Jack 2015). The new policy continues along this reform path, moving from product to producer support and now to a more land-based approach. This translates into three long-term CAP objectives: viable food production, sustainable management of natural resources and climate action and balanced territorial development. The most innovative aspect of the 2013 CAP reform measures was the introduction in the pillar 1 (Direct Payments) of payments for agricultural practices beneficial for the climate and the environment that is

Box 6.9 Stepping Stones for a Multifunctional Agriculture Policy in Europe (Source: DG Agriculture and Rural Development, European Union 2013)

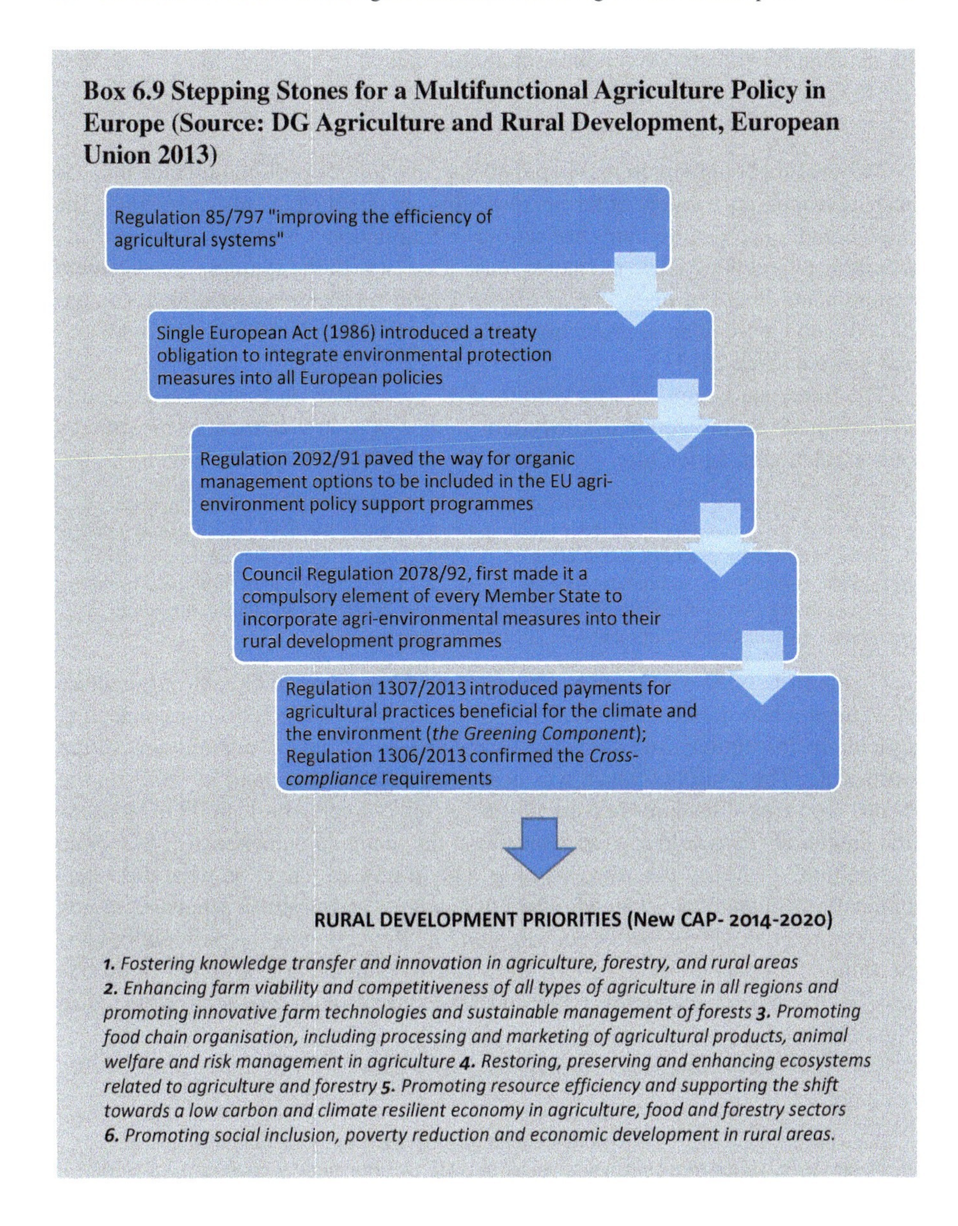

generally referred to as "the greening component". It introduces a mandatory requirement that all farmers receiving basic payments (decoupled "green" payment per hectare) should engage in environmental beneficial practices. It sets out three particular obligations, with farmers being required to comply with each that is relevant to their farm:

(a) to comply with crop diversification requirements;
(b) to maintain permanent grassland;
(c) to have ecological focus areas on farmland.

The greening component makes payments beyond the cross-compliance that has been a compulsory element of farmer eligibility for direct payments since 2005. The cross-compliance requirements are set out in Regulation 1306/2013. They provide that farmers receiving basic payments must comply with the statutory management requirements imposed under 13 European Union regulations and directives concerned with environmental protection, public animal and plant health and with animal welfare (Jack 2015).

The European Union, as a party to the UN Convention on Biological Diversity, has also made the protection of ecosystem services a central element of his strategy and, in relation to agriculture, its strategy sets the following specific targets:

> By 2020, maximise areas under agriculture across grasslands, arable land, and permanent crops that are covered by biodiversity-related measures under the CAP so as to ensure the conservation of biodiversity and to bring about a measurable improvement in the conservation status of species and habitat that depend on or are affected by agriculture and in the provision of ecosystem services as compared to the EU 2010 baseline, thus contributing to enhance sustainable development (EC 2011).

On the front of NGOs, IFOAM (International Federation of Organic Agriculture Movements) had started in the early 1970s its international activity for connecting agriculture institutions and farmers sharing the paradigm of organic agriculture word wide. The first conference was held in Sissach, Switzerland in 1977, on the theme "Towards a Sustainable Agriculture". After 20 years, the 11th IFOAM scientific conference was held in Copenhagen, on the theme "Fundamentals of Organic Agriculture" – "nearly twenty years on is a good time to reflect on what this international movement has achieved, where it has been and where it seems set to go" (Woodward et al. 1996). With a contribution on "Reflection on the past, outlook for the future", Woodward et al. (1996) were able to establish a framework of awareness for the movement membership and for external observers as represented in Fig. 6.39.

The four key words are articulated as follows (Woodward et al. 1996, pp. 259–261):

> But what is a *movement*? The synonyms are more relevant than the definition: action, activity, advance, agitation, campaign, change, crusade, development, faction, ground-swell, grouping, operation, organisation, party, progress, stirring. It implies agreement between people to move towards change; it implies a dynamic and it implies a goal.
>
> There can be no doubt that today an informed and coherent *alternative* to conventional agricultural "dogma" does exist theoretically, technically and practically in all parts of the world. That it does, is to a large extent due to the efforts and contribution made by individuals and organisations who are part of IFOAM.
>
> "The *health* of soil, plant, animal and man is one and indivisible" (Balfour 1975) [...] By recycling nutrients through this chain productivity could be maintained over time and health could be enhanced at all stages.

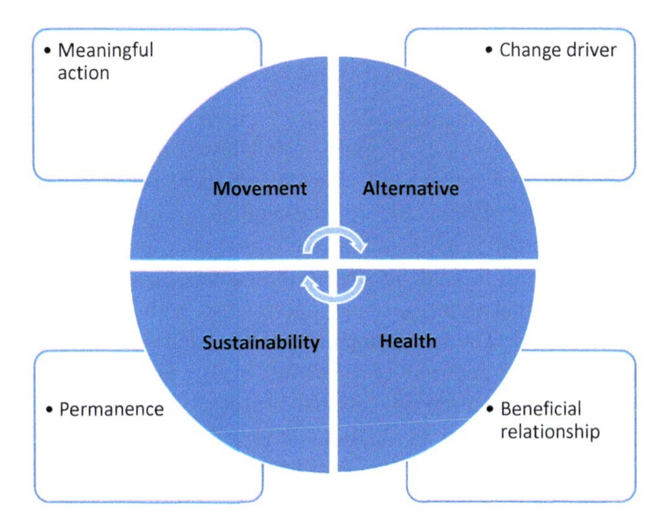

Fig. 6.39 Identity key words of the International Federation of Organic Agriculture Movements. (Adapted from Woodward et al. 1996)

> The criteria for a *sustainable agriculture* can be summed up in one world *permanence*, which means adopting techniques that maintain soil fertility indefinitely, that utilise, as far as possible, only renewable resources; that do not grossly pollute the environment, and that foster life-energy (or if preferred biological activity within the soil and throughout the cycles of all the involved food-chains).

The first two key words (movement and alternative) identify the efficient, material and formal causes of IFOAM, the other two words (health and sustainability) the final causes, or goals, to achieve through IFOAM in agriculture and society. With the acceptance of ecology as a principle for organic farming in 2005, a scientific consolidation and legitimation of the other synergic principles (health, fairness and care) has occurred, with a more potential of social credit and trust. For the future, according to a recent position paper (IFOAM EU Group 2019), the movement perspective should be guided by an ascendant tendency to blend technological innovation with institutional innovation in a way that, starting from external input substitution as regulated by law, a system redesign would follow, leading to agroecological landscapes and food systems. This prospective synergy between organic farming as a practice, and agroecology as a science, is a process of ecological intensification, at the same time, knowledge and labour intensive. The inclusive vision of the organic movement is as follows:

> To summarize, **organic farming should be strengthened as a practical and certified approach of agroecological farming**. Agroecology is a way to express the four principles of organic farming and can thus be seen as a free-thinking space to create concepts and practices for organic farming to develop beyond the organic regulation. In other words, **the development of organic agriculture and its principles should be seen within the agroecological production model** (IFOAM EU GROUP 2019, p. 9).

6.5.1 Agricultural Education and Research

The process of ecological intensification in agricultural education has moved the first steps under the framework of Magna Charta Universitatum, the new deal of the European agricultural policy and the international new paradigm of sustainable development. Since the 1990s, the institutionalisation of the "Inter-university Cooperation Programme" (ICP) under the "Erasmus programme" has been extremely beneficial both to students and to the cause of ecological agriculture to promote collaboration between European universities with research and teaching interests within the field. The institutional, curricular and personal benefits of this new European knowledge community are shown in Box 6.10. A degree curriculum can be regarded an input/output educational process. The decision to establish a degree course is taken by an institutional body, usually a Department, which employs human and material resources in order to give structure and functioning to the degree course. The process of making a decision is a step that stems from both external and internal inputs. The external context is given by a combination of information coming from international, national and local sources, altogether being regarded as the general framework of reference by the decisional institutional body. The internal context has its own composition as a group of people dedicated to an innovative project. Different feelings and prospects can fuel the initiative and, therefore, a purposeful internal context probably emerges in response to meaningful internal inputs, coming from the University, the other Departments and the personal attitudes of the people involved. Structure and functioning of a new degree curriculum probably comply with both the external and the internal context and, therefore, a new curriculum shows coherence among its internal components and correspondence with external expectations (Roling 2003).

Box 6.10 Benefits of Design and Implementation of Common Curricula in "Ecological Agriculture" (BSc Level) and "Agroecology" (MSc Level) (Adapted from ICP Internal Reports)

INSTITUTIONAL BENEFITS

To promote international awareness, cooperation, and knowledge amongst students, by students exchanges, inter-university intensive courses and staff exchanges.

To foster collaboration and information exchange between academic and research staff, facilitated by the increased student and teaching linkages.

To increase the profile of ecological agriculture within the agriculture departments of the participating universities due to the attainment of the benefits stated above.

CURRICULAR BENEFITS

The identification and development of common modules to create a specialisation, firstly at BSc and then at MSc degree levels.

By ensuring that the necessary courses are available at as many participating universities as possible; the credit points allocated to each module are common to all, and that the problem of course recognition and workload will be reduced.

PERSONAL BENEFITS

To provide students with the educational background required to meet the demands of both the agricultural sector and of society, in the context of increasing public interest in ecological food products and production methods.

To facilitate student mobility allowing for maximum choice in universities at which part of their chosen specialisation can be undertaken.

The outputs of the degree course can be estimated in terms of achievements in cultural education, professional skills and job opportunities of students and graduates. Individual and institutional achievements can be estimated in terms of improved attitude towards pedagogical and didactic tools for both students and teachers. More general societal benefits for public and private institutions derived from improved networking, sustainable development local strategies, education at local, regional, national and international levels, are altogether conditions that deserve attention, monitoring and evaluation.

The ICP programme in "Ecological agriculture" as a "Common European Degree Level Specialisation" started with an application for approval in 1994/95 academic year with the participation of seven universities in Europe from seven different countries (Fig. 6.40). The draft module descriptors and the necessary pre-requisite subjects required by the students for the proposed modules were discussed in depth at the group meetings in Lyon, 21–22 July 1995, and Wageningen, 26–27 January 1996. The final structure provided 1 year of specialisation in "Ecological agriculture" based on 13 modules, distributed in two semesters, with a package of six credits in between as an intensive summer course taught in English language. The experience of the summer course (2–3weeks) with the participation of all students and teachers in one site, and with one language(English), was really exciting for both learning outcomes and social relationships.

Integrating sustainability in agricultural higher education was one of the main outcomes of the EU Socrates Thematic Network for Agriculture, Forestry and the environment (AFANet).[6] Between 1997 and 2000, AFANet developed a number of

[6]Thematic networks are multi-actor projects that collect existing knowledge and best practices on a given theme to make it available in easily understandable formats for end users such as farmers, foresters, advisers and others.

Fig. 6.40 Universities involved in the specialisation course in "Ecological Agriculture"

initiatives to promote the integration of the concept and praxis of sustainability into the curricula of higher education institutions. A position paper of Wals and Bawden (2000) accounts for the AFANet aim to develop both a European dimension to education and cooperation between universities and colleges under the new paradigm of sustainability:

> As a concept it provides a focus for the building of bridges between different disciplines and between divergent interests and values. It also presents opportunities for fundamental reforms of curricula that involve the exploration of non-conventional epistemologies and ontologies, as well as non-traditional pedagogical practices that include more experiential or issue-based strategies, more interdisciplinary studies, and more applied practices [...] One such approach, derived from work on 'systemic learning' and participatory environmental education in Australia and presented here as a framework for considering new strategies both for sustainable rural development in practice, and for pedagogical approaches that facilitate the acquisition of competencies relevant to that.

The position paper of Wals and Bawden (2000) provides a range of anchors (epistemological, didactical, processual, and content anchors) for evaluating and establishing sustainability in agricultural higher education, a brief sketch of which is in Table 6.10.

The aim of presenting four different worldviews as epistemological anchors is not to replace one perspective with another, but to use each to inform the others. Each of these four perspectives on sustainability is legitimate and has potential to reveal insights in the process of enquiry or critical reflection: "we need not to see these dimensions as incommensurable philosophical dichotomies, but as heuristic devices to inform practical rationality or praxis" (Wals and Bawden 2000). According to these authors, the concept of sustainability is so complex, involving

Table 6.10 Kind of anchors for evaluating and establishing sustainability into agricultural education (adapted from Wals and Bawden 2000)

Epistemological anchors (worldviews)	Didactical anchors (learning)	Processual anchors	Content anchors
1. *Egocentric* Satisfaction of individual needs and wants 2. *Technocentric* Objective knowledge and reductionist perspective 3. *Ecocentric* Objective knowledge and systemic perspective 4. *Holocentric* Recognition of different worldviews	1. Learner-centred 2. Collaborative 3. Praxis-oriented 4. Problem solving 5. Self-regulative 6. With and from outsiders 7. Higher level cognitive 8. Emphasising both cognitive, affective and skill-related objectives	1. Total immersion 2. Diversity in learning styles 3. Active participation 4. System of values and valuing 5. Balancing the far and near 6. A case-study approach 7. The social dimension of learning 8. Learning for action	1. Interdisciplinarity 2. Environmental impact 3. Eye for alternatives 4. Socio-scientific disputes 5. Dealing with complexity

ethical, aesthetic, and spiritual issues as well as the more conventional technical, economic, social and cultural ones, to provide a new great opportunity for improving education and training in agriculture and integrated rural development.

AFANet continued the initiative to promote sustainability into higher agricultural education establishing a working group under the general item of "Education and Training for Integrated Rural Development" (ETIRD). Integrated Rural Development (IRD) was an emerging concept for raising the level of rural economic performances, promoting the shaping of viable communities, maintaining the native cultures, protecting the environment, and conserving the traditional features of the landscape. On-line quick surveys and a three-year journey of department visits by faculty members belonging to different European countries were organised in a project with the following objectives (Wals et al. 2004b):

1. make an inventory of current curriculum responses to changes in rural land use within European Institutions for Higher Agricultural Education;
2. investigate Communication, Education and Training (CE&T) programmes that were suitable for developing the notions of IRD among students, that is, problem-based learning, systems thinking, interdisciplinary learning, social learning;
3. describe four case studies of CE&T programmes for IRD in Europe (focusing on learning goals, learning and instruction methods, contents and learning outcomes);
4. share and reflect on participants' experiences with alternative learning and instruction methods for teaching and learning for IRD;
5. provide general guidelines for the development of CE&T programmes for IRD in Europe, complemented with examples of the contextual application of these guidelines.

One the four case study selected was the specialisation course in "Ecological Agriculture" held at the University of Tuscia, Viterbo (Italy), as one of the partners of the Common European Degree Level of Specialisation in "Ecological Agriculture". Detailed description of the four case study is in Wals et al. (2004a).

The ICP European working group on joint curriculum adopted a denomination ENOAT (European Network of Organic Agriculture Teachers) and developed next a plan for design and implementation of a MSc degree course in Agroecology still running in Europe, as an academic field for describing, analysing and improving farming and food systems according to the sustainability paradigm (Caporali 2000; Francis et al. 2003). The ICP group, or ENOAT group, recognised that sustainability of people and communities, stewardship of natural resources, and security of future food supplies are joint topics to develop in systems education for tomorrow's professionals. Agroecology has a positive role to play in society in order to create new culture and responsibility for an integrated rural development, and for a healthy and secure food system. Sustainable agroecosystems, both at the farm and landscape level, can help in providing practical and meaningful example of good social, economic and environmental practices and in generating an educational context for ecological literacy and behaviour. Figure 6.41 summarises the main criteria for the organisation of the MSc degree course in Agroecology.

Agroecology, as a major transdisciplinary paradigm, links together the biological, physical and social sciences in a systemic view. Curricula are "learning systems" representing the real systems, physical or abstract, to which they refer. The first facet of integration to be considered is the correspondence between the representation system (curriculum) and what is represented (real system), which is properly a matter of ontology and epistemology. The former refers to what is, the latter to our understanding of it. Epistemological, ontological and methodological tools inspired by the systems paradigm can be helpful in bettering connections between a curriculum as a whole and its context (external tools) and among the curriculum components themselves (internal tools).

On the front of theory, epistemological and ontological tools have a practical application on developing sustainability indicators suitable for monitoring, judging and bettering agroecosystem performances at different scale, with the final intent of connecting them in a coherent framework of reference. On the front of praxis, the involvement of agroecosystem stakeholders in research and didactic initiatives is essential to grasp the complexity of agriculture as a human activity systems, to understand the linkages between different scales of operation, and to build up trust in mitigating and eventually resolving conflicts and contrasting opinions and interests.

Building on the concept of learning adopted by ENOAT European Nordic universities for the implementation of their MSc in Agroecology, Lieblein and Francis (2007) put emphasis on linking cognition and practice through a learning ladder metaphor that integrates a personal dimension into the learning landscape, in addition to cognitive elements (Fig. 6.42). Lower level skills are acquired through training and learning facts, while principles and theories can be memorised and internalised for applications. Moving up the ladder corresponds to the entry point

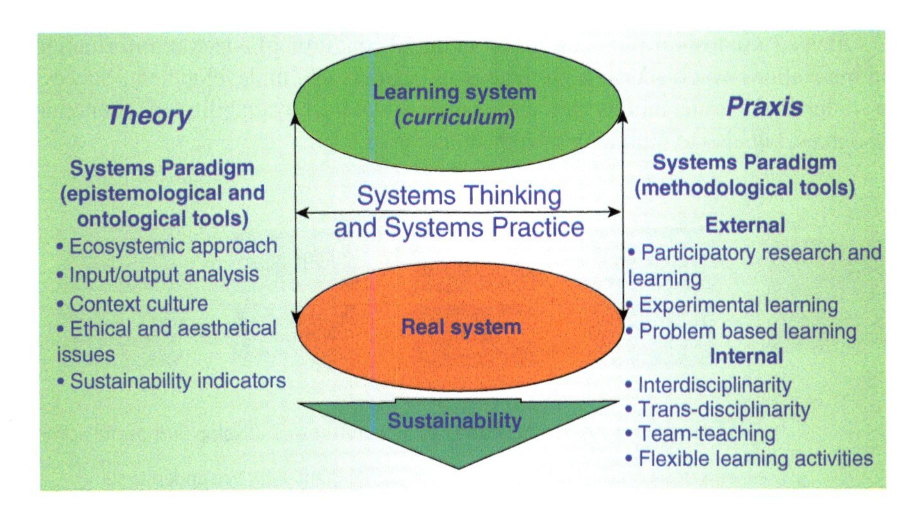

Fig. 6.41 Criteria for the organisation of the MSc in Agroecology (Wals et al. 2004a; Caporali 2007; Caporali et al. 2007)

for the MSc students in Agroecology, where they begin at the level of the farming or food systems and use their past knowledge to link practise with theory. When there is enough confidence to move toward action, students (often in teams) begin to envision and eventually responsibly suggest future options that can improve the farm or food system according to the goals of the farmer or the key client in the community.

In agroecological education, experiential based approach of farming and food systems integrates with relevant theory on concepts and methods, mixing cognitive and personal learning as necessary steps towards responsible actions as professionals (Lieblein and Francis 2007).

In view of enhancing knowledge building in rural areas, experiences from a partnership between an education-research centre of the University of Tuscia (Central Italy) and a rural SME (small-medium enterprise) within the framework of the BSc in Ecological Agriculture and the MSc in Agroecology have been investigated and documented under the sociological profile (Cannarella and Piccioni 2005). The focus was on the stable and long-term collaboration that qualifies entrepreneur-researcher relations and determines their success or failure. The possibility of establishing a scientific partnership presupposes the presence of pioneers and the emergence of innovation demand. This is a positive precondition for a flexible mentality that, together with a spirit of initiative and creativity, can be translated into a 'culture of innovation'. Three key interconnected factors contribute to this kind of cultural turn:

(a) time for adjustments (benefits from the implementation of innovation tend to decrease when the time required for adjustment increases);
(b) sensibility to problems (problems are or are not considered as operative impulses or as incentives with sense of responsibility);
(c) involvement (more the human resources involved, more is the effectiveness of innovative actions).

Often a rural firm or farm can suffer an initial condition of a lack of information on innovation, which acts as a bottleneck in business and in development. Success or failure of partnership is not exclusively linked to availability of economic resources, but also to mutual involvement and trust.

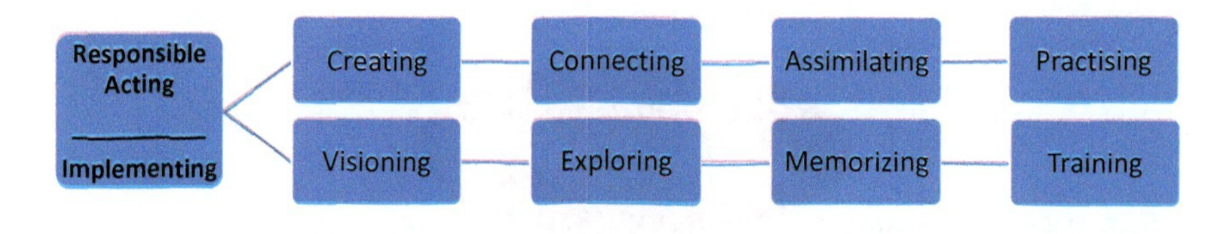

Fig. 6.42 Steps on the dual learning ladder towards responsible action as agroecologists. (Adapted from Lieblein and Francis 2007)

A constant topic of research and didactics within the MSc in Agroecology at the University of Tuscia has been the measurement of sustainability through Agricultural Sustainability Indicators (ASIs) (Caporali et al. 1989, 2003; Tellarini and Caporali 2000; Caporali 2010). The necessity to stress the importance of the decision-making process in society is well documented by the chapter 8 of Agenda 21 "Integrating environment and development in decision making", where the recommendation is "countries could develop systems for monitoring and evaluation of progress towards achieving sustainable development by adopting indicators that measure changes across economic, social and environmental dimensions". With the development and use of ASIs, agroecological research is getting more and more integrated in the structure of civil society, improving its role of scientific service for public utility.

The aim to develop ASIs has both an epistemological and a practical meaning, representing, respectively, (a) an efficient instrument of enquiry for studying agroecosystems functioning and performances according to an input/output approach, and (b) a relevant knowledge basis for both the design of sustainable agroecosystems and the decision-making process. According to Tellarini and Caporali (2000), ASIs can be subdivided in two large categories: (a) structural indicators, and (b) functional indicators. The first category aims to describe the most relevant components of agroecosystems, and therefore, to illustrate the differences and similarities between agroecosystem components, while the second category aims to measure the efficiency of transformation processes in agroecosystems. Structural indicators can be calculated in terms of numerical expression of component occurrence, extension, and distribution in spatial and temporal scales, while functional indicators can be calculated in terms of flow of energy-matter and information, including monetary exchanges. By relating each type or combination of output to each type or combination of input, it is possible to obtain a considerable amount of data on both the circulation of energy-matter and information within an agroecosystem and between an agroecosystem and its context. In this way, an assessment of resource use efficiency, both internal and external to the systems, can be made and a judgement of sustainability formulated for both the whole agroecosystem and each of its components. In other words, the sustainability level of the agricultural activity can be evaluated through a process-oriented assessment.

A recent paper of Wezel et al. (2018) provides a mapping of the current state and development of agroecology in Europe in its diverse forms: research, education, collective action networks and alternative food systems. Research and education are major components to agroecology in Europe and offer a bridge between science and practice. Policy development will be crucial to amplify agroecology and its proponents "must join forces and work hand-in-hand with the many stakeholders engaged in initiative to develop more sustainable agriculture and food systems".

A strong ecological intensification of agriculture involves both reducing external inputs and promoting better use of ecosystem services at field, farm and landscape level (Bergez and Therond 2019). New management and governance are required for achieving these goals at both farm and territorial levels (Magrini et al. 2019). A well-documented example of best practices on these topics for local organisation of stakeholders at basin level is in Bergez et al. (2019).[7] The first part of the book presents issues and drivers of the agroecological transition process, e.g. a systemic transformation consisting in the ecologisation of agriculture and food, with a discussion of the values underpinning both agricultural production and food consumption choices. The agroecological transition process has the following characters (Magrini et al. 2019, p. 69):

> (i) transition takes place over time intervals that vary, depending on the analysis scale (the farm or the agri-food system as a whole); (ii) transition is complex, systemic and requires changes of the whole sociotechnical regime; (iii) transition implies strong connections between niche-innovations and the dominant sociotechnical regime; and (iv) changes in values and individuals' abilities are fundamental drivers.

The second part of the book concerns the methodology used for supporting the territorial agroecological transition (AET) design and feedback from the TATA-BOX Project Experience. The TATA-BOX project aimed at developing a participatory toolbox to support local stakeholders in the design of an agroecological transition at local level. The purpose of the TATA-BOX project was the operationalisation of the conceptual and methodological frameworks proposed by Duru et al. (2015) for designing an AET. A new configuration of stakeholders, resources in the farming systems, supply-chains, and natural resources management, to form the agroecological transition, was adapted and tested on two adjacent territories in south-western France (Audouin et al. 2019). The book provides a detailed description of the participatory methods and the multimodal intermediary tools used to support the collective design of AET, and a projected action plan for transition from the initial to the desired agriculture and associated governance structures.

6.5.2 *Indicators of Ecological Intensification in Contrasting Farming Systems*

In one of the seminal papers on ecological agriculture, Kiley-Worthington (1981) pointed out that ecological agriculture has as first requirement to be *self-sustaining* and as second one to be *diversified.* Moreover, he advanced that "one of the first thing for the correct running of the ecological farms is to assess the ratio of land for arable to that for livestock". Reporting data on organic farms he visited in Europe, he suggested that for this ratio "the self-sustaining character of the enterprise is around 30%", thus providing a basic *structural indicator* of sustainability as the percentage of arable area to pasture on self-sustaining organic farms in Europe. This

[7] This book is based on research funded by the Agence National de la Recherche [French National Research Agency], which was carried out under The AGROBIOSPHERE programme, as part of the project "Transition agroécologique des territoires: une boite à outil pour concevoir et oeuvr mettre en une transition agroécologique des territoires agricoles avec les acteurs locaux (TATA-BOX)" (Agroecological Transition of Territories, a toolbox for designing and implementing an agroecological transition of agricultural territories with local actors).

indicator sounds prophetical in advancing the solution for farming system "greening" incentives that characterises the current European policy for payments to farmers. Looking at the composition of the cropping systems, he noted that the organic farms he visited had complex rotations, of 7–12 years, including several crops to increase or maintain available nitrates, such as the legumes, and green manuring crops that were ploughed in, such as rape. Measures of crop diversification as quantity and quality of crops grown constitute another group of structural indicators for evaluating farming system sustainability in terms of biological complementarity in both production and protective agroecosystem performances. Crop diversification is currently another option for farmers' eligibility to European direct payments.

The farm level is the more appropriate one in the hierarchical scale of agroecosystems for doing research and making decisions in favour of sustainability. Indeed, the farm is the management unit of agriculture with a biological base, easily identifiable because of its boundaries, and which represents the meeting point between human interest and the natural environment (Caporali et al. 1989). The most advanced regulations for sustainable agriculture, like those concerning organic farming, provide a framework of legitimation based on the agreement between the civil society and the farmer. That means that there is an explicit recognition of the farm as the crucial level of organisation of resources, both biophysical and socio-economical. Indeed, the production process at the farm level is able to affect the profile and sustainability of the next upper level of the agricultural systems (landscape level) as well as the next lower level or field level, where interactions between crops, animals and microorganisms affect soil fertility, which is the basis for agricultural sustainability.

Research based on ASIs at the farm level is therefore of great importance to decision-making processes, especially when groups of farms of contrasting management are involved. Organic farming systems are being considered as long term benchmark for the evaluation of apparently environmental benign agricultural production (OECD 1999). Therefore, the aim of research often includes the comparison of organic farming systems with conventional ones through appropriate ASIs (Reganold et al. 2001; Mader et al. 2002; Caporali et al. 2003; Reganold and Wachter 2016). Results of this kind of research are easily shown graphically with the help of a sustainability polygon or web, which simultaneously displays scores for different indicators and avoids having to aggregate across different scales.

An example of such a research is provided by Caporali et al. (2003) with farm-level indicators mainly based on farm structural components associated with farming and cropping systems diversity, in the framework of a comparison between two groups of farms, 18 organic and 15 conventional, located in the Tuscia region (Central Italy). Under a Mediterranean climate, which is normally characterised by three dry months in summer, crop productivity is drought constrained in rain-fed conditions. Consequently, farming and cropping systems need adjustments to summer drought if agriculture sustainability is to pursue. In this area, the most common crops are both tree and herbaceous crops. Olive, vine and hazelnut are predominant among the tree crops, while winter cereal crops, mostly hard wheat, and summer crops, like sunflower, are predominant in rain-fed conditions. In irrigation conditions, maize is the predominant summer crop. Crops are grown either as sole-crop or in mixed cropping systems with forage crops, such as Lucerne. In that paper (Caporali et al. 2003), ecological intensification within farm is meant as increase of *ecotopes* (i.e. kind of land use, like cropland or woodland, both as number of plots and size) and increasing *diversification*, both as kind of crops (number and size) and kind of animal reared. A total number of 15 indicators was used (Table 6.11), whose basic data were collected through direct interview with the farmers using an appropriate questionnaire. Farmers joined on a voluntary basis with the involvement of local farm union representatives. By summing up all the scores concerning 15 indicators, it was possible to get a global index (GI) that numerically summarises the performance on ecological intensification, diversity and sustainability of each farm or group of farms. The GI of a farm could reach the maximum score of 150 (15 × 10) in the case that a farm scores best in each kind of indicator, having weighted between 0 and 10 the relative score of each indicator.[8]

Organic farms scored better than conventional ones in 13 out of 15 indicators. The GI of the two farm clusters, 77.19 vs. 56.06, for organic and conventional farms respectively, confirms significantly that organic farms scored 1.38 times better than conventional ones.

As to the assessment of the ecological intensification gradient, the evaluation concerned:

(a) the farming system level, as *ecotopes* or land use diversity;
(b) the cropping system level, as number of crops grown, field size and density, kind and size of crops that can fix atmospheric nitrogen, such as legumes, which enhance the sustainability of cropping systems through a complementary use of natural resources in the rotation;
(c) the livestock rearing system, in that the number of animal species reared indicates the level of ecological intensification due to the integration of grazing and detritus chains, indispensable to enhance soil fertility through the increment of poly-annual forage crops, as Lucerne, and farmyard production and fertilisation.

As to the ecological intensification at farming system level, three indicators – wood area/farm area, number and wood plot size – are particularly meaningful. Their values show that organic farms scored 1.94, 1.95 and 1.57 times better than conventional ones. Organic farms had a larger (18.7% vs. 7.0%) and more dispersed wood area in the arable land, with more benefit for the biological diversity and the associated ecosystem services due to the ecotone effects.

As to the ecological intensification at cropping system level, biodiversity assessment is best expressed by the occurrence and spread of legumes, especially of Lucerne, which is the most important forage crop in a Mediterranean environment.

[8] The relative values in the range 0–10 for each spine or indicator of the web of sustainability was allocated to each farm or group of farms by taking into account that the value 0 and the value 10, respectively, correspond to the minimum and maximum absolute values recorded in the whole group of 33 farms checked.

Table 6.11 Indicators of structural biodiversity in organic and conventional farms grouped according to their gradient of ecological intensification (adapted from Caporali et al. 2003)

Farm indicators	Measure units	Organic farms (average size 54.7 ha)	Conventional farms (average size 70.4 ha)
Ecotope diversity			
Wood area/farm area	(%)	18.72	7.08
Number of wood plots	(n)	1.89	0.73
Wood plot size	(ha)	3.74	3.14
Wood area	(ha)	8.65	5.35
Cropping system diversity			
Arable area/farm area	(%)	84.26	92.37
Number of crops	(n)	4.44	4.27
Number of tree crops	(n)	1.67	1.20
Number of herbaceous crops	(n)	2.78	3.07
Arable area sown to legumes	(ha)	24.39	14.28
Arable area sown to Lucerne	(ha)	14.14	7.20
Fields density	($n\ ha^{-1}$)	0.47	0.25
Tree crop field density	($n\ ha^{-1}$)	0.84	0.83
Herbaceous crop field density	($n\ ha^{-1}$)	0.48	0.15
Occurrence of animal husbandry			
Occurrence of mixed farms	(%)	60.00	40.00
Animal species reared	(n)	1.11	0.53

The occurrence of forage legumes is crucial for both biological diversity and sustainability, since they involve the occurrence of animal husbandry as an enterprise in the farm system and since they supply biologically fixed N to the succeeding crops (Caporali and Onnis 1992). Organic farms scored 1.74 and 1.19 times higher than conventional ones for the area sown with legumes and the area sown with Lucerne, respectively. As an average, the percentage of arable area with legumes was 53% in the organic farms, while it was only 22% in the conventional ones. Moreover, Lucerne was grown on 31% of arable land in the organic farms, while it covered only 11% of arable land in the conventional ones. As to animal husbandry, the indicator of livestock occurrence scored 1.5 times higher for the organic farms (60 vs 40%), showing that the predominant typology of the conventional farm was not that of a mixed farm.

Other indicators useful in describing the cropping diversity conditions, especially the non-planned biodiversity, are those that concern the field density (number per hectare) of both tree and herbaceous crops. The higher the field density (or the smaller the field size), the greater the possibility of having field margins available for biological colonisation by plant and animal communities. In terms of landscape ecology science, a farm with a higher field density provides more ecotones and a more fine-scale diversity, which is functionally and aesthetically an important component of ecological intensification (Haber 1990; Kuiper 2000; Elliot 2017). Two out of three indicators of field density used in this survey – density of fields and density of herbaceous crops – scored higher in the organic farms than in the conventional ones (1.49 and 1.40), respectively.

From this survey emerges a picture of a typical organic farm of Tuscia region, where the average farm size is around 54 ha with diversified land use shared between cropland (81.3%) and wood land (18.7%), with diversified, ecologically intensified cropping systems, and with integration between crop and animal husbandry.

6.5.3 Ecological Intensification in Animal Husbandry

Livestock rearing is itself a means of ecological intensification of farming systems. Major principles that are at the basis of *standards* for organic farming (Woodward et al. 1996) and refer, directly or indirectly, to the importance of animal breeding are listed as follow:

⇒ work as much as possible within a closed system and rely on local resources;
⇒ maintain soil fertility;
⇒ avoid any form of pollution that is likely to result from agricultural techniques;
⇒ produce high nutritional value products in sufficient quantities;
⇒ minimize the use of fossil energy in agricultural practices;
⇒ guarantee livestock life conditions consistent with physiological needs and humanitarian principles;
⇒ allow farmers to earn their living and develop their potential skills;
⇒ use and develop technologies based on the understanding of organic systems;
⇒ use decentralized systems for product processing, distribution and marketing;
⇒ set up a system that is aesthetically pleasing for those who live inside and outside the system;
⇒ maintain and preserve natural forms of life and their habitat.

All aforementioned principles represent interdependent aspects of an integrated agroecosystem framework oriented to sustainability through ecological intensification.

Ecological intensification is to work as much as possible within a closed system, to rely on local resources, and to integrate nutrient chains on-farm, which guarantees a high degree of autonomy to the system thus making it productive and sustainable at the same time. Livestock simultaneously plays a dual role as herbivores (grazing chain) and activators of the chain of detritus and soil fertility through their excreta released on the field or following its distribution after proper composting. For animal husbandry, local resources are above all fodder crops that cover, at the farm level, a surface suitable for the maintenance of the raised animals. To achieve animal nutritional autonomy through on-farm grown fodders and grains is a condition of ecological intensification that allows an on-farm circular economy, distinguishing organic from conventional agriculture, where animals are most fed concentrates produced from off-farm resources.

Ecological intensification of animal husbandry can happen through *multispecies grazing*, i.e. the use of more than one species of large herbivores to graze a common forage resource (Walker 1994):

> These benefits are the result of different dietary habits of the animals because plants avoided by one kind of livestock may be relished by another. Differences in dietary habits are related to the physical limitation on the ability to select and the physiological limitation on the ability to detoxify forage phytochemicals. Compared to cattle, sheep diets usually have more forbs and less grass. Sheep can graze lower in the forage canopy, have a greater ability to select from a fine-scale mixture and have a more varied diet than cattle.

The ecological benefit results in more efficient utilization of forage resources and increases sustainable production. Multispecies grazing on small areas can be performed with small herbivores as rabbit as well (Fig. 6.43).

Soil fertility is at its apex when maximum integration of the grazing and detritus chains occurs at farm level. There are two main recommended ways of ecological intensification to maintain and increase soil fertility. The first consists of rotating crops conveniently so as to have poly-annual fodder crops suitable for enhancing organic matter content belowground and providing basic nourishment for livestock aboveground. The second way is the distribution of animal excreta on the field, as such or composted, for supplying decomposers with energy-matter and promoting humification. These two ways integrate and are strengthened in a mixed farm as an effective spot of ecological intensification.

Productivity of conventional husbandry is mainly based on the use of animal feedstuff and veterinary drugs that can easily contaminate the nearby and far-off environment. Within the framework of organic agriculture, the use of these synthetic compounds to feed and protect both plants and animals is not allowed. Therefore, the farm organisation must count on prevention of infestations from weeds, plant pathogens, phytophagous insects and animal parasites.

Providing livestock with life conditions consistent with their physiological and behavioural needs is an ethical rule that is emphasised in organic farming. Within the framework of an ecological culture, animals are entitled to live in accordance with their physiological and behavioural needs. In particular, they are entitled to live in a healthy environment, even if confined in a shed, that guarantees sufficient capability of movement; and to follow a healthy diet consistent with structural and functional features of their digestive tract (Kamra 2005). In particular, ruminant herbivores – that account for most animals reared – must follow a diet high in fibre (grass, hay, silage, and straw), given their complex digestive tract (rumen) which is able to digest cellulose due to its symbiosis with micro-organisms:

> Rumen, the four-chambered stomach of grazing animals, harbours a complex bionetwork, here all forms of primitive starting from archaea to protozoa exist in close proximity. Some of these microbes interact with each other in a synergistic relationship to extract energy while producing highly active lignocellulolytic enzymes supporting digestion of the host (Choudhury et al. 2015).

Fig. 6.43 Rabbit grazing on an organic dairy farm meadow in Po valley (Northern Italy)

Unfortunately in conventional agriculture, where entrepreneurial activity is often carried out only in accordance with the market rules, livestock are considered as machines turning a good (fodder) into a more valuable good (meat, milk, wool, eggs) and therefore are fed high levels of imported high protein foods in order to increase growth rate or milk yield. The structural and functional conditions of breeding are only optimized for the economic function, where concentration and processing efficiencies are dominant. As a result, animals are stacked in smaller and smaller spaces, and protected against diseases due to their high density. Only through a constant administration of antibiotics, and obliged to non-stop production cycles through the administration of hormones, they are subject to diets inconsistent with their anatomic and physiologic traits but administered only to encourage the most efficient energy and protein conversion from fodder of different origin into animal biomass and its by-products. Health-related problems that have been tragically experienced in Europe following the onset of mad cow disease (Bovine Spongiform Encephalopathy-BSE) which has also been transmitted to man, result from these stressful conditions, especially because ruminant animals (specialized herbivores) have been forced to feed on food derived from animal carcasses through a distortion in the food chain.

Industrial intensification and ecological intensification are the two poles on the axis of agriculture organisation and their conflicting goals and means reflect heavily on the animal husbandry conditions, affecting not only animal health and welfare, but also the environmental health and the economy of rural lands. The industrial organisation of agriculture tends to detach the animal husbandry from its sustaining local fodder resources, replacing them with fodder resources as concentrates coming even from other continents. The consequences at the local level are the concentration of farms on more accessible land, especially located in the valley floor, and the abandonment of mountain farms. Usually, the trend is a reduction of the farm number, the increase of heads per farm, and more land abandonment. Data relative to animal husbandry at the regional level of the Italian Alps are shown in Table 6.12. Losses of meadows and pastures (−26.6%), number of farms (−51.5%), and cattle heads (−22.8%) in only 20 years of recent history, reveal the dramatic change undergone as result of agriculture industrialisation. This process undermines at the roots the traditional rural organisation based on the ecosystem services provided by nature as availability of land and biological integrated resources as natural grasslands and local well adapted variety of livestock. The contrast between the modern tendency toward industrialisation and the traditional rural organisation as described by Battaglini et al. (2014) is summarised in Fig. 6.44.

The balance between highlands and lowlands in terms of human settlements and density of population, including livestock population, is undoubtedly the first condition to regain for a rational ecological intensification of agricultural land at the catchment level. There is a gloomy feeling around the past landscape planning that only urban areas has counted for economic investment and development, while the rest of land has been only "waste". Ecological intensification calls for a repopulation of less favourable lands, which are too far for easy and rapid global trade through standard air and sea communication networks. Mountain and hilly areas are spatial and temporal repository of natural goods and services, such as water and biomass, both as forests and grasslands that are renewable resources under a rational

Table 6.12 Evolution of livestock farming systems in the Italian Alps (adapted from Battaglini et al. 2014)

Year	1990	2010	Variation in 1990–2010, %
Meadows and pastures, ha	1,109,367	812,236	−26.6
Cattle, n			
Farms	43,774	21,221	−51.5
Heads	578,484	446,531	−22.8
Heads/farm	13.2	21.0	+59.2
Dairy cows	275,605	194,440	−29.4
Dairy farms	37,803	15,157	−59.9
Dairy cows/dairy farm	7.3	12.8	+76.0
Sheep, n			
Farms	7901	4402	−44.3
Heads	175,274	191,713	+9.4
Heads/farm	22.2	43.6	+72.5
Goats, n			
Farms	7221	4442	−38.5
Heads	84,455	89,625	+6.1
Heads/farm	11.7	20.2	+72.5

agroforestry use. The mosaic of woodland, grassland and farmland that is crown to each river catchment in every part of the world is a product of coevolution between man and nature that is proof of a past synergy operating for health, wealth, and beauty. To "waste" grassland pasture at home, and consequently, to enlarge one's ecological footprint upon other continents in order to feed one's own animals with grain protein is good only for trade but not for local and national economy, and for the global environment. Nevertheless, it happens in Europe, as Teixeira et al. (2018) recently report on a chapter titled "Livestock fodder production overseas: soy":

> The EU imports about 35 million tonnes of soy (*Glycine max*) every year, mainly from South America, which corresponds to about 35% of the global soy trade [...] About 95% of the soy produced in South Amerika is genetically modified (GMO) [...] Soy production has been one of the main drivers causing the loss of primary forests, areas of *cerrado* and unique wetlands in the Amazon, Pantanal, and Mato Grosso regions [...] Considering that the European legislation does not apply abroad, the production of fodder in Europe is generally advantageous to imports from South America, with respect to biodiversity and additional environmental concern.

The abandonment of less favoured but productive agricultural land would be really short-sighted (Marton et al. 2016), both for Europe and other continents.

Transhumance is the traditional movements of livestock from lowlands to highlands in Mediterranean areas to adjust nutrition requirements to seasonal grassland production. Transhumance is an ancient tool of ecological intensification for the use of local nutritional resources by creating a network of pasture steps located at different altitudes in order to furnish grazing areas spread over time according to the availability of plant biomass. During late spring the herd movements are upwards and during autumn downwards. "Transhumance shapes relations among people, animals and ecosystems. It involves shared rituals and social practices, caring for

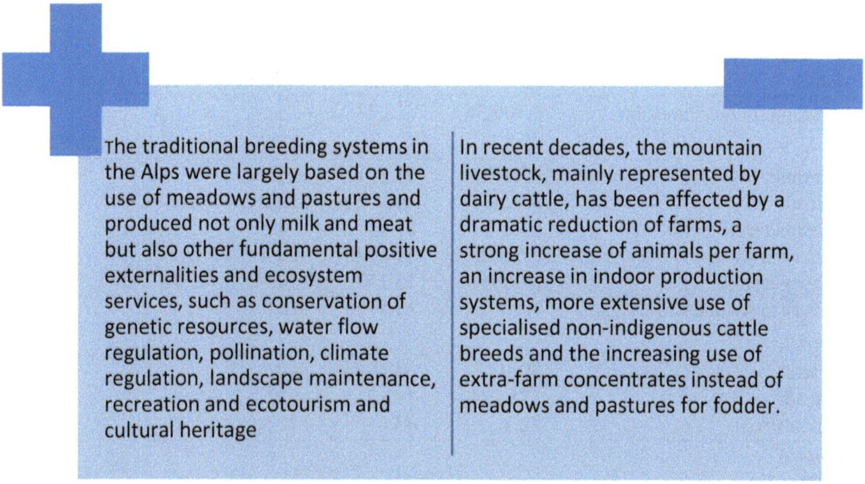

The traditional breeding systems in the Alps were largely based on the use of meadows and pastures and produced not only milk and meat but also other fundamental positive externalities and ecosystem services, such as conservation of genetic resources, water flow regulation, pollination, climate regulation, landscape maintenance, recreation and ecotourism and cultural heritage

In recent decades, the mountain livestock, mainly represented by dairy cattle, has been affected by a dramatic reduction of farms, a strong increase of animals per farm, an increase in indoor production systems, more extensive use of specialised non-indigenous cattle breeds and the increasing use of extra-farm concentrates instead of meadows and pastures for fodder.

Fig. 6.44 Pros and cons of traditional (ecological) intensification and industrial intensification of agriculture in the Italian Alps. (Adapted from Battaglini et al. 2014)

and breeding animals, managing land, forests and water resources, and dealing with natural hazards. Transhumant herders have in-depth knowledge of the environment, ecological balance and climate change, as this is one of the most sustainable, efficient livestock farming methods" (UNESCO 2019). For these reasons, UNESCO (14.COM 10.b.2) inscribed transhumance in 2019 on the Representative List of the Intangible Cultural Heritage of Humanity.

The traditional Alps management system for dairy cows (*Malga* system, see Box 6.11) consists of animals kept indoors from autumn to spring, mostly in tie-stalls, and moved to mountain pasture in summer. In general, studies have suggested that pasture is beneficial for cows' welfare because it allows the reduction of hock damage, lameness and claw disorders (Corrazzini et al. 2010). Therefore, grazing seems to be advantageous for cows' welfare and authorities and consumers show an increasing interest in animal welfare themes. Moreover, pastures and meadows are also one of the most important subjects of "greening" in European agricultural

Box 6.11 The Malga System in the Alps as a Leverage for Contrasting land Abandonment and Depopulation

In the mountain environment of Alps, the climatic constraint of seasonality for biomass production has been crucial in inspiring solutions for livestock husbandry through an integration of forage resources between lowlands (in winter) and uplands (in summer). This integration practice is called *alpeggio* in Italian, *alpage* in French, and *Bergweide* in German, and consists in moving livestock from lowlands to upland pastures in late spring and coming back at the end of summer. In this way, forage resources collection happens according to a basin level of organisation, on the base of the ecological principle of bringing livestock to forage instead of forage to livestock. This traditional alpine pastoral model is referred to as a "Malga system":

> A Malga system comprises a community-owned highland grassland area, with pasture or meadow use, that is associated with a mountain hut, cheese production facilities, and cattle sheds. The land is traditionally managed by a cattle herder and/or dairyman under community rules on behalf of cattle owners (Soane et al. 2012).

> A Malga system is an alpine *decentralised* farming system operating at different altitude and at different time in the whole river basin, which unveils appropriate adaptation of farming organisation and best practices of agro-forestry for animal husbandry to availability of biomass resources. For this reasons, it is paradigmatic example of ecological intensification for the best land use and production of ecosystems services in appreciated pastoral landscapes (Ianni et al. 2015; Krauss and Olwig 2018).

Points of strength for revitalising mountain communities in the Alps (after Ianni et al. 2015, modified)

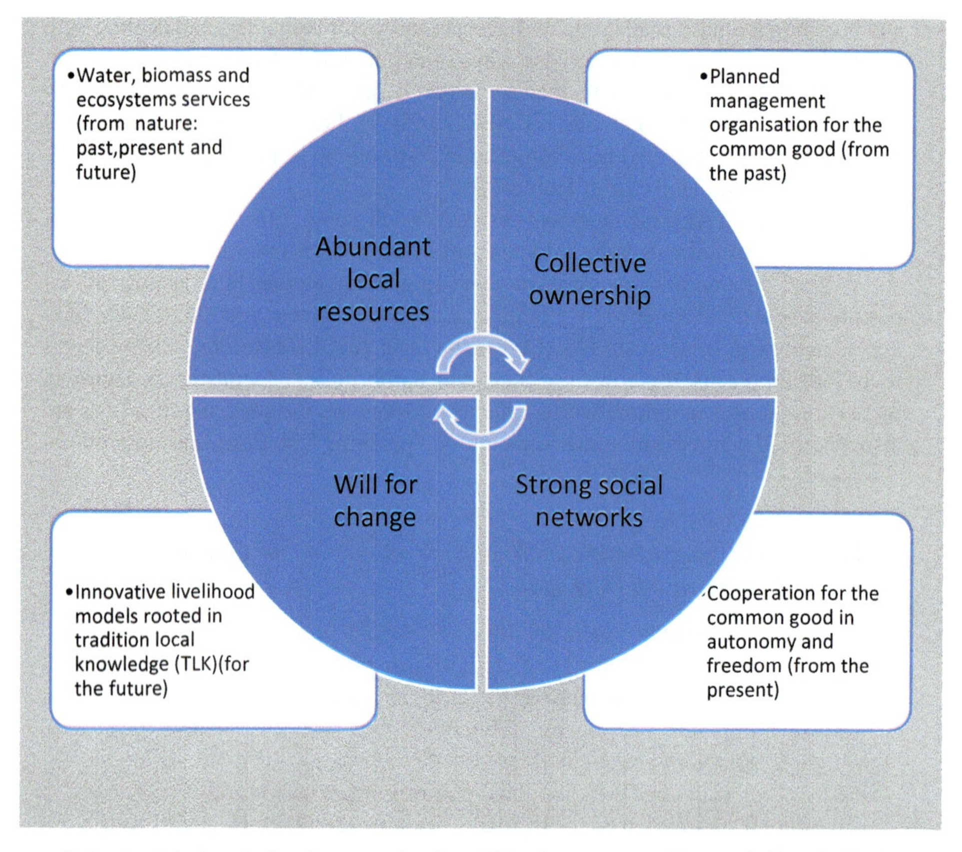

policies and their rule implementation is still in the process (Corrazzini et al. 2010; Krauß and Olwig 2018).

Systemic description of the Malga social-ecological system, showing the key players and processes for maintaining dairy production but also local biodiversity and related ecosystem services, can support better and more effective understanding of the dynamics involved as a source of inspiration for further development:

> The process of constructing and discussing system models can itself help groups of researchers, policy makers and land managers to obtain a greatly enhanced understanding of complexities involved (Gretter et al. 2018).

When a strong sense of local participation prevails and a degree of political and economic autonomy exists, a political ecology deeply rooted in democratic processes and cultural understanding can help find strategies for supporting both sustainable revitalization processes and the conservation of TEK (Traditional Local Knowledge) (Krauss and Olwig 2018). Modern systems of integration between highlands and lowlands are stepping-stones on the scale of ecological intensification of both agriculture and socioecological systems in general. They can also emerge from regional farmers' agreements, such as those mentioned by Marton et al. (2016) in Swiss:

> Farmers developed a collaborative dairy production scheme, where they take advantage of the specific environment of the two regions. In this contract rearing system, the young stock is reared on a mountain farm and the more intensive milk production is performed in the lowlands. This system is an example for the principle of comparative advantage, where each region focusses on the activity where it has the lowest opportunity costs.

In that system, the less requiring phase in the life cycle of a dairy cow is shifted to the less favoured highlands, while the most requiring phase is maintained in the favoured lowlands.

In the concept of ecological intensification, the ethics of farm animal husbandry is an important component. Appropriate animal breeding criteria, such as those suggested by Kiley-Worthington (1981), provide the basis for a careful livestock farming. They constitute the prerequisites for an ethics of care for animal husbandry and a series of self-explaining guidelines as summarised below:

(i) the farm must be economic, but the prime reason for its existence is to maximise production to feed people and animals, at as little environmental and social costs as possible;
(ii) the animals must be able to grow on low concentrate diets rather than selected for high performance on high protein diets;
(iii) the animals must be adapted to the local environment in that local varieties are likely to be more successful than imported ones;
(iv) the animals must be diseases resistant, and require low levels of medicaments;
(v) the animals must be behaviourally suited for such a system(for instance, cows usually suckling and looking after their calves, and having appropriate maternal behaviour);
(vi) the main aim should be towards high quality of animal product (for instance, milk with a high butter-fat is more appropriate when processed on the farm);
(vii) the selection for breeding animals with longer working lives is more important for the small producer because his profit per animal may be greater (due to lower input, less problems of diseases, lower capital expenditure, and direct marketing to the public).

On the base of the criteria listed above, Kiley-Worthington (1981) advances an important suggestion for promoting an ecological intensification of farming systems at the global level that has both a huge political and ethical meaning:

> The important agricultural development in the next few decades must inevitably be towards increasing the food self-sufficiency of developing countries, rather than encouraging them to grow cash crops off high input systems in order to earn currency to buy food from the west, which is produced at great energy costs. The latter is also, biologically unsound.

Industrialised countries are now far from these criteria of ecological intensification because the agriculture and animal husbandry development criteria have followed the opposite paradigm of industrial intensification, where external inputs constraint the rationale of farming organisation at any level, from the farm to the international structure and functioning of the human food system. Nevertheless, a turning point is required to face increasingly deteriorating environmental, social, and ethical conditions that call for global justice and care. An interesting systemic perspective for tackling this current human predicament, that involves also the animal breeding, comes from Nijland et al. (2013). The paper frames the topic of careful livestock farming into a systems view that extends from the self to the biosphere and examines the state and the prospect of animal husbandry in a context that links the particular with the global. They claim that the transition towards careful livestock farming systems calls for a joint and balanced effort of all stakeholder groups involved: the livestock sector, government, NGOs, retail, and consumers. All these stakeholders exist as a community of leaving beings that renew themselves in an evolving environment that is home to the present and future generations. Cultural evolution requires now a quality change that reduces the *psychological distance* between the ontological distances represented in Fig. 6.45, that links in a unique spatial, temporal and relational scenario the particular (the self) and the global (the biosphere and the future generations).

In this framework, the private (the farmer and the other single stakeholders) and the public (people, in the name of animals and the whole biosphere as well) interests must negotiate to find some points of agreement. Nijland et al. (2013) recognise that in current-day politics, production, trade and consumption of livestock, the main emphasis lies on bestowing care to the own group and the self in the form of profit maximization and food safety. On the other hand, the current societal debate on livestock production systems has brought impacts on animal welfare and longer term societal levels more to the forefront, not only at the level of consumers, but across all stakeholders in the livestock production network. Therefore, it is important to raise questions, such as those reported in Table 6.13, as an enquiring methodology for a participatory dialogue, balancing conflicts and seeking solutions

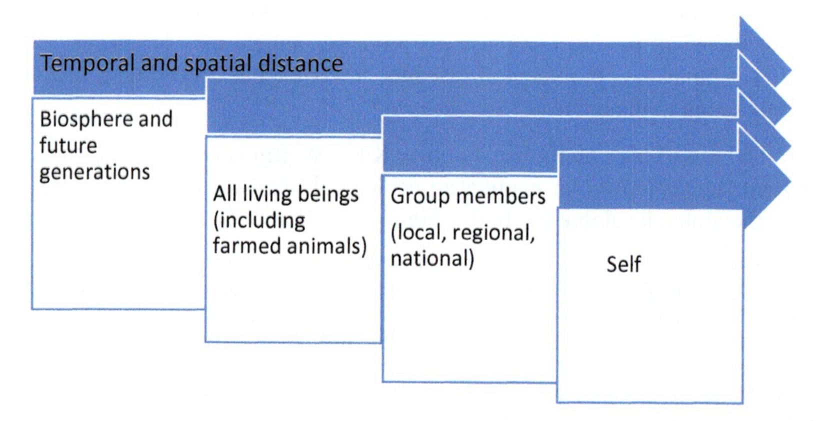

Fig. 6.45 Levels of ontological distances that relates the self to the other components of the extant universe on temporal and spatial scales. (Adapted for Nijland et al. 2013)

Table 6.13 Methodology for formulating and acceptable theory of careful livestock farming in the framework of the levels of ontological distances (adapted from Nijland et al. 2013)

Questions	Alternative motivations	Balancing
What levels of psychological distance are taken into consideration when designing careful livestock farming practices? What concrete farming system (product) features intended to help achieve carefulness are formulated? How are the benefits distributed over the levels?	(A) In regular current-day politics, production, trade and consumption of livestock, the main emphasis lies on best owing care to the own group and the self in the form of profit maximization and food safety (B) Inclusion of attention for animal welfare, environmental impact and/or long term global food security– whether by extending care directly to other living beings, the biosphere and/or future generations, or indirectly, by extending care to citizens and the taking into account of their concerns	Social objectives are next to immediate personal satisfaction The more levels are allowed to play a role and the better the accompanying benefits are balanced, the more complete and nuanced the dialogue on careful livestock farming becomes The closer to the self the selected levels are, the closer to home relationally, the less collective-minded, and the shorter-term oriented the approach is

shared by policy, stakeholders and the public, in order to attain consensus through increasing *perceived value* and *regulation of supply*. *Perceived value* is "increasing the demand through the recognition and valuation of the societal benefits of a system and its products. This is the route of communication, awareness raising, and education". *Regulation of supply* "entails the sector as a whole taking responsibility to achieve a certain standard of livestock production practice beyond current minimum legislation, for example through signing covenants; or the government imposing new legislation".

According to the authors, "taking care of the self and making a profit can be done not only in spite of but by taking care of larger circles, and that allocation of carefulness beyond the self and profit can indeed be defined and institutionalised in a combined effort of consumers, producers, retail and government".

Synthetic indicators of ecological intensification vs industrial production for livestock farming systems can be helpful in understanding the problem and finding agreements. In this respect, data on *Total milk production/cows lifetime* as reported in Fig. 6.46 are very illuminating on how the "niche" concept can operate in contrasting conditions of animal husbandry according to Kiley-Worthington's description:

> The appropriate way to design the conditions under which cattle live on an ecological farm is to keep them at grass with a bull and eating cellulose for as much of the year as possible [...] Designing and keeping cattle in environment suited to what they evolved to live in has the advantage that it reduces physiological stress and pay off in a longer life expectancy, an increase in fertility and ease management, although growth rates and milk yield may not be as high as on concentrate diets (Kiley-Worthington 1981).

In the organic farms (Fig. 6.46), where the conditions of ecological intensification are better satisfied, the dairy cows had a 2.8 time greater life expectancy, expressed as number of lactations, and a 2.2 time greater total milk yield than the

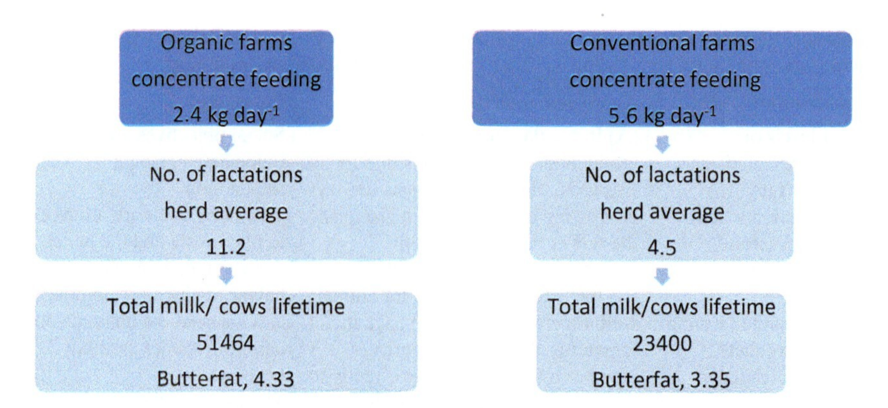

Fig. 6.46 Milk production from organic and conventional farms in similar localities and diversity of concentrate inputs. (Adapted from Kiley-Worthington 1981)

conventional concentrate fed cattle, even if the heard average milk yield for lactation was 4595 vs. 5200 l.

According to FAO (2014), achieving sustainable diets in animal husbandry through a reduction of concentrates is both an ecological intensification strategy and an ethical step toward an agroecological transition with the following implications:

> The concept of Sustainable Animal Diets[…] integrates the importance of efficient use of natural resources, protection of the environment, socio-cultural benefits, ethical integrity and sensitivity, in addition to currently recognized nutrition-based criteria for delivering economically viable, yet safe, animal products by producing safe feed (p. vii).

> Sustainable Animal Diets are expected to be beneficial for the animal, the environment and society, and are likely to generate socio-economic benefits, furthering poverty alleviation and food security efforts. This requires active participation of researchers, extension workers, science managers, policy-makers, industry and farmers (p. viii).

Minimising the use of fossil energy in agricultural practices is a hard task considering that in conventional agriculture such energy comes into the agricultural system in the form of machinery, fuels, industrial fertilizers and pesticides. This energy consumption can be even higher than that obtained in the form of agricultural products. The integration between animal and plant breeding offers opportunities for great savings in terms of fossil energy. Solar energy fixation through biomass and atmospheric nitrogen fixation through leguminous crops, and the subsequent farm use of C, N and other macro and microelements by grazing and detritus chains, are enough to guarantee productivity and sustainability of the system. Animal breeding guarantees the attainment of valuable products from both the nutritional and economical point of view. Furthermore, farm complexity related to the adoption of mixed breeding, undoubtedly improves professional, entrepreneurial and ethical skills with essential benefits in terms of personal satisfaction and social acknowledgment. Using and developing technologies for organic systems is a professional skill better achieved if based on the integration between animal and plant breeding. Using decentralized systems for product processing, distribution and marketing is an opportunity for resourceful farmers who want to increase added value of organic products by processing them at farm level, setting out shops on the farm or even door-to-door selling. Animal and plant breeding integration holds good prospects in this respect, especially when the product to process on-farm is milk. The residue of the processing of dairy products (whey) lends itself remarkably to act as food integrator when feeding livestock. The farm or door-to-door selling of cheeses, meat and eggs is a shorter way in the food market and it contributes to the reinforcement of the producer-consumer trust relationship.

Creating an aesthetically pleasing farmscape or landscape for those who live in and out of the system is a need suitable for the progressive development of the ecological intensification in post-modern society. Beauty and goodness are often associated as positive attributes of the same asset. Ecological intensification aspires to be structurally beautiful and functionally good at the same time. Animal and plant breeding integration meets the two conditions. Mixed breeding presupposes different crops covering various farm plots in space and time thus diversifying the landscape and not making it monotonous. Even in summer, when agricultural areas are generally dry in the Mediterranean environment, the presence of leguminous fodder crops like Lucerne is able to mark the agriculture landscape through unusual green islands into a yellowish matrix, thus arousing aesthetic pleasure and curiosity. Such a curiosity can be satisfied by consulting a text on herbaceous crops in which it will be clarified that the depth of Lucerne roots is such (a few meters) to guarantee, where the soil profile makes it possible, water-supply from water-tables even during summer drought. Lucerne will be able to provide animals with fresh grass through mowing, in a period that notoriously lacks green fodder.

6.5.4 *Ecological Intensification of Human Diet*

Inevitably, there will always be some trade-offs between biodiversity protection and human needs, even considering human diet:

> It may not be possible simultaneously to achieve high [animal] welfare, good environmental outcomes, and at the same time, enjoy current diets which are high in meat and dairy products: something may have to give (Godfray and Garnett 2014).

In compliance with the ecological intensification paradigm, even human diet needs a re-orientation framework, which currently takes the name of "environmental nutrition" (Sabaté et al. 2016; Sabaté 2019). Starting from the legitimate ecological assumption that even food systems must operate within environmental constraints to avoid disastrous consequences to the biosphere, Sabaté et al. (2016) propose a further step of ecological intensification of human mindset with envisaging a new discipline, "environmental nutrition", as a tool for public education, suitable for comprehensively addressing the sustainability of food systems. Such constraints must also take in account challenges such as food security and safety, and nutritional quality and health outcomes. In an ecological framework, diet is exactly the relationship which links the human species with its context of life (Tilman and

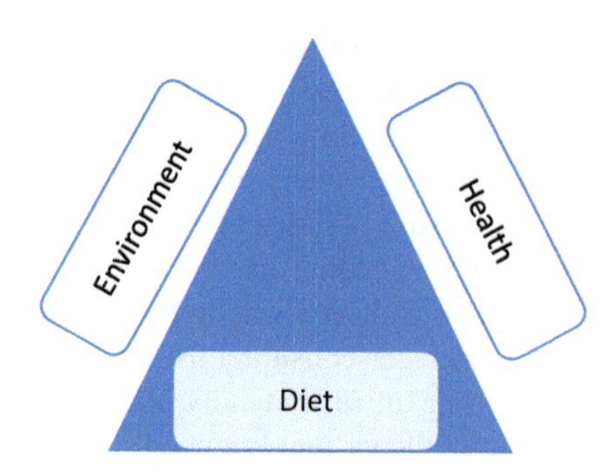

Fig. 6.47 Human diet as the basic trophic relationship between man and his/her environment

Clark 2014), and this relationship (Fig. 6.47) takes the form of agriculture for almost the whole humanity.

Human food preferences determine the type of diet and, consequently, the kind of agriculture that must provide the basic sources of food.

Since the early book of Ancel Keys on the benefits of Mediterranean diet (Keys and Keys 1963), the interest in comparing the implications of different types of diet has grown. This happened in parallel with the increasing concerns for human health, environmental problems and economic performances associated with agriculture and food system development (Sofi et al. 2008; Tilman and Clark 2014; Godfray and Garnett 2014; Leroy and Praet 2015; Sabaté et al. 2016; Sabaté 2019; Chen et al. 2019).

Tilman and Clark (2014) report on data comparing disease incidence rates of individuals who consumed typical omnivorous diets with those who had diets classified as Mediterranean, pescetarian or vegetarian.[9] Relative to conventional omnivorous diets, across the three alternative diets incidence rates of type II diabetes were reduced by 16–41% and of cancer by 7–13%, while relative mortality rates from coronary heart disease were 20–26% lower and overall mortality rates for all causes combined were 0–18% lower. Sofi et al. (2008) encourage a Mediterranean-like dietary pattern for primary prevention of major chronic diseases. Greater adherence to a Mediterranean diet is associated with a significant improvement in health status, as seen by a significant reduction in overall mortality (9%), mortality from cardiovascular diseases (9%), incidence of mortality from cancer (6%), and Alzheimer's disease (13%). In general, the alternative diets tend to have higher consumption of fruits, vegetables, nuts and pulses and lower empty calories[10] and meat consumption. A shift to alternative diets has also positive implications for the reduction of GHG emissions. From 2009 to 2050 global population is projected to increase by 36%, with an estimated 80% increase in global GHG emissions from

[9] A vegetarian diet consists of grains, vegetables, fruits, sugars, oils, eggs and dairy, and generally not more than one serving permonth of meat or seafood. A pescetarian diet is a vegetarian diet that includes seafood. A Mediterranean diet is rich in vegetables, fruit and seafood and includes grains, sugars, oils, eggs, dairy and moderate amounts of poultry, pork, lamb and beef. Omnivorous diets, such as the 2009 global-average diet and the income-dependent 2050 diet, include all food groups.

[10] 'Empty calories', defined as calories from refined fats, refined sugars, alcohols and oil.

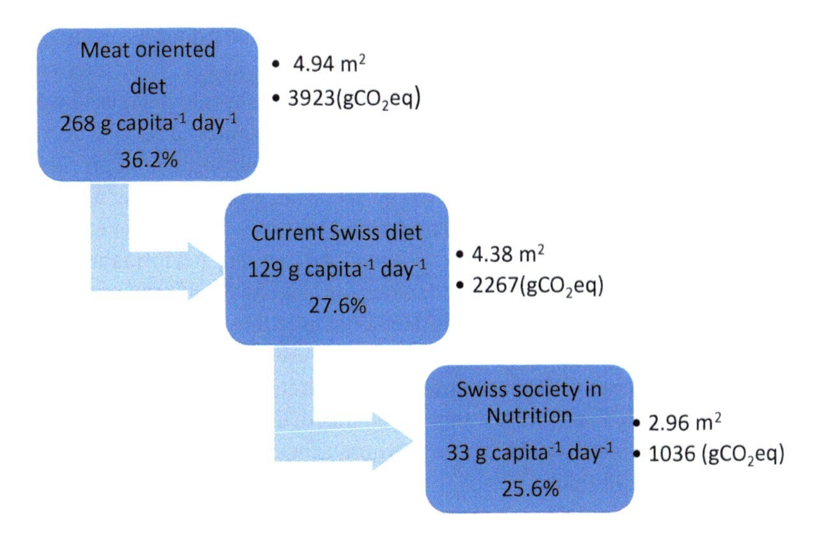

Fig. 6.48 Meat product average consumption (g capita^{-1} day^{-1} and percentage of total diet), cropland use (m^2 per capita), and GHG emissions (gCO_2eq per capita) of three types of diet in Switzerland. (Adapted from Chen et al. 2019)

food production (from 2.27 to 4.1 Gt year^{-1} of CO_2-Ceq global emission) due to a sharp increase in per capita daily demand for meat protein and 'empty calories'. In contrast, there would be no net increase in food production emissions if by 2050 the global diet had become the average of the Mediterranean, pescetarian and vegetarian diets (Tilman and Clark 2014). From these considerations, the authors formulate the following conclusion:

> The dietary choices that individuals make are influenced by culture, nutritional knowledge, price, availability, taste and convenience, all of which must be considered if the dietary transition [more meat protein and 'empty calories'] that is taking place is to be counteracted. The evaluation and implementation of dietary solutions to the tightly linked diet–environment–health trilemma is a global challenge, and opportunity, of great environmental and public health importance.

A recent investigation in Switzerland (Chen et al. 2019) focuses on multi-dimension and multi-indicator analysis in order to identify the potential trade-offs between nine alternative dietary scenarios. They found that transition towards a healthy diet following the guidelines of Swiss society of nutrition is the most sustainable option and is projected to result in 36% lesser environmental footprint, 33% lesser expenditure and 2.67% lower adverse health outcome (DALYs)[11] compared with the current diet. Data reported in Fig. 6.48 show a comparison of three different types of diet (meat oriented, current Swiss diet, and diet recommended by Swiss Society in Nutrition). The reduction of meat consumption according to the diet

[11] DALYs (Reduced Disability-Adjusted Life Years) measure the quality-adjusted life years, which is by definition the sum of Years of Life Lost (YLL) and the Years Lived with Disability (YLD).

recommended by the Swiss Society in Nutrition would have a double advantage, the cropland use would be 32.4% and 41.9% less than the current Swiss diet and the meat oriented diet, respectively, while the GHG emissions would decrease by 54.3% and 73.6% for the same types of diet. The ecological footprint would be further reduced for freshwater, nitrogen and phosphorus use. Shifting from current to the meat oriented diet would increase the cost by 10–20%, while adopting a diet based on Swiss nutrition guidelines would decrease the cost by ~35% (Chen et al. 2019). All types of indicators of ecological, economic and sanitary performances used in this enquiry show how a healthy and nutritious diet that has low impact on the environment needs not be expensive. The authors conclude that the current average diet in Switzerland has higher than acceptable greenhouse gas emissions and nitrogen footprints, threatening to transgress their global planetary boundaries and therefore points towards needs for policies that can reduce the embodied environmental impacts in the food items or reduce the consumption of items with high footprints per kg, such as meat products.

Large-scale quantitative assessment of the effect on land use and GHG emissions of human diet with reduced meat content agree on the expected benefits from such a change in consumer preferences (Stehfest et al. 2009; Wirsenius et al. 2010; Havilk et al. 2014). By using an integrated assessment model, Stehfest et al. (2009) found a global food transition to less meat, or even a complete switch to plant-based protein food to have a dramatic effect on land use:

> Up to 2700 Mha of pasture and 100 Mha of cropland could be abandoned, resulting in a large carbon uptake from regrowing vegetation. Additionally, methane and nitrous oxide emission would be reduced substantially. A global transition to a low meat-diet as recommended for health reasons would reduce the mitigation costs to achieve a 450 ppm CO_2-eq. stabilisation target by about 50% in 2050 compared to the reference case. Dietary changes could therefore not only create substantial benefits for human health and global land use, but can also play an important role in future climate change mitigation policies.

Concerning *animal footprint* for land, energy use and global warming potential in OECD (Organization for Economic Cooperation and Development) countries, a review paper (de Vries and de Boer 2010) of studies adopting a LCA (Life Cycle Assessment) methodology[12] has shown important comparative differences among the animal components of human diet. Production of 1 kg of beef used most land (27–49 m^2) and energy (34–52 MJ), followed by production of 1 kg of pork (8.9–12.1 m^2; 18–45 MJ), chicken (8.1–9.9 m^2; 15–29 MJ, eggs (4.5–6.2 m^2; 10–15 MJ), and milk (1.1–2.0 m^2; 1–8 MJ). If expressed as land use per kg of protein, milk production required 33–59 m^2, which overlapped with pork (47–64 m^2), chicken (42–52 m^2), and eggs (35–48 m^2), whereas beef production required 144–258 m^2. Production of milk and eggs have largely better performances in terms of environmental impact compared with meat production (pork, chicken and beef).

[12] *Life cycle assessment* is a systems method to evaluate the environmental impact during the entire life cycle of a product. Two types of environmental impact are considered during the life cycle of a product: use of resources such as land or fossil fuels, and emission of pollutants such as CO_2, ammonia or methane (Guinée et al. 2002).

As to the global warming potential for livestock products per average daily intake of each product, the major impact of the OECD consumption pattern resulted from consumption of beef; followed by pork, chicken and milk; and then by eggs.

Moreover, most of grain crops grown in the world are not for food but for feed. Therefore, as Foley et al. (2005) explain:

> more food can be de delivered by changing our agricultural and dietary preferences. Simply put, we can increase food availability (in terms of calories, protein and critical nutrients) by shifting crop production away from livestock feed, bioenergy crops and other non-food applications.

Quantitatively, the 1 billion tonnes of cereals (wheat, barley, oats, rye, corn, sorghum and millet) annually fed to livestock could feed some 3.5 billion humans (Eisler et al. 2014). These considerations, once let known through more education in environmental nutrition, could convince consumers to change their mind and shift to a safer diet suitable for man, livestock and the environment. Indeed, an environmental nutrition that shows how human diet, agriculture, livestock, health, and environment are connected together in an inescapable treadmill (Fig. 6.49), sheds lights on the hidden role of consumers in the agro-food-system, where traditionally they are considered as passive clients and unconscious components. The challenge for the future is just to intensify ecologically the consumers' mind set in order to make them responsible for what to eat, how to grow it, and which goals to achieve with the agro-food-systems locally and globally.

Changing human diets towards less animal products means potentially reducing their density of population, especially at the farm and landscape levels and consequently, reducing the pressure on both planned and associated biodiversity, and enhancing the animal welfare and health conditions. Welfare definitions tend to

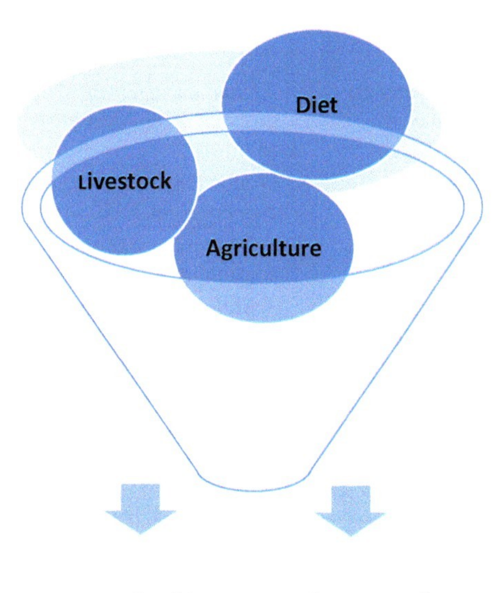

Fig. 6.49 The treadmill of a responsible human diet affecting human health and environmental care

include the requirement not only that animals are in good health, but also that they are somehow experiencing a 'life worth living', performing activities and behaviours that are instinctive and that may cause distress if they cannot be performed (Godfray and Garnett 2014). If the heavy burden to depend on concentrates for nutrition and maximum yield is relieved, and more possibility of grazing is allowed, more diversified crop rotations can be implemented and more pasture areas provided, with benefits for both animal health and welfare. Between human health, animal health and welfare, and environmental care, there are not basic, insuperable trade-offs, it is only a question of balancing multiple expectations and goals.

As transition steps of ecological intensification to sustainable livestock, Eisler et al. (2014) highlight eight strategies of animal husbandry (Table 6.14) suitable for cutting both environmental and economic costs, while ensuring quantity and quality of the food livestock produce. The first four strategies concern care for animal needs in terms of nutrition, welfare and health; the others put emphasis on promoting changes of human dietary patterns, management practices, and economic and cultural evaluations in order to achieve more systemic awareness and to adjust human goals to livestock requirements. To explore these multidisciplinary strategies, the authors suggest that building up a global network of research farms is appropriate in order to produce tangible evidence and convince farmers to adopt them.

Table 6.14 Strategies and ecological motivations to sustainable livestock husbandry (adapted from Eisler et al. 2014)

Strategies	Motivations
Feed animals less human food	Ruminants graze pastures and can eat hay, silage and high-fibre crop residues
Raise regionally appropriate animals	More can be done to encourage farmers to realize the advantages adapted to local areas
Keep animals healthy	Animal management should include measures to contain transmissible diseases. Keeping animals at high densities spreads infectious diseases far and fast
Adopt smart supplements	Productivity of ruminants can be boosted with supplements which encourage microbes in the rumen to provide better nutrition
Eat quality not quantity	Eating less, better quality meat. There are nutritional advantages to consuming small amounts of high-quality animal foods
Tailor practice to local culture	Traditional animal husbandry supplies more than just food; it provides wealth and status. Many of these benefits are disrupted when traditional grazing and mixed farming practices are replaced with industrial systems that maximize short-term production
Track costs and benefits	The livestock sector accounts for 14.5% of human-induced greenhouse-gas emissions, but sustainably managed grazing can increase biodiversity, maintain ecosystem services and improve carbon sequestration by plant and soil. Tune livestock policies to socioeconomic and geographic environments
Study best practices	There will be no one-size-fits-all solutions. Optimize the use of livestock in different regions, using local resources, breeds and feedstuffs

However, there are structural barriers to ecological intensification of human diet in that the same model of industrial intensification applied to animal husbandry applies to human settlements. Urbanisation is the pattern to concentrate humans as CAFOs (Concentrated animal feeding operations) do to concentrate animals. As Thorntorn (2010) claims, historical changes in the demand for livestock products have followed human population growth, income growth and urbanisation, all factors that intensified in the recent past. The production response in different livestock systems has been associated with science and technology, and 'industrial intensification', which is synonymous with a particular form of livestock production in which animals are kept in highly artificial environments and provided with high-input diets very different from their natural food. Breeding of animals aims at narrow yield goals (Godfray and Garnett 2014). How urbanisation has brought about a push towards nutrition transition in terms of industrial intensification is topic of many papers (Hawkes et al. 2017; IFPRI 2017) that can be summarised in the following points:

- Diets are changing with rising incomes and urbanization – people are consuming more animal-source foods, sugar, fats and oils, refined grains, and processed foods.
- This "nutrition transition" is causing increases in overweight and obesity and diet-related diseases such as diabetes and heart disease.
- Urban residents are making the nutrition transition fastest – but it is occurring in rural areas too.
- Urban food environments – with supermarkets, food vendors, and restaurants – facilitate access to unhealthy diets, although they can also improve access to nutritious foods for people who can afford them.
- For the urban poor, the most easily available and affordable diets are often unhealthy.

The process of coevolution of concentrated operations for both humans and animal husbandry is not sustainable and demand for livestock products should decrease due to socio-economic factors such as human health concerns and changing socio-cultural values. In the future, "production will increasingly be affected by competition for natural resources, particularly land and water, competition between food and feed and by the need to operate in a carbon-constrained economy" (Thorntorn 2010).

In the framework of "The European Green Deal", which is an integral part of the European Commission's strategy to implement the United Nation's 2030 Agenda and the sustainable development goals (European Commission 2019), one important element is the food value chain, defined under the title "From Farm to Fork: designing a fair, healthy and environmentally-friendly food system" (European Commission 2020a). The Farm to Fork strategy operates under the heading "healthy people, healthy society and a healthy planet" which subsumes "a new comprehensive approach to how Europeans value food sustainability". It appears an opportunity to improve lifestyles, health, and the environment as follows:

> The creation of a favourable environment that makes it easier to choose healthy and sustainable diets will benefit consumers' health and quality of life, and reduce health-related costs for society. People pay increasing attention to environmental health, social and ethical issues and they seek value in food more than ever before. Even as societies become more urbanised, they want to feel closer to their food. They want food that is fresh, less processed and sustainably sourced. And the calls for shorter supply chains have intensified during the current outbreak [COVID-19 pandemic]. Consumers should be empowered to choose sustainable food and all actors in the food chain should see this as their responsibility and opportunity (European Commission 2020a, p. 4).

In May 2020, the European Commission has adopted the new *EU Biodiversity Strategy for 2030* (European Commission 2020b) and an associated plan – a comprehensive, ambitious, long-term plan for protecting nature and reversing the degradation of ecosystems by 2030 with benefits for people, the climate and the planet. The *Biodiversity Strategy* addresses the main drivers of biodiversity loss, in line with the 2030 Agenda for Sustainable Development and with the objectives of the Paris Agreement on Climate Change:

> to support the long-term sustainability of both nature and farming, this strategy will work in tandem with the new Farm to Fork Strategy and the new Common Agricultural Policy (CAP), including by promoting eco-schemes and result-based payment schemes.

As to "bringing nature back to agricultural land", it advances recommendations such as:

- to provide space for wild animals, plants, pollinators and natural pest regulators, there is an urgent need to bring back at least 10% of agricultural area under high-diversity landscape features. These include, *inter alia*, buffer strips, rotational or non-rotational fallow land, hedges, non-productive trees, terrace walls, and ponds. These help enhance carbon sequestration, prevent soil erosion and depletion, filter air and water, and support climate adaptation;
- to make the most of this potential, at least 25% of the EU's agricultural land must be organically farmed by 2030;
- the decline of genetic diversity must also be reversed, including by facilitating the use of traditional varieties of crops and breeds. This would also bring health benefits through more varied and nutritious diets. The Commission is considering the revision of marketing rules for traditional crop varieties in order to contribute to their conservation and sustainable use;
- protecting and restoring nature will need more than regulation alone. It will require action by citizens, businesses, social partners and the research and knowledge community, as well as strong partnerships between local, regional, national and European level.

In recent papers, questions have arisen as to why more ecologically based agriculture and sustainable food systems have not been widely adopted in the USA (Fernandez et al. 2013; DeLonge et al. 2016). Miles et al. (2017) consider that "there is a profound gap between the potential of agroecology to resolve agrifood problems and the current levels of federal funding to support advances in this field". In their opinion, "closing this gap may be a key lever for triggering a positive

Table 6.15 Stepping stones for converting the entire global food systems toward sustainability (after Gliesmann 2016, modified)

Levels of change	Goals
1. Increase the efficiency of industrial and conventional practices	Reduce the use and consumption of costly, scarce, or environmentally damaging inputs
2. Substitute alternative practices for industrial/conventional inputs and practices	Replace external input with those that are more renewable, based on natural products, and more environmentally sound
3. Redesign the agroecosystem so that it functions on the base of a new set of ecological processes	Increase diversity in farm structure and management through ecologically-based rotations, multiple cropping, agroforestry, and the integration of animals with crops
4. Re-establish a more direct connection between those who grow our food and those who consume it	Ground a "food citizenship", where communities of growers and eaters can form alternative food networks around the world
5.Build a new global food system, based on equity, participation, democracy, and justice	Restore and protect earth's life-support systems upon which we all depend

feedback cycle in which agroecological research facilitates greater adoption, which in turn can encourage additional research investment". Finally, they propose a set of "key policies to be implemented at local to federal scales to successfully scale agroecology and achieve a more ecologically sustainable and socially equitable society" (Miles et al. 2017). These proposals ground their basis on Gliesmann's (2016) five levels of agroecological transition, as reported in Table 6.15.

Gliesmann (2016) claims that "agroecology is a way of redesigning food systems, from the farm to the table, with a goal of achieving ecological, economic, and social sustainability". The first level is about using industrial inputs more efficiently so that fewer inputs will be needed and the negative impacts of their use will also be reduced. According to Gliessman et al. (2019), "so-called *precision agriculture* is a recent focus of research at Level 1. Although this kind of research has reduced some of the negative impacts of industrial agriculture, it does not help break its dependence on external material inputs and monoculture practices". While the first three levels describe the first steps farmers can actually take for converting their farms from industrial or conventional to more sustainable agroecosystems, the other two levels connect farms to the broader food system and the societies which they belong to. Globally, "all five levels taken together can serve as a roadmap that outlines in an almost stepwise manner a process for transforming the entire global food system" (Gliesmann 2016). The transformational process entails "basic beliefs, values and ethical system change" and "involves change that is global in scope and reaches beyond the food system to the nature of human culture, civilization, progress, and development".

Cultural improvements due to scientifically proved new ontology of life and human individuals, as shown in Box 6.12, can help in re-framing our membership in nature biodiversity and our food preferences:

> Recognizing the "holobiont" – the multicellular eukaryote plus its colonies of persistent symbionts – as a crirically important unit of anatomy, development, physiology, immunology, and evolution opens up new investigative avenues and conceptually challenges the

ways in which the biological subdisciplines have heretofore characterized living entities (Gilbert et al. 2012).

Ethically and politically, the notion of community offers rich possibilities for guiding future approaches to research, to medicine, and, more broadly, to all of our encounters with the living world. Acknowledging that all living beings are individuals who are paradoxically nested collectives within larger collectives challenges us to rethink our place in the complex order of things (Schneider and Winslow 2014).

The above-mentioned insights suggest that scientific knowledge can change our way of thinking, speaking, understanding and acting in a new horizon of significant *membership* where "we do not exist outside ecosystems which can then choose to enter at freely chosen times. We are always already inside the ecosystem, just as the

Box 6.12 Human Gut Microbiota Potential (HMP)
A Symbiotic View of Life: We Have Never Been Individuals (Gilbert et al. 2012)

We acquire our consortia of microbes from the environment, beginning at the time of birth. By adulthood, the number of bacteria that colonize our epidermal and mucosal surfaces is thought to exceed our population of human cells. The largest community resides in our gut where 500–1000 species are assembled at densities estimated to reach 1011 per gram of luminal contents in the proximal colon, creating a bacterial 'nation' of 10–100 trillion citizens. In this sense, we should not view ourselves as individuals, but rather as a highly diversified co-evolving collection of members of Eukarya, Bacteria and Archaea (Xu et al. 2004).

EVIDENCES

The human distal gut is home to a densely populated microbial community (microbiota) that plays key roles in health and nutrition. The microbial symbionts that occupy this habitat produce an arsenal of enzymes that degrade dietary complex carbohydrates (glycans) that cannot be hydrolyzed by host enzymes. The simple sugars generated are fermented into host-absorbable end products, including short-chain fatty acids, that can contribute as much as 10% of the calories extracted from the human diet and are thought to play a role in preventing colorectal cancer (Martens et al. 2011).

The HMP [Human Microbiome Project] now estimates that the commensal bacteria that reside on and in that individual's body incorporate a total of some 3 million different genes, suggesting a ratio of microbial to human genes of at least 130 to 1. These parallel genomes represent an intriguing partnership (Schneider and Winslow 2014).

The gut microbiota exhibits many important physiological functions that include regulation of energy levels and metabolism, neutralization of drugs and carcinogens, modulation of intestinal motility, regulation of immunity, barrier effects, and protection against pathogens. Host behavior and cognitive functions such as learning, memory, and decision-making are also believed to be affected by the gut microbiota. In a broad sense, the gut microbiota appears to be critical to maintain host homeostasis and health order to maintain health (Chuan-Sheng Lin et al. 2016).

PROSPECTS

The path forward to addressing basic and applied questions about the gut microbiome is so inherently interdisciplinary that it provides a splendid opportunity to craft new alliances, spawn new fields, and transform how we educate ourselves (Gordon 2012).

The gut microbiota could be manipulated using diet, prebiotics, and probiotics in order to maintain health [...] Changes in the diet of the host could be used to modulate the gut microbiota and restore homeostasis [...] Currently, protein and animal fat consumption appears to be more closely linked with disease than the intake of carbohydrates (Chuan-Sheng Lin et al. 2016).

The long-term goals of the entire HMP involve trying to understand the role of human-associated microbial communities in health and disease (Schneider and Winslow 2014).

ecosystem is always already inside us in the form of microbes, food, water and air" (Wirzba 2019, p. 101).

6.6 Ecological Intensification in Theology of Creation and Food Ethics

Ecology is a contaminating science and its process of contamination has encompassed the field of theology at the crossroads point that scientists and philosophers call 'nature' and theologians call 'creation'. The most known document that reveals this on-going cross-fertilisation process is the papal encyclical letter *Laudato si'* (Pope Francis 2015, 76):

> In the Judaeo-Christian tradition, the word "creation" has a broader meaning than "nature", for it has to do with God's loving plan in which every creature has its own value and significance. Nature is usually seen as a system which can be studied, understood and controlled, whereas creation can only be understood as a gift from the outstretched hand of the Father of all, and as a reality illuminated by the love which calls us together into universal communion.

According to Pope Francis, creation is nature plus God's love, a kind of equation that marks the difference between ecology, or the science of nature organisation, and eco-theology, the transdisciplinary new field of inquiry that emerges from the dialogue between ecology and theology (Cobb 1972; Edwards 2006; Habel and Trudinger 2008; Henning 2015).

Many authors, including Bakken (2000) and Henning (2015), credit the publication of the historian Lynn White Jr.'s essay, "The Historical Roots of Our Ecological

Crisis" in 1967, to be the spark that fired the new field of study connecting ecology with theology (eco-theology). The paper charged Christianity of bearing a "huge burden of guild" for having blinding followed the biblical commandment to have "dominion" over the earth. After comparing contrasting interpretations of man/nature relationship from different sources of Christianity (ancient, the medieval of St. Francis, and the modern ones) as briefly summarised in Fig. 6.50, White's conclusion sounds alarmistic, involving an ethical revision of thinking and practice that involves both religion and science.

Since this start, a process of ecological intensification in theology has flourished and is still advancing (see Sect. 4.3.2). In 1979, Pope John Paul II did in fact make St. Francis the patron saint of Ecology (papal bull *Inter Sanctos*, 29 November) and, in 1990, he addressed the ecological crisis as a moral issue with the world peace day speech "The ecological Crisis. A Common responsibility". After the publication of Pope Francis's *Laudato si'*, scholars' field of interest has enlarged to include in a recent book the relationship between religion and sustainable agriculture as a matter of world spiritual traditions and food ethics (LeVasseur et al. 2016). Vandana Shiva (2016) explains in the foreword of her book "the Dharma of food" how agriculture and theology link as follows:

> My own approach to the sacredness of food and farming has been informed by my culture. *Dharma* is the unique gift of Indian civilisation to humanity. It has provided the compass for right action and right livelihood. There is no equivalent word in English. Dharma is not reducible to religion, as has often been erroneously done. It is the "right way of living" and "path of righteousness". All religions that grew from Indian soil – Hinduism, Buddhism, Sikhism, and Jianism – refer to dharma (p. viii).

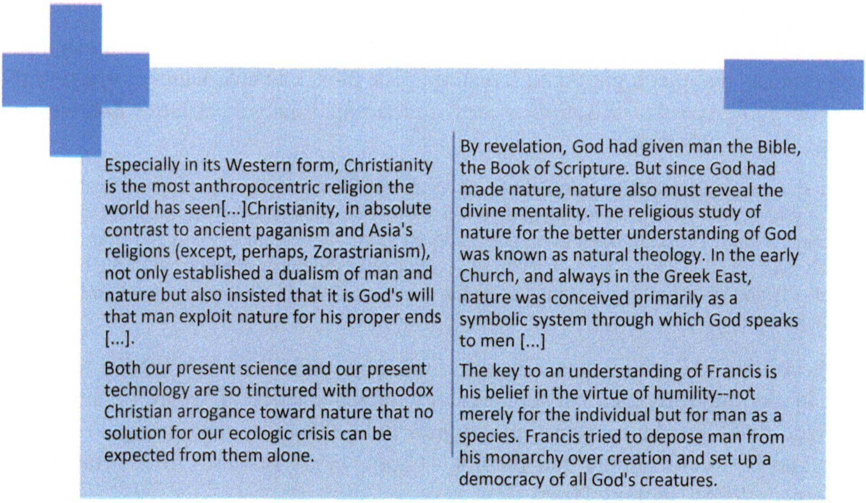

Fig. 6.50 Contrasting visions of human dominion of nature in the history of Christianity. (Adapted from White 1967)

Shiva makes clear that the etymological root of the word *dharma* is *dhr*, "to hold, bear, support, maintain, keep, carry" and its meaning is "that which holds", "that which sustain":

> Because Dharma holds and sustains the earth community and all of humanity embodies the principle of unity – of humans with the rest of nature, and of humans across our diversities. Dharma arises from the interconnectedness of all life and from our duty to care for all humans and all species (p. viii).

In *Hinduisms*, a strict correlation exists between *dharma,* "the right way of living", and *rta*, "the right order" of reality, as represented in Fig. 6.51.

How *dharma*, the righteous conduct in society, and *rta*, the moral order in the universe, correspond to each other, is confirmed as follows by a scholar of Hindu morality like Saral Jhingran (1989, p. 37):

> In the Vedas *dharma*, like *rta*, is given a near-ontological status. Both are equated to *satya* (truth) and truth is what eternally is. They are also opposed to *anrta* (falsehood) and false is both ontologically less real and morally reprehensible. *Rta* or *dharma* signifies the moral order of the universe that governs and determines the course of events, so as to ensure that truth or *dharma* is always victorious and that untruth or *adharma* (morally wrong or unjust) is finally defeated.

Considering that "the growing and giving the food in abundance is the highest dharma", Shiva (2016, p. ix) concludes that "the web of life is the food web. Food is life; therefore it is sacred. It is not a commodity".

How the relationship between human conduct and natural order appears in *Laudato si'* (Pope Francis 2015) and inspires an "ecological conversion" is briefly reported in Box 6.13. In chapter 6, III-"Ecological conversion", *Laudato si'* offers Christians "suggestions for an ecological spirituality grounded in the convictions of our faith", in that "the life of the spirit is not dissociated from the body or from nature or from worldly realities, but lived in and with them, in communion with all that surrounds us" (LS, 216). In LS (235), it is stated that through the Sacraments

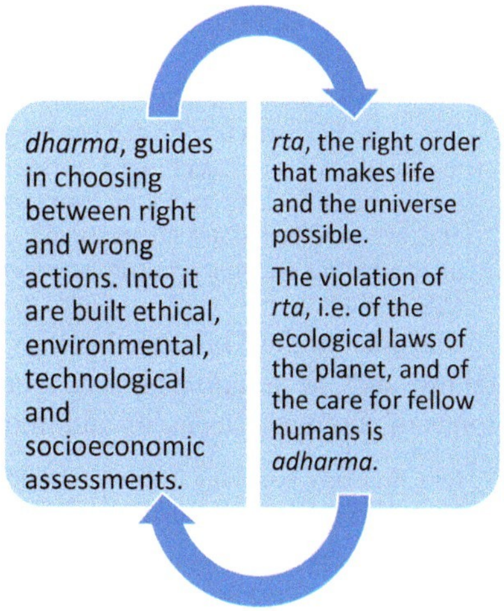

Fig. 6.51 Correspondence between *dharma* and *rta* in Hinduism. (Adapted by Shiva 2016)

"we are invited to embrace the world on a different plane" and that "for Christians, all the creatures of the material universe find their true meaning in the incarnate Word":

Box 6.13 Key Words for an Ecological Conversion According to Laudato si' (220–221)

"Ecological conversion calls for a number of attitudes which together foster a spirit of generous care, full of tenderness" (LS, 220).

GRATITUDE

Ecological conversion "entails gratitude and gratuitousness, a recognition that the world is God's loving gift" (LS, 220).

AWARENESS

Ecological conversion entails "a loving awareness that we are not disconnected from the rest of creatures, but joined in a splendid universal communion [...]we do not look at the world from without but from within, conscious of the bonds with which the Father has linked us to all beings" (LS,220).

Ecological conversion entails "the awareness that each creature reflects something of god and has a message to convey to us, and the security that Christ has taken unto himself this material world and now, risen, is intimately present to each being, surrounding it with his affection and penetrating it with his light" (LS, 221).

RESPONSIBILITY

Responsibility stems from "the recognition that God created the world, writing into it an order and a dynamism that human beings have no right to ignore" (LS, 221).

"We do not understand our superiority as a reason for personal glory or irresponsible dominion, but rather as a different capacity which, in its turn, entails a serious responsibility stemming from our faith" (LS, 220).

> It is in the Eucharist that all that has been created finds its greatest exaltation [...] Joined to the incarnate Son, present in the Eucharist, the whole cosmos gives thanks to God. Indeed the Eucharist is itself an act of cosmic love [...] The Eucharist joins heaven and earth; it embraces and penetrates all creation [...] Thus, the Eucharist is also a source of light and motivation for our concerns for the environment, directing us to be stewards of all creation (LS, 236).

The apex of the Christian theology of creation is achieved in chapter 6, VII-"The Trinity and the relationship between creatures" (LS, 238–241), of which some points of ecological meaning and implications are reported below:

> for Christians, believing in one God who is trinitarian communion suggests that the Trinity has left its mark on all creation;

(a) the divine Persons are subsistent relations, and the world, created according to the divine model, is a web of relationships;

(b) the human person grows more, matures more and is sanctified more to the extent that he or she enters into relationships, going out from themselves to live in communion with God, with others and with all creatures;
(c) everything is interconnected, and this invites us to develop a spirituality of that global solidarity which flows from the mystery of the Trinity.

The functional role of food in maintaining life through trophic chains and webs is not only a scientific notion of ecological organisation, but also unveils the ontology of existence of all and each living beings, and therefore has a metaphysical weigh, which is even celebrated in Christian liturgy. Norman Wirzba (2019) summarizes the reasons for an ecotheology of food as follows:

> In our global economy food is a commodity much like any other, serving the business need for profit, the consumer desire for cheapness, and the political quest for power. In this context, food ceases to speak as the grace of God. Eating ceases to be the occasion through which we experience life as a membership of belonging, responsibility, and gratitude (p. xiv).

Yet, this plentiful role of food is not fully recognised in the so called "developed" society. This is a cultural void that needs to be filled. More dialogue, even between different religions and theological positions, can help give new value to agriculture and food system, and to farmers that drive the process. Positive responses in this regard have arisen after the *Laudato si'*, such as the vision of a *regenerative agri-culture* advanced by Freudenberger and Freudenberger (2015) which is briefly reported in Box 6.14.

The vision of a *regenerative agri-culture* operates in the framework of agriculture as a 'human activity systems' and strictly complies with "the ethics of eco-poiesis" (self-nutrition, self-development, and self-maintenance) advanced by Arran Gare (2017, p. 193) for an ecological civilisation on the base of the following Aldo Leopold's land ethic premises (Leopold 1949):

(a) the extension of ethics is a process in ecological evolution;
(b) an ethic, ecologically, is a limitation on freedom of action in the struggle for existence;
(c) an ethic, philosophically, is a differentiation of social from anti-social conduct;
(d) the process has its origin in the tendency of interdependent individuals or groups to evolve modes of co-operation (symbioses);
(e) politics and economics are advanced symbioses in which the original free-for-all competition has been replaced, in part, by co-operative mechanism with an ethical content;
(f) all ethics so far evolved rest upon a single premise: the individual is a member of a community of interdependent parts;
(g) the land ethic simply enlarges the boundaries of the community to include soils, waters, plants, and animals, or collectively: the land. The extension of ethics is a process in ecological evolution;

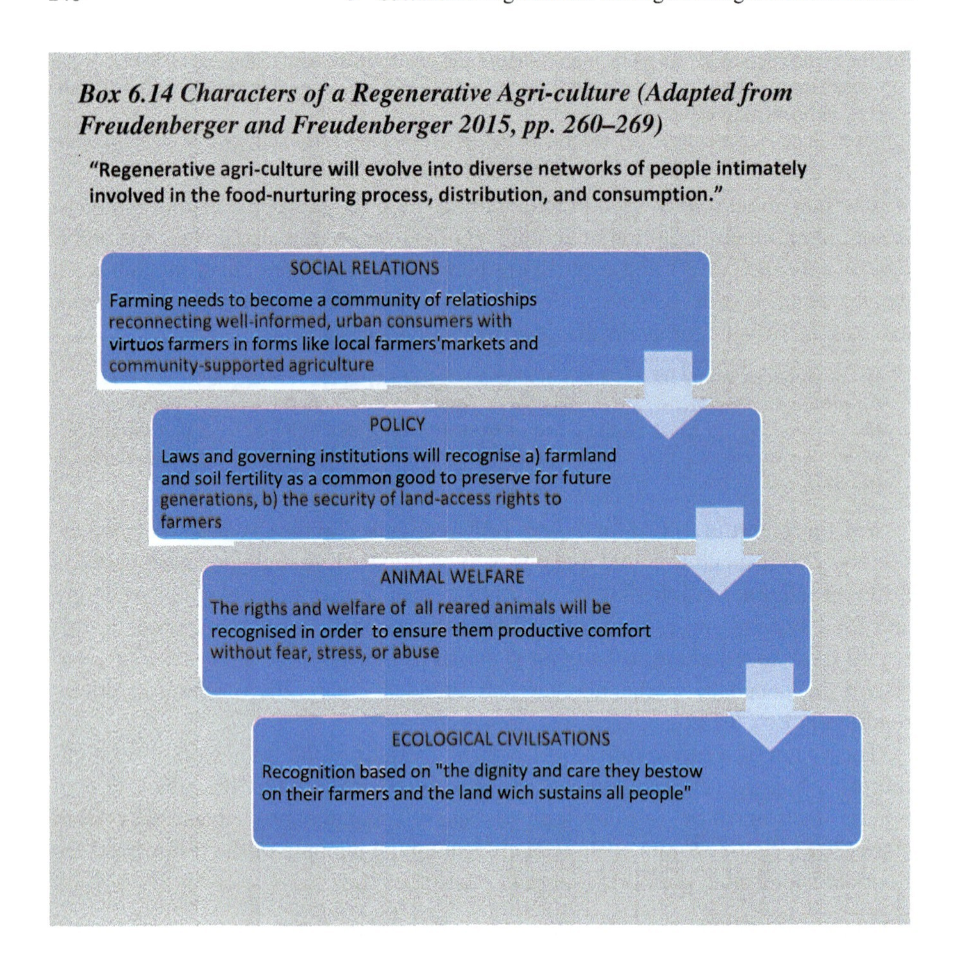

Box 6.14 Characters of a Regenerative Agri-culture (Adapted from Freudenberger and Freudenberger 2015, pp. 260–269)

Food (or feed) is 'mother' to all relationships within the earth living community and compels humanity to "relearn how to eat within the biosphere limitations and gift" (Freudenberger and Freudenberger 2015). Pope Francis appeals for "a healthy relationship with Creation" (LS, 218) that cannot do without agriculture. We should take a responsible stance against animal suffering caused by animal husbandry as McMahan (2019) advocates:

> Even if animal suffering does not matter as much as the equivalent suffering of persons, it still matters very much. And it is ubiquitous (see Box 6.15).

Box 6.15 Eating Animals: From Causing to Alleviating Suffering (Elaborated from McMahan 2019)

EVIDENCE FROM FACTS

"Most of the meat consumed in economically developed societies is from animals raised in factory farms. In these 'farms', the animals are tightly packed into filthy, stifling, indoor spaces, and thus suffer more or less continuous physical and psychological torment throughout their entire lives. When they reach full size, pumped with antibiotics because of the unsanitary conditions in which they are kept, they suffer a final period of panic and terror as they are mass-slaughtered, often in mechanized, assembly-line fashion" (p 292)

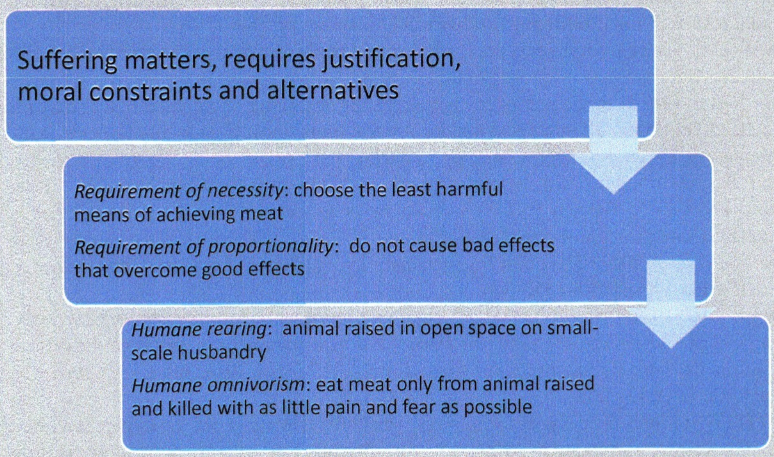

Suffering matters, requires justification, moral constraints and alternatives

Requirement of necessity: choose the least harmful means of achieving meat

Requirement of proportionality: do not cause bad effects that overcome good effects

Humane rearing: animal raised in open space on small-scale husbandry

Humane omnivorism: eat meat only from animal raised and killed with as little pain and fear as possible

FINAL SUGGESTIONS

"We ought not to be indifferent to the great suffering that many billions of animals endure every day. When our science becomes sufficiently advanced to enable us to ameliorate their suffering without doing more harm than good, it will become our duty to do that" (p.300)

References

Agnoletti M et al (2019) Terraced landscapes and hydrogeological risk. Effects of land abandonment in Cinque Terre (Italy) during severe rainfall events. Sustainability 11:235. https://doi.org/10.3390/su11010235

Akkari C et al (2019) Multifunctionality of farmland and farm activities & multi-actor involvement in agricultural development planning. MOJ Food Process Technol 7(4):132–134

Altieri MA, Toledo VM (2011) The agroecological revolution in Latin America: rescuing nature, ensuring food sovereignty and empowering peasants. J Peasant Stud 38(3):587–612

Anderson CR, Anderson MD (2020) Resources to inspire a transformative agroecology: a curated guide. In: Herren H, Haerlin B et al (eds) Transformation of our food systems: the making of a paradigm shift. Zukunftsstiftung Landwirtschaft, Berlin, pp 169–179

Antrop M (2000) Background concepts for integrated landscape analysis. Agric Ecosyst Environ 77:17–28

Audouin E et al (2019) Participatory methodology for designing an agroecological transition at local level. In: Bergez JE et al (eds) Agroecological transitions: from theory to practice in local participatory design. Springer, Cham, pp 177–206

Bakken PW (2000) Nature as a theatre of grace: the ecological theology of Joseph Sittler. In: Bouma-Prediger S, Bakken P (eds) Evocations of grace. William B. Eerdmans Publishing Company, Cambridge, pp 1–19

Bakshi BR, Fiksel J (2003) The quest for sustainability: challenges for process systems engineering. AICHE J 49(6):1350–1358

Balfour EB (1975) The living soil and the Haughley experiment. Universe Books, New York

Barbera G et al (2018) The "jardinu" of Pantelleria as a paradigm of resource efficient horticulture in the built-up environment. Acta Hortic 1215:ISHS 2018. https://doi.org/10.17660/ActaHortic.2018.1215.65

Barbour I (1993) Ethics in an age of technology, The Gifford lectures, vol 2. Harper, San Francisco

Barrow CJ (1999) Alternative irrigation: the promise of runoff agriculture. Earthscan, Routledge, London

Battaglini L et al (2014) Environmental sustainability of alpine livestock farms. Ital J Anim Sci 13:431–443

Bellah RN et al (1991) The good society. Alfred A. Knopf, New York

Bergez JE, Therond O (2019) Introduction. In: Agroecological transition: from theory to practice in local participatory design. Springer, Cham

Bergez JE et al (2019) Agroecological transition: from theory to practice in local participatory design. Springer, Cham

Bianchi FJJA et al (2006) Sustainable pest regulation in agricultural landscapes: a review on landscape composition, biodiversity and natural pest control. Proc R Soc B 273:1715–1727

Bommarco R et al (2013) Ecological intensification: harnessing ecosystem services for food security. Trends Ecol Evol 28(4):230–238

Borin M et al (2010) Multiple functions of buffer strips in farming areas. Eur J Agron 32:103–111

Bradshaw AD (1996) Underlying principles of restoration. Can J Fish Aquat Sci 53(Suppl. 1):3–9

Bu CF et al (2008) Effects of hedgerows on sediment erosion in Three Gorges Dam Area, China. Int J Sediment Res 23:119–129

Caborn JM (1971) The agronomic and biological significance of hedgerows. Outlooks Agric 6:279–284

Campi P et al (2009) Effects of tree windbreak on microclimate and wheat productivity in a Mediterranean environment. Eur J Agron 30:220–227

Cannarella C, Piccioni V (2005) Knowledge building in rural areas: experiences from a research centre-rural SME scientific partnership in Central Italy. Int J Rural Manag 1(1):25–43

Caporali F (2000) Ecosystems controlled by man. In: Frontiers of life, vol 4. Academic, New York, pp 519–533

Caporali F (2007) Agroecology as a science of integration for sustainability in agriculture. Ital J Agron/Rivista di Agronomia 2(2):73–82

Caporali F (2010) Agroecology as a transdisciplinary science for a sustainable agriculture. In: Lichtfouse E (ed) Biodiversity, biofuels, agroforestry and conservation agriculture. Springer, Dordrecht, pp 1–71

Caporali F (2015a) History and development of agroecology and theory of agroecosystems. In: Monteduro M et al (eds) Law and agroecology. A transdisciplinary dialogue. Springer, Berlin, pp 3–29

Caporali F (2015b) Pietro Cuppari precursore dell'agroecologia e del governo sostenibile del territorio. Edizioni ETS, Pisa

Caporali F, Campiglia E (2001) Increasing sustainability in Mediterranean cropping systems with self-reseeding annual legumes. In: Gliessman SR (ed) Agroecosystem sustainability. Developing practical strategies. CRC Press, Boca Raton, pp 15–27

Caporali F, Onnis A (1992) Validity of rotation as an effective agroecological principle for a sustainable agriculture. Agric Ecosyst Environ 41:101–113

Caporali F et al (1989) Concepts to sustain a change in farm performance evaluation. Agric Ecosyst Environ 27:579–595

Caporali F et al (2003) Indicators of cropping systems diversity in organic and conventional farms in Central Italy. Int J Agric Sustain 1:67–72

Caporali F et al (2007) Teaching and research in agroecology and organic farming: challenges and perspective. ENOAT meeting, Pieve Tesino, Università degli Studi della Tuscia

Caron P et al (2014) Making transition towards ecological intensification of agriculture a reality: the gaps in and the role of scientific knowledge. Curr Opin Environ Sustain 8:44–52

Castonguay A et al (2016) Resilience and adaptability of rice terrace social-ecological systems: a case study of a local community's perception in Banaue, Philippines. Ecol Soc 21(2):15. https://doi.org/10.5751/ES-08348-210215

Ceccarelli S (1996) Adaptation to low/high input cultivation. Euphytica 92:203–214

Chapagain T, Raizada MN (2017) Agronomic challenges and opportunities for smallholder terrace agriculture in developing countries. Front Plant Sci 8:331. https://doi.org/10.3389/fpls.2017.00331

Chen C et al (2019) Dietary change scenarios and implications for environmental, nutrition, human health and economic dimensions of food sustainability. Nutrients 11:856. https://doi.org/10.3390/nu11040856

Choudhury PK et al (2015) Rumen microbiology: an overview. In: Puniya AK, Singh R, Kamra DN (eds) Rumen microbiology: from evolution to revolution. Springer. https://doi.org/10.1007/978-81-322-2401-3_1

Chuan-Sheng Lin et al (2016) Impact of the gut microbiota, prebiotics, and probiotics on human health and disease. Biom J 37:259–268

Cobb JB Jr (1972) Is it too late? A theology of ecology. Bruce, Beverly Hills

Corrazzini M et al (2010) Effect of summer grazing on welfare of dairy cows reared in mountain tie-stall barns. Ital J Anim Sci 9(3):e59. https://doi.org/10.4081/ijas.2010.e59

Couëdel A et al (2018) Cover crop crucifer-legume mixtures provide effective nitrate catch crop and nitrogen green manure ecosystem services. Agric Ecosyst Environ 254:50–59

Creanza N et al (2017) Cultural evolutionary theory: how culture evolves and why it matters. PNAS 114(3):7782–7789

Dainese M et al (2016) High cover of hedgerows in the landscape supports multiple ecosystem services in Mediterranean cereal fields. J Appl Ecol. https://doi.org/10.1111/1365-2664.12747

Darnhofer I et al (2010) Conventionalisation of organic farming practices: from structural criteria towards an assessment based on organic principles. A review. Agron Sustain Dev 30:67–81

De Schutter O (2012) Agroecology, a tool for the realization of the right to food. In: Lichtfouse E (ed) Agroecology and strategies for climate change. Springer, pp 1–16. http://hdl.handle.net/2078.1/105949. https://doi.org/10.1007/978-94-007-1905-7_1

de Vries M, de Boer IJM (2010) Comparing environmental impacts for livestock products: a review of life cycle assessments. Livestock Sci 128:1–11

DeLonge MS et al (2016) Investing in the transition to sustainable agriculture. Environ Sci Pol 55:266–273. https://doi.org/10.1016/j.envsci.2015.09.013

Diamond J (1987) Reflexions on goals and on the relationship between theory and practice. In: Jordan WR, Gilpin ME, Aber JD (eds) Restoration ecology: a synthetic approach to ecological research. Cambridge University Press, Cambridge, pp 329–336

Diamond J (2002) Evolution, consequences and future of plant and animal domestication. Nature 418:700–707

Diaz S et al (2015) The IPBES conceptual framework-connecting nature and people. Curr Opin Environ Sustain 14:1–16

Duru M et al (2015) Designing agroecological transitions: a review. Agron Sustain Dev 35:1237–1257. https://doi.org/10.1007/s13593-015-0318-x

Edwards D (2006) Ecology at the hearth of faith. Orbis Book, New York

Eisler MC et al (2014) Steps to sustainable livestock. Nature 507:32–34

Elliot GP (2017) Treeline as ecotone. In: Richardson D, Castree N, Goodchild NF et al (eds) The international encyclopedia of geography. Wiley, pp 1–10. https://doi.org/10.1002/9781118786352.wbieg0539

European Commission (2011) Our life insurance, our natural capital: an EU Biodiversity Strategy to 2020. COM 82011,244 final

European Commission (2019) The European Green Deal. COM(2019) 640 final

European Commission (2020a) Farm to fork strategy. For a fair, healthy and environmentally-friendly food system. EU Green Deal

European Commission (2020b) EU biodiversity strategy for 2030. Bringing nature back into our lives. COM(2020) 380 final

European Parliament (2017) Report on the state of play of farmland concentration in the EU: how to facilitate the access to land for farmers (2016/2141(INI)) Committee on Agriculture and Rural Development, A8-0119/2017

European Union (EU) (2013) Overview of CAP reform 2014–2020. Agricultural Policy Perspectives Brief, N°5

FAO (1976) Soil conservation and management in developing countries, FAO soils bulletin, 33. FAO, Rome

FAO (1995) Necessità e Risorse. Atlante dell'Alimentazione e dell'Agricoltura. FAO, Rome

FAO (2011) Save and grow. A policymaker's guide to the sustainable intensification of smallholder crop production. FAO, Rome

FAO (2014) Towards a concept of sustainable animal diets. In: Harinder P, Makkar S, Ankers P (eds) FAO animal production and health report. No. 7. FAO, Rome

FAO (2015) The 5 principles of sustainable food and agriculture. FAO, Rome

FAO (2018a) The 10 elements of agroecology guiding the transition to sustainable food and agricultural systems. FAO, Rome

FAO (2018b) Scaling up agroecology initiative: transforming food and agricultural systems in support of the SDGs. http://www.fao.org/3/I9049EN/i9049en.pdf

FAO (2019) TAPE tool for agroecology performance evaluation 2019 – process of development and guidelines for application. Test version. FAO, Rome

FAO and IFAD (2019) United Nations decade of family farming 2019–2028. Global Action Plan. FAO, Rome

Fedoroff NV (2010) The past, present and future of crop genetic modification. New Biotechnol 27(5):461–465

Ferguson BG et al (2019) Special issue editorial: what do we mean by agroecological scaling? Agroecol Sustain Food Syst 43(7–8):722–723. https://doi.org/10.1080/21683565.2019.1591565

Fernández González C et al (2020) Transdisciplinarity in agroecology: practices and perspectives in Europe. Agroecol Sustain Food Syst. https://doi.org/10.1080/21683565.2020.1842285

Fernandez MK et al (2013) Agroecology and alternative agri-food movements in the United States: toward a sustainable agri-food system. Agroecol Sustain Food Syst 37(1):115–126
Flannery KV (1972) The cultural evolution of civilizations. Annu Rev Ecol Syst 3:399–426
Foley JA et al (2005) Global consequences of land use. Science 309:570–574
Forman RTT, Baudry J (1984) Hedgerows and hedgerows networks in landscape ecology. Environ Manage 8:495–510
Forman RTT, Godron M (1986) Landscape ecology. Wiley, New York
Francis AC (1989) Biological efficiencies in multiple-cropping systems. Adv Agron 42:1–42
Francis CA et al (2003) Agroecology: the ecology of food systems. J Sustain Agric 22(3):99–118
Freudenberger D, Freudenberger CD (2015) Principles of regenerative agriculture for ecological civilizations. In: Cobb JB, Castuera I (eds) For our common home. Process Century Press, Anoka, pp 260–269
Frison EA et al (2011) Agricultural biodiversity is essential for a sustainable improvement in food and nutrition security. Sustainability 3:238–253
Frissel MJ (1977) Cycling of mineral nutrients in agricultural ecosystems. Agro-Ecosystems 4:7–16
Gallardo-López F et al (2018) Development of the concept of agroecology in Europe: a review. Sustainability 10:1210. https://doi.org/10.3390/su10041210
Garcia-Ruiz JM, Lana-Renault N (2011) Hydrological and erosive consequences of farmland abandonment in Europe, with special reference to the Mediterranean region – a review. Agric Ecosyst Environ 140:317–338
Gare A (2017) The philosophical foundation of ecological civilization. Earthscan Routledge, London
Garibaldi LA et al (2017) Farming approaches for greater biodiversity, livelihoods, and food security. Trends Ecol Evol 32(1):68–80
Garratt MPD et al (2017) The benefits of hedgerows for pollinators and natural enemies depends on hedge quality and landscape context. Agric Ecosyst Environ 247:363–370
Gepts P (2004) Crop domestication as long-term selection experiment. Plant Breed Rev 24(2):1–42
Gilbert SF et al (2012) A symbiotic view of life: we have never been individuals. Q Rev Biol 87(4):325–341
Gliesmann SR (2016) Transforming food systems with agroecology. Agroecol Sustain Food Syst 40(3):187–189. https://doi.org/10.1080/21683565.2015.1130765
Gliessman SR (2015) Agroecology: the ecology of sustainable food systems, 3rd edn. CRC Press/ Taylor & Francis, Boca Raton
Gliessman SR (2018) Defining agroecology. Agroecol Sustain Food Syst 42(6):599–600. https://doi.org/10.1080/21683565.2018.1432329
Gliessman SR et al (2019) Agroecology and food sovereignty. IDS Bull Polit Econ Food 50(2):91–109
Godfray HCJ, Garnett T (2014) Food security and sustainable intensification. Philos Trans R. Soc B 369:20120273. https://doi.org/10.1098/rstb.2012.0273
Golley FB (1993) A history of the ecosystem concept in ecology. Yale University Press, New Haven
Gonzales TA (2000) The culture of the seed in the Peruvian Andes. In: Brush SB (ed) Genes in the field. IPGRI, Rome, pp 193–216
Gonzales de Molina M et al (2020) Political agroecology: advancing the transition to sustainable food systems. CRC Press, Boca Raton
Gordon JI (2012) Honor. Thy gut symbionts redux. Science 336:1251. https://doi.org/10.1126/science.1224686
Gretter A et al (2018) Governing mountain landscapes collectively: local responses to emerging challenges within a systems thinking perspective. Landsc Res. https://doi.org/10.1080/01426397.2018.1503239
Grove R et al (2020) Pastoral stone enclosures as biological cultural heritage: Galician and Cornish examples of community conservation. Land 9:9. https://doi.org/10.3390/land9010009
Guinée JB et al (2002) Handbook on life cycle assessment; operational guide to the ISO standards. Institute for Environmental Sciences, Leiden

Gurr GM et al (2016) Multi-country evidence that crop diversification promotes ecological intensification of agriculture. Nat Plants 2:16014. https://doi.org/10.1038/nplants.2016.14
Habel NC, Trudinger P (2008) Exploring ecological hermeneutics. Society of Biblical Literature, Atlanta
Haber W (1989) Using landscape ecology in planning and management. In: Zonneveld IS, Forman RTT (eds) Changing landscapes: an ecological perspective. Springer, New York, pp 217–232
Haber W (1990) Basic concepts of landscape ecology and their application in land management. In: Kawanabe H, Ohgushi T, Higashi M (eds) Ecology for tomorrow, Physiology and ecology Japan, (special number) 27. Kyoto University, Kyoto, pp 131–146
Harlan JR (1995) The living fields: our agricultural heritage. Cambridge University Press, Cambridge
Harper JL (1982) After description. In: Newman EI (ed) The plant community as a working mechanism, British ecological society. Special publication, N° 1. Blackwell Scientific, Oxford, pp 11–25
Havilk P et al (2014) Climate change mitigation through livestock system transitions. PNAS 111(10):3709–3714
Hawkes C et al (2017) Urbanization and the nutrition transition. In: IFPRI (International Food Policy Research Institute) (ed) Global food policy report. International Food Policy Research Institute, Washington, DC, pp 34–41
Heinrich Boll Foundation, Rosa Luxemburg Foundation, Friends of the Earth Europe (2017) Agrifood Atlas. Facts and figures about the corporations that control what we eat. Heinrich Foundation, Berlin
Hellerstein D et al (2002) Farmland protection: the role of public preferences for rural amenities, Agricultural economic report no. 815. Boll Foundation, Berlin
Henning B (2015) Stewardship and the roots of the ecological crisis. In: Cobb JB, Castuera I (eds) For our common home. Process Century Press, Anoka, pp 41–45
Hudson N (1976) Research needs for soil conservation in developing countries. In: Soil conservation and management in developing countries, FAO soils bulletin, 33. FAO, Rome, pp 169–181
Huyck L, Francis CA (1995) Designing a diversified farmscape. In: Exploring the role of diversity in sustainable agriculture. American Society of Agronomy, Madison, pp 95–120
IAASTD (International Assessment of Agricultural Knowledge, Science and Technology for Development) (2006) Agriculture at a crossroads. Global report of the International Assessment of Agriculture Knowledge, Science and Technology for Development. Island Press, Washington, DC
Ianni E et al (2015) Revitalizing traditional ecological knowledge: a study in an alpine rural community. Environ Manag. https://doi.org/10.1007/s00267-015-0479-z
IFOAM (International Federation of Organic Agriculture Movements) (2005) Principles of organic agriculture. Bonn: IFOAM. Available on-line at: http://www.ifoam.org/
IFOAM EU GROUP (2019) Position paper on agroecology. Organic and agroecology: working to transform our food system. Available on line at: https://www.organicseurope.bio/library-type/position-papers/
IFPRI (International Food Policy Research Institute) (2017) Global food policy report. International Food Policy Research Institute, Washington, DC. https://doi.org/10.2499/9780896292529
IPES FOOD (International Panel of Experts on Sustainable Food Systems) (2016) From uniformity to diversity. A paradigm shift from industrial agriculture to diversified agroecological systems. www.ipes-food.org
Jack B (2015) Ecosystem services: European agricultural law and rural development. In: Monteduro M et al (eds) Law and agroecology. A transdisciplinary dialogue. Springer, Berlin, pp 127–150
Jackson L et al (2010) Biodiversity and agricultural sustainagility: from assessment to adaptive management. Curr Opin Environ Sustain 2:80–87
Jagger S (2015) What does your garden show? Exploration of the semiotics of the garden. In: Trifonas PP (ed) International handbook of semiotics. Springer, Berlin, pp 629–646

Janssens MJJ et al (1990) Low-input ideotypes. In: Gliessmann SR (ed) Agroecology, Ecological studies 78. Springer, Berlin, pp 130–145
Jarvis DI et al (2008) A global perspective of the richness and evenness of traditional crop-variety diversity maintained by farming communities. PNAS 105:5326–5331
Jhingran S (1989) Aspects of Hindu morality. Motilal Banarsidass Publishers, Delhi
Jordan WR III et al (1987) Restoration ecology: ecological restoration as a technique for basic research. In: Jordan WR III, Gilpin ME, Aber JD (eds) Restoration ecology. University Press, Cambridge, pp 3–21
Kamra DN (2005) Rumen microbial ecosystem. Curr Sci 89(1):124–134
Karlen DL et al (1994) Crop rotations for the 21st century. Adv Agron 53:2–45
Keys A, Keys M (1963) Eat well and stay well. Doubleday, Garden City
Kiley-Worthington M (1981) Ecological agriculture, what is and how it works. Agric Environ 6:349–381
Kleijn D et al (2019) Ecological intensification: bridging the gap between science and practice. Trends Ecol Evol 34(2):154–166
Koohafkan P, Altieri M (2011) Global important agricultural heritage systems. A legacy for the future. FAO, Rome
Krauß W, Olwig KR (2018) Special issue on pastoral landscapes caught between abandonment, rewilding and agro-environmental management. Is there an alternative future. Landsc Res 43(8):1015–1020. https://doi.org/10.1080/01426397.2018.1503844
Kremen C, Miles A (2012) Ecosystem services in biologically diversified versus conventional farming systems: benefits, externality, and trade-offs. Ecol Soc 17(4):40. https://doi.org/10.5751/ES-05035-170440
Kuiper J (2000) A checklist approach to evaluate the contribution of organic farming to landscape quality. Agric Ecosyst Environ 77:143–156
LaFevor MC (2014) Restoration of degraded agricultural terraces: rebuilding landscape structure and process. J Environ Manag 138:32–42
Lal R (2014) Soil conservation and ecosystem services. Int Soil Water Conserv Res 2(3):36–47
Landis DA (2017) Designing agricultural landscapes for biodiversity-based ecosystem services. Basic Appl Ecol 18:1–12
Larson G et al (2014) Current perspective and the future of domestication studies. PNAS 111(17):6139–6146
Leopold A (1949) A Sand County almanac and sketches here and there. Oxford University Press, Oxford
Leroy F, Praet I (2015) Meat traditions: the coevolution of humans and meat. Appetite 90:200–211
Lescourret F et al (2015) A social-ecological approach to managing multiple agro-ecosystem services. Curr Opin Environ Sustain 14:68–75
LeVasseur T et al (2016) Religion and sustainable agriculture. World spiritual traditions and food ethics. University Press of Kentucky, Lexington
Lieblein G, Francis C (2007) Towards responsible action through agroecological education. Ital J Agron/Rivista di Agronomia 2(2):83–90
Liebman M, Dyck E (1993) Crop rotation and intercropping strategies for weed management. Ecol Appl 3(1):92–122
Lin BB et al (2011) Effects of industrial agriculture on climate change and the mitigation potential of small-scale agro-ecological farms. CAB Rev 6(020). https://doi.org/10.1079/PAVSNNR20116020
Loomis RS, Connor DJ (1992) Crop ecology. Cambridge University Press, Cambridge
Lowder SK et al (2016) The number, size, and distributions of farms, smallholder farms, and family farms worldwide. World Dev 87:16–29
Lowrance RR (1998) Riparian forest ecosystems as filters for non-point source pollution. In: Pace ML, Groffman PM (eds) Successes, limitations and frontiers in ecosystem science. Springer, New York, pp 113–141

MacRae RJ et al (1990) Farm-scale agronomic and economic conversion from conventional to sustainable agriculture. Adv Agron 43:155–198
Mader P et al (2002) Soil fertility and biodiversity in organic farming. Science 296:1694–1697.
Magrini MB et al (2019) Agroecological transition from farms to territorialised agri-food systems: issues and drivers. In: Bergez JE et al (eds) Agroecological transition: from theory to practice in local participatory design. Springer, Cham, pp 69–98
Malèzieux E (2012) Design cropping systems from nature. Agron Sustain Dev 32:15–29
Marshall EJP, Moonen AC (2002) Field margins in northern Europe: their functions and interactions with agriculture. Agric Ecosyst Environ 89:5–21
Martens EC et al (2011) Recognition and degradation of plant cell wall polysaccharides by two human gut symbionts. PLoS Biol 9(12):e1001221
Martinez-Alier J (2011) The EROI of agriculture and its use by the Via Campesina. J Peasant Stud 38(1):145–160
Martinez-Torres ME, Rosset PM (2010) La Via Campesina: the birth and evolution of a transnational social movement. J Peasant Stud 37(1):149–175
Marton SMRR et al (2016) Environmental and socioeconomic benefits of a division of labour between lowland and mountain farms in milk production systems. Agric Syst 149:1–10
McMahan J (2019) Eating meat. In: Edmonds D (ed) Ethics and contemporary world. Routledge, New York, pp 291–300
McMichael P (2009) A food regime genealogy. J Peasant Stud 36:139–169
MEA (Millennium Ecosystem Assessment) (2005) Ecosystems and human well-being: biodiversity synthesis. World Resources Institute, Washington, DC
Mendez VE et al (2013) Agroecology as a transdisciplinary, participatory, and action-oriented approach. Agroecol Sustain Food Syst 37:3–18
Mendez VE et al (2020) Agroecology, a transdisciplinary, participatory, and action-oriented approach, 2nd edn. CRC Press, Boca Raton
Miles A et al (2017) Triggering a positive research and policy feedback cycle to support a transition to agroecology and sustainable food systems. Agroecol Sustain Food Syst 41(7):855–879. https://doi.org/10.1080/21683565.2017.1331179
Milla R et al (2017) Looking at past domestication to secure ecosystem services of future croplands. J Ecol 105:885–889
Miñarro M, Prida E (2013) Hedgerows surrounding organic apple orchards in north-west Spain: potential to conserve beneficial insects. Agric For Entomol 15:382–390
Mooney P (2018) Blocking the chain. Industrial food chain concentration. Big data platforms and food sovereignty solutions. ETC Group, Val David
Morandin LA, Kremen C (2013) Hedgerow restoration promotes pollinator populations and exports native bees to adjacent fields. Ecol Appl 23(4):829–839
Murdoch W (1990) World hunger and population. In: Carrol CR, Vandermeer JH, Rosset PM (eds) Agroecology, Wayne M Getz Series Editor. University of California, Berkeley, pp 3–20
Myskja BK (2006) The moral difference between intragenic and transgenic modifications of plants. J Agric Environ Ethics 19:225–238
National Research Council (1989) Alternative agriculture. National Academy Press, Washington, DC
Nielsen KM (2003) Transgenic organisms – times for conceptual diversification? Nat Biotechnol 21:227–228
Nijland HJ et al (2013) What is careful livestock farming? Substantiating the layered meaning of the term 'careful' and drawing implications for the stakeholder dialogue. NJAS 66:23–31
Norgaard RB (1984a) Coevolutionary agricultural development. Econ Dev Cult Chang 32(31):525–546
Norgaard RB (1984b) Coevolutionary development potential. Land Econ 60(2):160–173
Norgaard RB (1988) Sustainable development: a co-evolutionary view. Futures 20(6):606–620
Novak MA (2006) Five rules for the evolution of cooperation. Science 314:1560–1563

Nyéléni (2007) Declaration of Nyéléni, Declaration of the Forum for Food sovereignty. Full text available at: https://nyeleni.org/IMG/pdf/DeclNyeleni-en.pdf
OECD (Organisation for Economic Cooperation and Development) (1999) Environmental indicators for agriculture: issues and design, vol 2. OECD, Paris
Oreszczyn S, Lane A (2000) The meaning of hedgerows in the English landscape: different stakeholder perspectives and the implications for future hedge management. J Environ Manag 60(1):101–118
Packham R, Srikandarajah N (2005) Systemic action research for postgraduate education in agriculture and rural development. Syst Res 22:119–130
Parker KA (1996) Pragmatism and environmental thought. In: Light A, Katz E (eds) Environmental pragmatism. Routledge, London, pp 21–37
Pasqual U et al (2017) Valuing nature's contributions to people: the IPBES approach. Curr Opin Environ Sustain 26:7–16.
Pimentel D, Kounang N (1998) Ecology of soil erosion in ecosystems. Ecosystems 1:416–426
Pollard E et al (1974) Hedges. Collins, London
Pope Francis (2015) Laudato si'. Encyclical letter on Care for our Common Home. www.vatican.va
Post WM, Kwon KC (2000) Soil carbon sequestration and land-use change: processes and potential. Glob Chang Biol 6:317–328
Quist DA et al (2013) Hungry for innovation: pathways from GM crops to agroecology. In: Late lessons from early warnings, European Environmmental Agency (EEA) report no 1, vol 2. Publications Office, Luxembourg, pp 458–485
Ratnadass A et al (2012) Plant species diversity for sustainable management of crop pests and diseases in agroecosystems: a review. Agron Sustain Dev 32:273–303
Raybould A, Poppy GM (2012) Commercializing genetically modified crops under EU regulations. GM Crops Food 3(1):9–20. https://doi.org/10.4161/gmcr.18961
Re F (1806) Elementi D'Agricoltura. Stamperia Vitarelli, Venezia
Reganold JP, Wachter JM (2016) Organic agriculture in the twenty-first century. Nat Plants 2:15221. https://doi.org/10.1038/NPLANTS.2015.221
Reganold JP et al (2001) Sustainability of three apple production systems. Nature 410:26–930
Ridolfi C (1838) Istituto Agrario di Meleto. Cont Atti dei Georgofili 16:557
Robson MC et al (2002) The agronomic and economic potential of break crops for ley/arable rotations in temperate organic agriculture. Adv Agron 77:369–427
Rockstrom J et al (2009) Planetary boundaries: exploring the safe operating space for humanity. Ecol Soc 14(2):32 [online] URL: http://www.ecologyandsociety.org/vol14/iss2/art32/
Roger-Estrade J et al (2010) Tillage and soil ecology: partners for sustainable agriculture. Soil Tillage Res 111:33–40
Roling N (2003) From causes to reasons: the human dimension of agricultural sustainability. Int J Agric Sustain 1(1):73–88
Rosenzweig ST, Schipanski ME (2019) Landscape-scale cropping changes in the High Plains: economic and environmental implications. Environ Res Lett 14:124088
Sabaté J (ed) (2019) Environmental nutrition. Connecting health and nutrition with environmentally sustainable diets. Academic Press, London
Sabaté J, Hanvatt H, Soret S (2016) Environmental nutrition: a new frontier for public health. Am J Public Health 106:815–821
Savo V et al (2014) Combining environmental factors and agriculturalists' observations of environmental changes in the traditional terrace system of the Amalfi coast (Southern Italy). Ambio 43:297–310
Schade JD et al (2002) Sources of nitrogen to the riparian zone of a desert stream: implications for riparian vegetation and nitrogen retention. Ecosystems 5:68–79
Schneider GW, Winslow R (2014) Parts and wholes: the human microbiome, ecological ontology, and the challenges of community. Perspect Biol Med 57(2):208–223

Shiva V (2016) The dharma of food. In: LeVasseur T, Parajuli P, Wirzba N (eds) Religion and sustainable agriculture. World spiritual traditions and food ethics. University Press of Kentucky, Lexington, pp VII–X
Smith LM et al (2011) Ecosystem services provided by playas in the High Plains. Ecol Appl 21(3, Supplement):S82–S92
Soane ID et al (2012) Exploring panarchy in alpine grasslands: an application of adaptive cycle concepts to the conservation of a cultural landscape. Ecol Soc 17(3):18. https://doi.org/10.5751/ES-05085-170318
Socci P et al (2019) Terracing: from agriculture to multiple ecosystem services. Agric Environ. https://doi.org/10.1093/acrefore/9780199389414.013.206
Sofi F et al (2008) Adherence to Mediterranean diet and health status: meta-analysis. BMJ 337:a1344
Stehfest E et al (2009) Climate benefits of changing diets. Clim Chang 95(1–2):83–102
Stetter MG et al (2017) How to make a domesticate. Curr Biol 27:R853–R909
Struik PC, Kuyper TW (2017) Sustainable intensification in agriculture: the richer shade of green. A review. Agron Sustain Dev 37:39. https://doi.org/10.1007/s13593-017-0445-7
Swift MJ, Anderson JM (1994) Biodiversity and ecosystem function in agricultural systems. In: Schulze ED, Mooney HA (eds) Biodiversity and ecosystem function. Springer, Berlin, pp 15–41
Tamburini G et al (2020) Agricultural diversification promotes multiple ecosystem services without compromising yield. Sci Adv 6:eaba1715
Tarolli P et al (2014) Terraced landscapes: from an old best practice to a potential hazard for soil degradation due to land abandonment. Anthropocene 6:10–25
Teixeira CMGL, et al. (2018) Animal husbandry. Biodiversity Fact Sheet. Global Nature Fund
Tellarini V, Caporali F (2000) An input/output methodology to evaluate farms as sustainable agroecosystems. An application of indicators to farms in Central Italy. Agric Ecosyst Environ 77:111–123
Thiele LP (1999) Evolutionary narratives and ecological ethics. Political Theory 27(1):6–38
Thomas BA (2003) Removal of hedgerows: proper management of land. Councils: hedgerow regulation (Sl 1997, NO. 11 60). Environ Law Rev 5:279–287
Thorntorn PK (2010) Livestock production: recent trends, future prospects. Philos Trans R Soc B 365:2853–2867
Tilman D (2000) Causes, consequences and ethics of biodiversity. Nature 405:208–211
Tilman D, Clark M (2014) Global diets link environmental sustainability and human health. Nature 515:518–522
Tilman D et al (2011) Global food demand and the sustainable intensification of agriculture. PNAS 108(50):20260–20264
Tittonell P (2014) Ecological intensification of agriculture—sustainable by nature. Curr Opin Environ Sustain 8:53–61
Tittonell P, Giller KE (2013) When yield gap are poverty traps: the paradigm of ecological intensification in African smallholder agriculture. Field Crop Res 143:76–90
Tomich TP et al (2011) Agroecology: a review from a global-change perspective. Annu Rev Environ Resour 36:193–222
Treacy JM (1987) Building and rebuilding agricultural terraces in the Colca Valley of Peru. Yearb Conf Lat Am Geogr 13(1987):51–57. https://www.jstor.org/stable/25765680
Trevelyan RC (1941) The infinity. In: Singh G (ed) Canti di Giacomo Leopardi nelle traduzioni inglesi. Centro nazionale di Studi Leopardiani, Recanati, p 163
Tsatsakis AM et al (2017) Environmental impacts of genetically modified plants: a review. Environ Res 156:818–833
Tscharntke T et al (2005) Landscape perspectives on agricultural intensification and biodiversity: ecosystem service management. Ecol Lett 8:857–874
Tscharntke T et al (2012) Global food security, biodiversity conservation and the future of agricultural intensification. Biol Conserv 151:53–59

UN (2015) Transforming our world: the 2030 agenda for sustainable development. UN General Assembly, 25 September 2015

UNESCO (2014) Intangible Cultural Heritage nomination file no. 00720 for inscription on the representative list of the intangible cultural heritage of humanity in 2014. http://www.unesco.org/culture/ich/doc/download.php?versionID=30503

UNESCO (2019) » Culture » Intangible Heritage » Lists » Transhumance, the seasonal droving of livestock along migratory routes in the Mediterranean and in the Alps. https://ich.unesco.org/en/RL/

Val V et al (2019) Agroecology and La Via Campesina I. The symbolic and material construction of agroecology through the dispositive of "peasant-to-peasant" processes. Agroecol Sustain Food Syst 43(7–8):872–894. https://doi.org/10.1080/21683565.2019.1600099

Van der Sluis T et al (2014) Landscape change in Mediterranean farmlands: impacts of land abandonment on cultivation: terraces in Portofino (Italy) and Lesvos (Greece). J Landsc Ecol 7(1):23–44

Vandermeer J (1989) The ecology of intercropping. Cambridge University Press, Cambridge, UK

Varotto M et al (eds) (2019) World terraced landscapes: history, environment, quality of life. Springer, New York

Walker JW (1994) Multispecies grazing: the ecological advantage. Sheep Research J (USA), Special issue, pp 52–64.

Wals AEI, Bawden R (2000) Integrating sustainability into agricultural education: dealing with complexity, uncertainty and diverging worldviews. AFANet, (ICA), University Gent, Gent

Wals AEI et al (2004a) Education and training for integrated rural development: stepping stones for curriculum development. Elsevier-Reed Business Information, The Hauge

Wals AEI et al (2004b) Education for integrated rural development: transformative learning in a complex and uncertain world. J Agric Educ Ext 10(2):89–100

WCED (World Commission on Environment and Development) (1987) Our Common Future (The Bruntland Report). Oxford University Press.

Walton RJ et al (2014) Regulatory services delivered by hedges: the evidence base. Report of Defra project LM0106. 99pp

Weal A (2010) Ethical arguments relevant to the use of GM crops. New Biotechnol 27(5):582

Webber GD et al (1976) Farming systems in South Australia. Department of Agriculture 587

Wei W et al (2016) Global synthesis of the classifications, distributions, benefits and issues of terracing. USDA Forest Service/UNL Faculty Publications. 312. http://digitalcommons.unl.edu/usdafsfacpub/312

Weißhuhn P et al (2017) Supporting agricultural ecosystem services through the integration of perennial polycultures into crop rotations. Sustainability 9:2267. https://doi.org/10.3390/su9122267

Wezel A et al (2015) The blurred boundaries of ecological, sustainable, and agroecological intensification: a review. Agron Sustain Dev 35:1283–1295

Wezel A et al (2018) Agroecology in Europe: research, education, collective action networks, and alternative food systems. Sustainability 10:1214. https://doi.org/10.3390/su100412142214

White L (1967) The historical roots of our ecological crisis. Science 155:1203–1207

White R (2009) Understanding vineyard soils. Oxford University Press, New York

Wirsenius S et al (2010) How much land is needed for global food production under scenarios of dietary changes and livestock productivity increases in 2030? Agric Syst 103(9):621–638

Wirzba N (2019) Food and faith. A theology of eating, 2nd edn. Cambridge University Press, Cambridge

Woodward L et al (1996) Reflections on the past, outlook for the future. In: Østergaard TV (ed) Fundamentals of organic agriculture. Proceedings volume I of the 11th IFOAM international scientific conference. IFOAM, Copenhagen, pp 259–270

Wratten SD et al (2012) Pollinator habitat enhancement: benefits to other ecosystem services. Agric Ecosyst Environ 159:112–122

Xu J et al (2004) Message from a human gut symbiont: sensitivity is a prerequisite for sharing. Trends Microbiol 12(1):21–28

Yehong S et al (2011) Tourism place: a discussion forum. Terraced landscapes as a cultural and natural heritage resource. Tour Geogr 13(2):328–331

Zimmerer KS, de Haan S (2017) Agrobiodiversity and a sustainable food future. Nat Plants 3:17047. https://doi.org/10.1038/nplants.2017.47

Zohary D, Spiegel-Roy P (1975) Beginnings of fruit growing in the Old World. Science 187:319–327

Chapter 7
Conclusions: A New Ecological Ethic for Grounding Sustainability in Agriculture and Society

> *Agriculture – perhaps the most crucial technology today because of the universal need for food and the continuing crises of hunger and environmental degradation*
>
> *(Barbour I. 1993, p. 85).*

Ethics is a process of cybernetic account that human beings in action use in their context of life. Ethics evolves according to changes induced, information retrieved and prospected new action. Rapid changes induced by the Anthropocene to the local and global environments require adequate responses of both mitigation and adaptation in order to keep balanced conditions suitable for a sustainable development of both human society and the larger planet leaving community. Maintaining life on this planet, as we know it, is becoming an emergent ethical question and must be tackled exclusively by human kind. Agriculture is the human activity system that induces major changes at the ecosystem level and brings about large effects on biodiversity and biogeochemical cycles, while producing the indispensable assets (food, feed, fibres, etc.) for human health, wealth and welfare. Rapid human population growth and industrial civilisation favoured by input of fossil energy sources have led agriculture to a crossroads that requires responsible decisions in order to re-establish conditions for a sustainable agriculture. Increasing urban concentration of people and facilities, and industrial agriculture instrumentally subservient to it are not more sustainable. Current agricultural policy at the international level (UN, FAO, EU, etc.) has produced important steps for orientating agriculture towards an agroecological transition that must be implemented at the level of single States, according to their national, regional and local conditions of climate, soil, water resources, and socioeconomic development. Agroecology, as the science of ecology applied to agriculture, has both epistemological and ontological bases for producing a change of paradigm in favour of sustainable agriculture. It reveals that sustainability is a property emerging spontaneously from the becoming of natural

F. Caporali, *Ethics and Sustainable Agriculture*,
https://doi.org/10.1007/978-3-030-76683-2_7

processes operating at the agroecosystem level, whether an adequate integration of production, consumption, decomposition, and recycling is ensured at both farm and landscape levels. The driving condition for implementing integration of ecological intensification processes is a viable rural society that allows human and livestock population to permeate the land fabric in a way that complementarity between natural processes and human operations occurs in congruence with the principles of eco-development. Maximisation of the solar radiation flux, matter re-cycling, and biodiversity potential through internal web of grazing and detritus chains are the conditions facilitating agroecosystem ecological intensification. Site-specificity of agroecosystem design and management is a major criterion of agriculture sustainability in that local cultural tradition and environmental constraints have already met and balanced through successful experiential learning of best practices by farmers' generations. Education toward local production and consumption of food and feed is a powerful means for building up knowledge and trust of agriculture stakeholders at the local level, benefitting a local circular economy and strengthening the sense of identity and belongingness to a rural community, which is guaranty of resilience and safe development. Appropriate agricultural regulations and norms for sustaining financially, technically, and culturally small and medium enterprises of farmers, transformers, and traders at local level, are indispensable to create an interdependent web of business relations that are also social relations of both mutual support and recognised development of skills, competences, and job creation and innovation. The maintenance of a balanced rural society is still an appropriate means of sustainable development for both national States and the leaving community of the planet Earth.

An ecological ethic emerges when agriculture is recognised a complex process of development involving at the same time socio-economic, biological and physical components at different levels of organisation (field, farm, and landscape). These components need to be synergistically organised in a way that natural processes are recognised hierarchical, dominant driving forces. Design and management are consequent arrangements of components that satisfy both their complementarity (or internal coherence) for the agroecosystem performances of production and protection, and their likely positive impact (or external correspondence) for the ecological services (production, support, regulation and culture) provided in the context of action. A robust multifunctional agriculture is the outcome of an ecological ethic responsibly pursued. Whether humanity has hope to achieve the goal of a sustainable agriculture will depend on what we think about our planet, land, agriculture, justice, peace, and solidarity. We must establish a new culture for a new citizenship, i.e. an earthbound citizenship, grounded on the facts and values recognised and professed by science, philosophy, and theology. These cultural values must forge ethical ideals to disseminate with an appropriate eco-linguistics, made up of frames, metaphors and narratives, which can change the old paradigm of unlimited growth into the new paradigm of balanced development within both the planetary and local limits. For example, concepts, metaphors, and narratives about food as the primary asset provided by agriculture, must insist more on the food as a process than as a product. Indeed, food is the basic relationship linking all the living beings in a unique food community, where abiotic and biotic components are joint in a sustainable, regenerative flow of ecosystem life. It may be useful to suggest a hierarchical organisation of values in order to strive for an ecological ethic and a sustainable agriculture (Table 7.1).

To achieve a sustainable agriculture would be necessary to climb the ladder of values and ethics which starts from the lowest step (anthropocentric), passes through the intermedium one (biocentric), and reaches the top one (ecological). It is a cultural ascension marking the passage from and egocentric or self-centred utilitarian stance to a progressive inclusiveness, which recognises values to the partnership of the agroecosystem components. A growth in ecological knowledge and practice, which also means an ethical progression, is the next challenge for humanity, and the first test bench for evaluating its effectiveness is the attainment of a sustainable agriculture.

Agriculture really has the potential to guide the ecological transition toward an integral sustainable development, at both local and global level, following the basic agroecological principles reported in Table 7.2.

The first two principles (*wisdom* and *life*) concern progress in systems knowledge and understanding for informing systems practice through the second two principles (*complementarity* and *multi-functionality*). Both kind of principles constitute indispensable immaterial values for grounding an ecological ethic more appropriate for a socio-ecological organisation of care, justice and peace. On the base of etymological, epistemological, ontological and ethical reasons and values, agroecology is both a transdisciplinary science and a paradigm for promoting sustainable agriculture at every scale of intervention, be it local or global. Social recognition and implementation of its potential through appropriate law and institutional change in academic and professional parties can pave the way for a humankind's future of *harmony-with-nature- development*.

Table 7.1 Hierarchy of values and kind of ethics for a sustainable agriculture

Values	Ethics	Agriculture
Instrumental	Anthropocentric	Industrial
Intrinsic	Biocentric	↓
Systemic	Ecological	
		Sustainable

Table 7.2 Basic agroecological principles for an ecological ethic or an ethic for sustainability

Category	Formulation	Explanation
Wisdom principle **or of "ontological respect"**	**Agriculture must not contradict nature.**	Categorical imperative**: agriculture must comply with ontology**, i.e. the organisation (structure and functioning) of the planet Earth. "Quo natura vergit, eo ducere oportet"[a]
Life principle **or of "autopoietic respect"**	The Earth is a photosynthetic planet which takes on a life of its own (**all beings are photosynthesis-dependent**), therefore agriculture must ground on the best use of solar radiation and the other natural resources.	Fields, farms and rural land make up potential **"solar plants"** that collect and storage energy as biomass useful for grazing and detritus chains, building up soil fertility, biological control, climatic regulation and system sustainability. All these agroecosystem services are released how much more ecological intensification proceeds.
Complementarity principle **or of "balanced land use"**	**Agriculture must occupy and defend its own fertile soils, without contributing to erosion of strategic natural ecosystems on both a local or global scale.**	**Fertile agricultural soils need protection from land consumption by other alternative use (urban, industrial, etc.).** **Tropical forest ecosystems need protection from agricultural land use.**
Multi-functionality principle **or of "socio-ecological sustainability"**	Agriculture is a human activity system that co-evolves with the natural environment and must yield **ecosystem services** (support, regulation, provision and culture).	Balanced demography of humans and livestock at the catchment scale for yielding **soil production and protection, water regulation, climatic mitigation, biomass production, and culture for sustainability** (for instance, **bio-districts** for local governance and **academic curricula in agroecology for education and training).**

[a]Where nature goes, it is better to go there (from the physician-philosopher Galeno)

Afterword

A scholarly review is not only a routine procedure letting your work published. In my case, it was also an enrichment process of spiritual sharing and confidential companionship. The process had a direction orchestrated by a publishing editor who first made a connection between the author and anonymous reviewers, and then filtered information and knowledge flowing reciprocally between them. This process was a fluid and dynamic event that led to emergent results – a creative process unveiling the modality through which spirit works- but flowing with words.

A first reviewer made comments and suggestions as follows:

"The subject is of intense current and future interest, as many scientists realize that the objectivity and absolute truths we have pursued for decades are in fact tempered by many social, cultural, historical, and especially personal values. **Current interest in the North with 'diversity, equity, and inclusiveness' [DEI] could be summarized**".

"This focus is now permeating our thinking as we navigate the pandemic, and realize even more than before how marginalized people are differentially affected by such disturbances. We are seeking a sustainable and equitable food system that is resilient in the face of climate and social change. **This perspective if added to the proposed book on ethics and sustainable agriculture would definitely expand and extend its relevance to the next generation. I appreciate that this is both painful and difficult for me and for the author, as our generation has dealt primarily with the geo-bio-physical aspects of farming and food systems**. In my opinion, this should be woven into the ethics discussions".

These inspiring comments and suggestions led me to document myself on the strength of DEI movement, looking first at the meaning of the component words of the acronym DEI and then at commenting on them. Consulting the "**Glossary of Terms for Diversity, Equity, & Inclusion**",[1] their definitions read as follows:

[1] Equity, Diversity, and Inclusion Glossary of Terms UNIVERSITY OF WASHINGTON – SCHOOL OF PUBLIC HEALTH DEPARTMENT OF EPIDEMIOLOGY – EQUITY, DIVERSITY, AND INCLUSION (EDI) COMMITTEE, Summer 2019.

F. Caporali, *Ethics and Sustainable Agriculture*,
https://doi.org/10.1007/978-3-030-76683-2

Diversity describes the myriad ways in which people differ, including the psychological, physical, and social differences that occur among all individuals, such as race, ethnicity, nationality, socioeconomic status, religion, economic class, education, age, gender, sexual orientation, marital status, mental and physical ability, and learning styles. Diversity is all-inclusive and supportive of the proposition that everyone and every group should be valued. It is about understanding these differences and moving beyond simple tolerance to embracing and celebrating the rich dimensions of our differences.

Equity ensures that individuals are provided the resources they need to have access to the same opportunities, as the general population. While equity represents impartiality, i.e. the distribution is made in such a way to even opportunities for all the people, conversely equality indicates uniformity, where everything is evenly distributed among people.

Inclusion/Inclusiveness. Authentically bringing traditionally excluded individuals and/or groups into processes, activities, and decision/policy making in a way that shares power.

All together, these three words build up an intriguing system of interdependence such as that represented in Fig. 1, where ecological meaning emerges through the confluence of convergent "values" of different nature: ontological, epistemological and normative.

In a systems view of reality, **diversity** is the structural component of ecosystems bearing an ontological value because sensitively recognised as *being and becoming*. **Inclusion** is the functional component of ecosystems that science epistemologically explains as *integration*, i.e. ecological complementarity among structural components in a dynamic and sustainable organisation. **Equity** is an anthropic component of ecosystems as a criterion with potential for deliberative power and normative value. Equity is a challenge that includes humanity and each man as subject/object of discrimination for rights and duties in enjoying ecosystem goods and services according to justice.

Equity is establishing right relationships among people and between people and the environment- in practice, the present and future challenge of the entire humanity or *the human predicament*. In an ecotheological perspective, humanity is not only the Anthropocene driver but also the responsible trustee of creation. The duty of humanity as a theological category is reaching out for taking care of creation. Peace among people, and between Earth and people, is the basic condition to start the

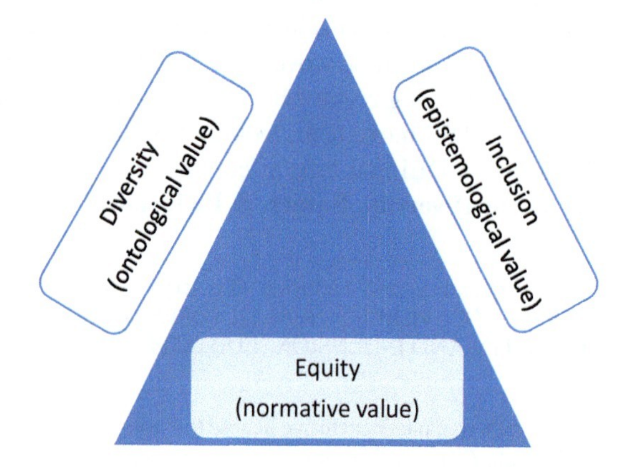

Fig. 1 DEI (Diversity, Equity, Inclusion) transdisciplinary field of concepts and values

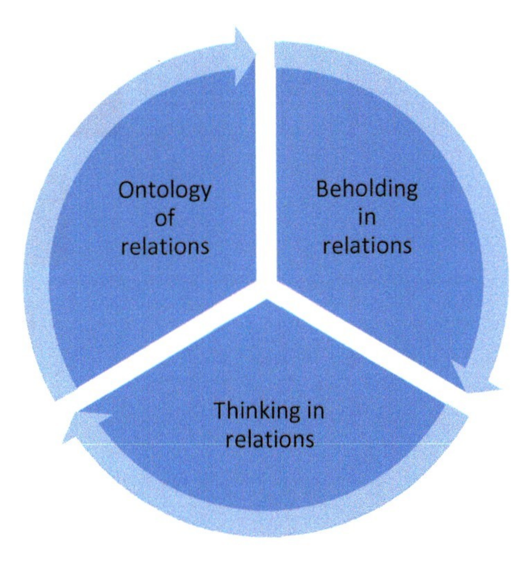

Fig. 2 "Reality is known only in relations". (Adapted from Sittler 1970)

process. Seeking for equity is the normative way for achieving this peaceful condition of ecological justice. Joseph Sittler (1970)[2] was one of the first ecotheologians to recommend systems thinking as an epistemological push toward an ecological commitment. He argues that "the question of reality is itself an ecological question. Because the question is ecological, reality itself must be spoken of ecologically. Reality is known in relations". Figure 2 shows what kind of relations are involved in the epistemology of reality and in grounding a theological responsibility for ecological commitment.

Sittler claims that "there is no ontology of isolated entities, or instances, of forms, of processes" and, consequently,

> the only adequate ontological structure we may utilize for thinking things Christianly is an ontology of community, communion, ecology – and all three words point conceptually to thought of a common kind. "Being itself" may be a relation, not an entitative thing.

The challenge of socio-ecological justice at global and local levels pertains to ethics as the most inclusive system of values for balancing relationships. Growing migrations of people from continent to continent, from State to State, and from countryside to cities, are currently one of the most evident outcomes of some kind of socio-ecological injustice which need to be healed. Most migrants are marginalized races, nationalities, genders and classes of people, such as peasants that have lost even their pour land due to lack of socio-ecological support. Recent pope Francis' encyclical letter "on Fraternity and social Friendship"[3] devotes Chapter 4

[2] Sittler J 1970. Ecological commitment as theological responsibility. In "Evocation of Grace. Writings on Ecology, Theology, and Ethics" Bouma-Prediger S and Bakken P (Eds), pp. 76–86,William B Eerdmans Publishing Company, Grand Rapid, Michigan.

[3] ENCYCLICAL LETTER "FRATELLI TUTTI" (FT) OF THE HOLY FATHER FRANCIS "ON FRATERNITY AND SOCIAL FRIENDSHIP" Given in Assisi, at the tomb of Saint Francis, on 3 October, Vigil of the Feast of the Saint, in the year 2020, the eighth of my Pontificate. Libreria Editrice Vaticana, Rome.

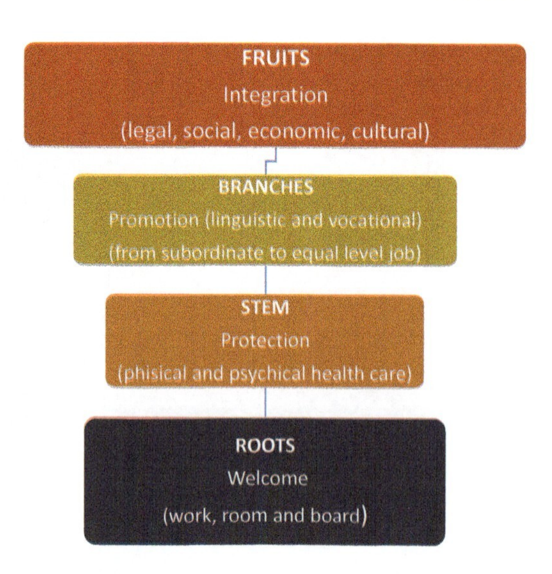

Fig. 3 A "migrant tree"

to the care of migrants, both people and individual, which recalls the elements of DEI strategy to adopt in both the country or place of origin and the country or place of welcome. This strategy is based on four key-words, *welcome*, *protect*, *promote,* and *integrate*" (FT 129). Using an agricultural metaphor, these four key words form a "migrant tree", where welcome is "the roots", protection is "the stem", promotion is "the branches", and integration or inclusion is "the fruits" (Fig. 3).

A migrant would develop in a "new land" like a transplanted cutting, first routing in the new soil, if it is fertile, then developing, if it is protected and promoted, until flourishing and giving fruits and seeds according to his/her integration in a welcoming context.

Moreover, an ecological aspect of DEI movement is emerging these days, specifically concerning the inclusion of biodiversity and the imbalance of the biotic community under the human control. There is mounting evidence that understanding interactions between humans and animal is critical in preventing outbreaks of zoonotic disease. At this time, when Covid-19 pandemics is still running and causing human suffering and death worldwide, it is urgently necessary to reconsider, as suggested in a paper title by Graham et al. (2008), "the animal-human interface and infectious disease in industrialised food animal production [by] rethinking biosecurity and biocontainment".[4]

It has been advanced that Confined Animal Feeding Operations (CAFOs) act as amplifiers of influenza:

> Influenza pandemics occur when a novel influenza strain, often of animal origin, becomes transmissible between humans. Domestic animal species such as poultry or swine in confined animal feeding operations (CAFOs) could serve as local amplifiers for such a new strain of influenza.[5]

[4] Graham et al. 2008. Public Health Reports/May–June 2008/Volume 123.

[5] Saenz et al. 2006 Confined Animal Feeding Operations as Amplifiers of Influenza. Vector Borne Zoonotic Dis. 6(4): 338–346.

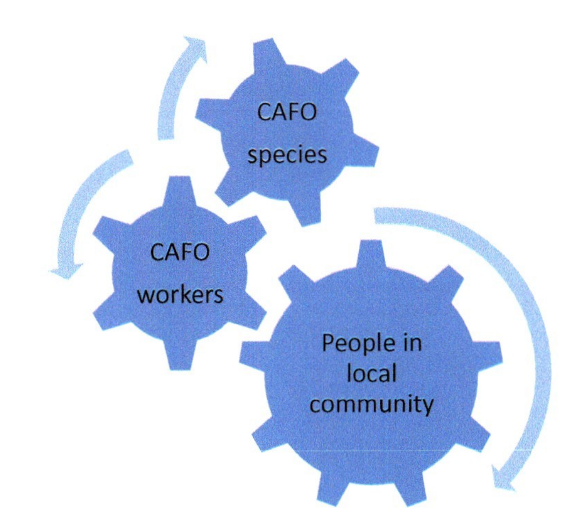

Fig. 4 Potential amplification of a new influenza virus by domestic birds or animals raised in CAFOs. Transmission dynamics between the CAFO species, CAFO workers, and the rest of the local community (after Saenz et al. 2006, modified)

Figure 4 shows how the crowding of swine and poultry in CAFOs increases the transmission of influenza viruses within the groups considered, the CAFO species, the CAFO workers (the bridging population), and the rest of the local human population.

According to Wallace (2016),[6] considerable modelling has been conducted around the relationship between pathogen epidemiology and virulence, including evidence of on-site transitions from low to high pathogenicity. Major factors in livestock production system that most likely select for virulence are:

> Increased population sizes, increased densities, declining genetic diversity, increasing throughput speed, ever-younger livestock, increasing geographic concentration, overlapping geographies of different livestock species, more extensive transport, and an encroachment on forest and wetlands expanding the interface between livestock and wildlife.[7]

Moreover, the emergence of a new, more virulent strain might be a viral evolutionary reaction to campaign of vaccination, literally emerging underneath the vaccine coverage.[8]

To reduce virulence,

> We must devolve much of the production to smaller, locally owned farms. Genetic monocultures of domesticated birds must be diversified back into heirloom varieties, as immunological firebreaks.[9]

Indeed, the Covid-19 pandemics might be more than a sanitary emergency and entail a radical change of humanity's industrial way of life.

---

[6] Wallace R 2016. Big farms make big flu. Dispatches on infectious disease, agribusiness, and the nature of science. Monthly review Press, New York.

[7] Ibidem, p.182.

[8] Ibidem, p.19.

[9] Ibidem, p.29.